Shem Wambugu Maingi
Vanessa GB Gowreesunkar
Maximiliano E Korstanje
Editors

Tourist Behaviour and the New Normal, Volume II

Implications for Sustainable Tourism Development

palgrave
macmillan

Editors
Shem Wambugu Maingi
Kenyatta University
Nairobi, Kenya

Vanessa GB Gowreesunkar
Indian Institute of Management
Sirmaur, Himachal Pradesh, India

Maximiliano E Korstanje
University of Palermo
Buenos Aires, Argentina

ISBN 978-3-031-45865-1 ISBN 978-3-031-45866-8 (eBook)
https://doi.org/10.1007/978-3-031-45866-8

This Palgrave Macmillan imprint is published by the registered company Springer Nature Switzerland AG.
The registered company address is: Gewerbestrasse 11, 6330 Cham, Switzerland

Paper in this product is recyclable.

Foreword

Sustainability is a significant issue for the tourism industry and cognate industries. Overtourism for instance was a phenomenon that really challenged the sustainability of the industry (experience of the visitors, locals, damage to buildings and natural heritage, etc.). The outbreak of COVID-19 did not address the issue. However, it provided to all stakeholders of the industry evidence of how destructive our behavior is for local heritage, the environment (fauna and flora), and the local population. Has the behavior of tourists changed after the outbreak of COVID-19? That is the question. Existing academic research and industry reports are not systematically considering the issue of the intended behavior gap (gap between what individuals are saying and their actual behavior) when collecting data. To evaluate changes in behavior (host, visitor, governance perspectives), a different, radical approach needs to be adopted regarding data collection.

Future research could be about:

(a) Developing strong arguments why covert research should not be considered inappropriate, but instead viewed as an additional triangulation tool. Lugosi (2006: 541), argued that: "concealment is sometimes necessary, and often unavoidable. Criticisms leveled against covert methods should not stop fieldworkers from engaging in research that involves covertness."

(b) Introduce covert research method as a "must" in the validity process when it comes to research in Leisure/tourism, due to the growing phenomenon of eco-guilt and eco-shame (guilt/shame individuals develop when they are aware their behavior is harmful to the environment) among consumers (Mkono & Hughes, 2020). Because of this growing trend, overt research (face-to-face survey) carried out in this area such as the one carried out by Casado-Diaz et al (2022) on tourists' attitudes with regard to water conservation might present some limitations due to the fact that respondents might present themselves more positively than they are to comply to what is expected from them, and fit within the norm (Ellison et al., 2006; Hancock & Tomas, 2009).
(c) Providing a portfolio of covert research method tools. Lugosi (2006), started his investigation under cover (covert) by just attending a bar as a customer, then started to interact with the staff and owner, then he got a job within the organization, and step by step disclosed his intentions, moving slowing from covert to semi-covert, to overt research. Providing guidelines/tools to collect data in a niche area is important otherwise, academics tend not to investigate this field, leading the field, the topic, or the research method to remain unexplored despite its importance for the advancement of understanding and practices. The topic of children in tourism is a living example (Canosa & Graham, 2016; Canosa, Graham & Wilson, 2018; Poria & Timothy, 2014).

Meanwhile, this book entitled "*Tourist Behavior in the New Normal: Implications for Sustainable Tourism Development*" (volume 2) is providing a discussion around the topic of sustainability from different perspectives that is, visitors, governance, and hosts. This book is also mapping how Sustainable Development Goals (SDGs) could be achieved through the different suggested strategies. Through this book, readers will discover how eco-guilt and eco-shame, in other words, guilt/shame individuals or organizations develop when they are aware their behavior is being harmful to the environment (Mkono & Hughes, 2020) and is translated into the new normal of the new world.

Oxford Brookes Business School
Oxford, UK
Hugues Séraphin

References

Canosa, A., & Graham, A. (2016). Ethical tourism research involving children. *Annals of Tourism Research, 61*(15), 1–6.

Canosa, A., Graham, A., & Wilson, E. (2018). Child-centered approaches in tourism and hospitality research: Methodological opportunities and ethical challenges. In *Handbook of research methods for tourism and hospitality management*. Edward Elgar Publishing.

Casado-Diaz, A. B., Sancho-Esper, F., Rodriguez-Sanchez, C., & Sellers-Rubio, R. (2022). Tourists' water conservation behavior in hotels: The role of gender. *Journal of Sustainable Tourism, 30*(7), 1518–1538.

Ellison, N., Heino, R., & Gibbs, J. (2006). Managing impressions online: Self-presentation processes in the online dating environment. *Journal of Computer-Mediated Communication, 11*, 415–441.

Hancock, J. T., & Tomas, C. L. (2009). Putting your best face forward: The accuracy of online dating photographs. *Journal of Communication, 59*, 367–386.

Lugosi, P. (2006). Between overt and covert research. Concealment and disclosure in an ethnographic study of commercial hospitality. *Qualitative Enquiry, 12*(5), 541–561.

Mkono, M., & Hughes, K. (2020). Eco-guilt and eco-shame in tourism consumption contexts: Understanding the triggers and responses. *Journal of Sustainable Tourism, 28*(8), 1223–1244.

Poria, Y., & Timothy, D. J. (2014). Where are the children in tourism research? *Annals of Tourism Research, 47*, 93–95.

Contents

Notes on Contributors

Eliyas Ebrahim Aman is a senior lecturer at the School of Business and Economics at Madda Walabu University, Ethiopia. Currently, he is a PhD researcher at István Széchenyi Economics and Management Doctoral School, University of Sopron, Hungary. He holds a Bachelor's degree in Marketing Management from Madda Walabu University and a Master's degree in Business Administration from Punjabi University, Patiala, India. His research area of interest is tourism marketing, digital marketing, and sustainable tourism development.

Raquel Arguedas-Sanz Full Professor of Financial Economics at the Faculty of Economics and Business Studies of the UNED. She is the author of 24 book chapters and has disseminated her research in various scientific articles and at recognized conferences in the area of finance.

Maria Criselda Badilla is an associate professor from the University of the Philippines, Diliman with research interest in tourism marketing, product development, digital media marketing, and destination branding.

Viktorija Baranauskienė is a researcher at the Lithuanian Centre for Social Sciences. Her research interests include geography of population, regional and urban studies, sparsely populated territories, socio-spatial exclusion, change of network of public service infrastructure, and tourism. She has published several articles and book chapters on socio-spatial exclusion, problems of unequal regional development, and, typology of the level of urbanization.

Sneha Bhattacharyya has completed her M.Phil. in Social Sciences and pursuing PhD and is currently working as Manager, Project, and Research at Contact Base (trading style: banglanatak dot com, www.banglanatak.com). Working in the domain of inclusive growth and sustainable development following the pathway of culture, at Contact Base she works toward ensuring rural empowerment and upliftment of indigenous communities. Her work is dedicated to using culture as a vehicle for development addressing SDGs—poverty alleviation, gender, wellbeing, inclusive development, global cooperation etc.

Francis Boadu has a Bachelor of Science (BSC) in Business Administration. He is currently doing a Master's in Tourism Organizations Management (TOM) at the Faculty of Economics, University of Algarve (Portugal). And he is doing his research on topics related to digital media/marketing and tourism business opportunities for Africa, especially in Ghana.

Nimit Chowdhary is a professor with Jamia Millia Islamia, New Delhi. He holds BE (Mech), MBA, and PhD (Management) and has more than 27 years of postgraduate teaching and research experience. He has been a full professor for close to 14 years at Mizoram University, Indian Institute of Tourism and Travel Management (Gwalior, Noida, and Nellore), and now Jamia Millia Islamia, New Delhi. He has also been a professor at ITESM-Mexico, Shaoxing University, China. He has also taught at the University of Girona, Spain, University of Gotland, Sweden, and GEA Academy Slovenia. Recently, Arizona State University accorded him the honor of Adjunct Professorship. He has received awards and accolades in and outside of India to recognize his academic contributions. He is a recipient of the AICTE career award for Young Teachers; SIDA fellowship, Sweden; Guest scholarship, Sweden; Linnaeus Palm Exchange Programme Grants, Sweden; PIMG Research Excellence Award; Scholars' Grant (EMTM) Erasmus Mundus, Europe, among others. Recently he was chosen for the prestigious LEAP programme at Oxford University. His research has focused on destination marketing, tourism entrepreneurship and guiding and interpretation, supervised 15 PhDs, authored eight books, edited four books, and contributed more than 125 papers. He has funded research support from UGC, ICSSR, AICTE, and MoT-GoI.

Alain Daou Associate Professor of Entrepreneurship at the Olayan School of Business and Director of the AUB Nature Conservation Center (AUB-NCC) at the American University of Beirut, in Lebanon.

Maraísa da Silva Soares Costa Postgraduate and MBA professor in Brazil. Doctoral student in Administration at PUC Minas. Member of the Research Group GEDI—Study Group in Management, Diversity and Inclusion of the Pontifical Catholic University of Minas Gerais. Research Focus: Diversity and Inclusion.

Leila El Zeenni is a project manager at the AUB-NCC. Leila holds an MSc in Ecosystem Management from the American University of Beirut.

Silvia Giralt Escobar PhD in Administrative Sciences. Professor of Financial Economics and Accounting and inspector attached to the Inspectorate of Services, Department of Economics and Business Management. Alcala University, Madrid, Spain. Research interest in administration and tourism.

Silvia Fernandes is Assistant Professor at the Faculty of Economics, University of Algarve (Portugal). She holds a PhD in Innovation Economics and is a member of CinTurs (Research Centre for Tourism, Sustainability, and Well-being). She lectures on information technology and innovation management in undergraduate and master courses. She has several publications in books and journals, in themes such as: enterprise information systems, smart tourism, innovation, mobile systems, etc.

Cláudia Aparecida Avelar Ferreira PhD in Business Administration by the Graduate Program in Business Administration at the Pontifical Catholic University of Minas Gerais—PUC MINAS. Member of the GEDI Research Group—Study Group in Management, Diversity, and Inclusion of the Pontifical Catholic University of Minas Gerais. Research Focus: Gender and Race Diversity and Inclusion.

Mahendar Reddy Gavinolla is working as Assistant Professor (Tourism) & Coordinator of International Relations at the National Institute of Tourism and Hospitality Management, Hyderabad, India, and Researcher at the Institute of Social, Economic, and Humanities Research (HESPI), Vidzeme University of Applied Sciences, Valmiera, Latvia. Teaching and research interest focus on sustainability and heritage aspects of tourism. He has published papers in Routledge, Emerald, Sage, and Springer. He has been a visiting/guest faculty at the National Institute of Fashion Technology (NIFT), Hyderabad, India, National Institute for Micro Small & Medium Enterprises (NIMSME), Hyderabad, Indi, and Vidzeme University of Applied Sciences, Latvia.

Vanessa GB Gowreesunkar is a citizen of the island of Mauritius. She is currently serving as Associate Professor at the Indian Institute of Management (Sirmaur, India). With over two decades of experience in teaching, training and research, Vanessa has brought her contributions to various international universities and educational institutions. She is an editorial board member of several scientific journals and has a number of publications in leading international journals to her credit. Outside her academic role, she holds the position of Associate Editor for the Emerald International Journal of Tourism Cities and Vice President of the International Tourism Studies Association (ITSA). Vanessa also assumes the role of social media director for both entities and she has the role of observer at the World Tourism Network.

Vanessa has varied interests in tourism and her areas of expertise include Tourism Management and Marketing, Tourism Education, Post-Colonial and Post-Conflict Destination Management, Sustainable Tourism, Island Tourism, Research Methodology, Employability Skills, Social Entrepreneurship and related topics.

Zeynep Gulen Hashmi with more than 20 years of experience in the field of hospitality and tourism, has capitalized on her work experience with multiple hotel brands and moved to the academic side of tourism and hospitality. Hashmi's academic research and published articles focus on tourism and hospitality sustainability strategies, sustainable luxury management, and organizational transformation for sustainability in tourism and hospitality. Hashmi is currently Assistant Professor at the

Emirates Academy of Hospitality Management in Dubai and Visiting Professor at the National University of Sciences and Technology in Islamabad.

Julián Alberto Álvarez Hernández holds a Doctorate in Administration. Professor/researcher of the Tourism Bachelor Program at the Autonomous University of Ciudad Juárez. Av. Universidad, Chamizal, Ciudad Juárez, México.

Manuel Ramón González Herrera has a Doctorate in Geography of Tourism from de Havana University and Alcala University, Madrid, Spain. Professor/researcher of the Tourism Bachelor Program at the Autonomous University of Ciudad Juárez. Professor of the Master's and Doctorate in Administration, Autonomous University of Ciudad Juarez. Visiting Professor of the Master and Doctorate in Administration at the University of Alcala, Madrid. Av. Universidad, Chamizal, Ciudad Juárez, México.

Anna Hourani is the Operations Manager at the AUB-NCC. Anna holds an MSc in Territory Engineering Landscape and Heritage from the Lebanese University.

Philip Ireri has more than 15 years of work experience in ecotourism, community participation, and environmental and social safeguards in the public, private, and NGO sector. He is presently the Principal Consultant at Essibility Africa, a Kenyan-based consultancy that provides an array of environmental and social sustainability services. His research interests cover ecotourism, sustainable tourism, community participation, nature-based enterprises, social justice, and environmental conservation.

Rami K. Isaac was born in Palestine. Rami Isaac did his undergraduate studies in The Netherlands, graduate studies in the UK, and earned his PhD from the University of Groningen, in Spatial Sciences, in The Netherlands. He is currently Senior Lecturer in Tourism Teaching at the undergraduate as well as postgraduate levels at the Academy for Tourism at the Breda University of Applied Sciences in The Netherlands. In addition, he is an assistant professor at the Faculty of Tourism and Hotel Management at Bethlehem University, Palestine. Currently, he is the

Vice-President of the Research Committee 50 on International Tourism, International Sociologist Association ISA (2018–2024). His research interests include tourism development and management, critical theory, and political aspects of tourism. He published numerous articles and book chapters on tourism and political (in)stability, occupation, tourism and war, dark tourism, violence, and transformational tourism.

Deborah Kangai is Assistant Lecturer in Tourism Management in the Department of Environmental Studies and Resource Development, at Chuka University, Kenya. She holds a Bachelor of Science Degree in Ecotourism and Hospitality Management and a Master's in Tourism Management from Chuka University. Currently, she is a PhD researcher at István Széchenyi Economics and Management Doctoral School, University of Sopron, Hungary. Her research interests are sustainable tourism, ecotourism, coastal and marine ecotourism, cultural tourism, tourism product development, and destination management.

Isaac Kimunio lectures in the School of Economics, Kenyatta University. Isaac Kimunio is also a researcher at the Kenyatta University Women Economic Empowerment Hub under a project funded by the Bill & Melinda Gates Foundation (BMGF), "Violence, Crisis, and Women's Work." The research Topic in the WEE hub is "Effectiveness of SGBV Response and Prevention Programmes in Enhancing Women's Work and Promoting Their Economic Empowerment." He holds a PhD in Economics, specializing in monetary economics from Kenyatta University. Isaac has a strong desire for research and teaching and his research interests are tourism, gender, monetary, financial economics, and econometrics. Isaac is also a member of The Econometrics Association of Kenya (EAK) where he is the head of research. Isaac's work experience includes having worked with Barclays Bank of Kenya for more than five years in various departments. He has also worked as a trainer of trainers with the Kenya National Bureau of Statistics and with Mukuru Slums Development Projects (MSDP). At MSDP, he worked on a project funded by OXFAM UK using mobile money transfer to help the urban poor.

Maximiliano E Korstanje is Editor in Chief of *International Journal of Safety and Security in Tourism* (UP Argentina) and Editor in Chief Emeritus of *International Journal of Cyber Warfare and Terrorism* (IGI-

Global US). Besides being Senior Researcher in the Department of Economics at University of Palermo, Argentina, he is a global affiliate of Tourism Crisis Management Institute (University of Florida US), Centre for Ethnicity and Racism Studies (University of Leeds), The Forge (University of Lancaster and University of Leeds UK) and The International Society for Philosopher, hosted in Sheffield UK, with more than 1200 published papers and 35 books. As author his recent works include, *Strategic Tool and Methods for Promoting Hospitality and Tourism Services*, *Virtual Traumascapes and Exploring the Roots of Dark Tourism*, *Research Practices and Innovations in Global Risk and Contingency Management* by IGI Global, *The Rise of Thana Capitalism and Tourism* by Routledge, *Terrorism, Tourism and the End of Hospitality in the West* by Springer Nature, *The Mobilities Paradox, a Critical Analysis* by Edward Elgar, *Risk and Safety Challenges for Religious Tourism and Events* by CABI, and *Terrorism in a Global Village* by Nova among others. Currently he co-edits almost 10 specialized journals and takes part of almost 30 journals associated to themes such as human rights, mobility, tourism, and terrorism. Korstanje was nominated with five honorary doctorates for his contribution to the study of the effects of terrorism in tourism. In 2015 he was awarded as Visiting Research Fellow at School of Sociology and Social Policy, University of Leeds, UK and the University of La Habana Cuba. In 2017 was elected as Foreign Faculty Member of AMIT, Mexican Academy in the study of Tourism, which is the most prominent institutions dedicated to tourism research in Mexico. He has vast experience in editorial projects working as advisory member of Elsevier, Routledge, Springer, IGI Global, and Cambridge Scholar Publishing. Korstanje has visited and given seminars in many important universities worldwide. He has also recently been selected to take part in the 2018 Albert Nelson Marquis Lifetime Achievement Award, a great distinction given by Marquis Who's Who in the world.

Edis Kriaučiūnas is a geographer and senior research fellow at the Lithuanian Centre for Social Sciences. His scientific interests are regional geography, recreational geography, rural development, and tourism. Author/co-author of scientific peer-reviewed publications and scientific monographs, participant, and manager of research projects.

Darius Liutikas is a senior scientific researcher at the Lithuanian Centre for Social Sciences (Vilnius, Lithuania). Research interests include the sociology of tourism, anthropology of tourism, modern and traditional pilgrimage, and forms of mobility. He has published around 50 articles and book chapters in Lithuanian and international academic journals. He has prepared and published six books (three in Lithuanian language and three in English). His most recent book, of which he is the editor, is *Pilgrims: Values and Identities* (CAB International, 2021). He is a leader of the scientific project "Transformations and perspectives of the tourism sector after the COVID-19 pandemic" (2021–2023).

Aapo Lundén is a PhD researcher at the Geography Research Unit at the University of Oulu and a member of Tourism Geographies and Sustainable Mobilities (TOGSUM) and Transformation and Transnationalization of state spaces: Geo-politics of Borders, Mobilities and Planning in the Arctic (SPARCTIC) research groups and an affiliate of the Biodiverse Anthropocenes (ANTS) research program. His research interests include nature-based tourism, governance and regulation, wilderness studies, conservation, development in Arctic areas, environmental policy, political ecology, and institutional theory.

Shem Wambugu Maingi is Lecturer and Researcher in Tourism Management at the Department of Hospitality and Tourism Management at Kenyatta University in Kenya. He has been a tourism lecturer and researcher from Kenyatta University for the last twelve (12) years. Shem Maingi has widely published in Scopus indexed journals and books internationally, one of which received a joint award as article of the year 2011. He has edited and authored books on the future and resilience of tourism, notably *Tourism Destination Management in a Post-Pandemic Context: Global Issues and Destination Management Solutions* and *Tourism through Troubled Times: Challenges and Opportunities of the Tourism Industry in 21st Century* published by Emerald Group Publishing (UK) as well as *Management of Tourism Ecosystem Services in a Post Pandemic Context: Global Perspectives* published by Routledge (UK). He also jointly developed an international policy brief entitled, "*Rebuilding Tourism Post Pandemic – Policy Recommendations from Global Case Studies*" published by Emerald Open

Research (UK). He is an expert member of the African Union Economic, Social and Cultural Council (AU-ECOSOCC) Tourism and Infrastructure Cluster. He is also an expert member of the International Scientific Committee on Cultural Tourism of ICOMOS (ICTC) as well as the IUCN WCPA Tourism and Protected Areas Specialist Group TAPAS. Shem Maingi is currently a member of an ICOMOS ICTC Advisory Committee developing a Session Proposal on "Polycentric and participatory governance in cultural tourism for community resilience" for the 2021 United Nations Climate Conference (COP26) Scientific Symposium on the theme Living Heritage and Climate Change. Shem Maingi has served as a member of the Advisory Panel, IFLA Africa – African Landscape Network/ ICOMOS: UNESCO Programme of Participation in the activities of Member States for 2020–2021. He is also a board member of Ecotourism Kenya's National Eco-rating and Standards Committee and has been involved in environmental conservation through Ecotourism Certifications assessments and Tourism Standards formulation.

Steve Makambi is a lecturer in the Department of Econometrics and Statistics at the School of Economics, Kenyatta University. He holds a PhD in Economics, specializing in econometrics, from Kenyatta University. With a keen interest in various fields, Steve Makambi focuses on applied econometrics, finance, time series econometrics, econometric analysis, econometric modeling, panel data econometrics, and financial econometrics. In addition to his lecturing role at the School of Economics, Kenyatta University, Makambi actively contributes to research as a member of the Kenyatta University Women Economic Empowerment Hub. He is involved in a project funded by the Bill & Melinda Gates Foundation (BMGF). Within this project, Makambi utilizes his expertise to further research and support women's economic empowerment. Steve Makambi is a self-directed and result-oriented professional with advanced knowledge in economics, econometrics, and statistical analysis. He possesses proficiency in conducting advanced scientific research, encompassing a diverse set of skills ranging from problem conceptualization and identification, research design formulation, sampling and data collection, statistical analysis, and report writing.

Adrián Mendieta-Aragón Research Associate of the Department of Economic Analysis at the National University of Distance Education (UNED). His research interests are in the areas of tourist behavior, digital tourism, and demand modeling.

Riza Mae Mirandilla is a tourism graduate from the University of the Philippines, Diliman with research interests on nature-based tourism, domestic tourism, and tourism development.

Sweety Mishra is a junior research fellow with the Department of Tourism and Hospitality Management in Jamia Millia Islamia, Delhi. She has worked in the tourism sector for one year. She graduated with a Bachelor of Computer Application from Jaipur National University and completed her Postgraduation in Master of Travel and Tourism Management from Banaras Hindu University. Her interest areas are entrepreneurship, social entrepreneurship, women entrepreneurship, and gender equality, women empowerment, and tourism mindfulness.

Joseph K. Muriithi is a senior lecturer at the Department of Environmental Studies and Community Development at Kenyatta University. His main areas of research interest include ecotourism, nature-based tourism, urban greening, participatory development, and environmental conservation. He has also published in the same fields of specialization.

Martin W. Nandelenga is Senior Country Economist in the Country Economics Department (ECCE) at the African Development Bank. He holds a Doctor of Philosophy (PhD) degree in Economics from the University of Cape Town, South Africa. Before joining the African Development Bank, he worked at the Kenya Institute for Public Policy Research and Analysis (KIPPRA) as Senior Policy Analyst. Nandelenga has also worked at Egerton University, Kenya, as a lecturer in the Department of Economics. Martin's research interests include public finance, particularly on the macroeconomic analysis of fiscal and monetary policy, financial markets with a focus on systemic spillovers, and portfolio diversification. He has also worked on fiscal sustainability evaluating debt dynamics in developing countries. He has published widely in leading peer-reviewed journals such as the *Journal of Economic*

Asymmetries. He has also authored several research works that have been published as working papers by the Economic Research Southern Africa (ERSA) and the Africa Development Bank. Martin has consulted with several international organizations such as the World Bank, the African Development Bank, and United Nations Development Programme.

Chloe Marie Novo is a tourism graduate from the University of the Philippines, Diliman with research interests in ecotourism, sustainable tourism, and urban planning.

Árpád Ferenc Papp-Váry is a senior research fellow and lecturer at the István Széchenyi Economics and Management Doctoral School, University of Sopron, Hungary. His area of research is branding marketing and marketing communication trends, product placement, branded entertainment, country branding, nation branding, and city branding.

Gintarė Pociūtė-Sereikienė is a geographer and research fellow at the Lithuanian Centre for Social Sciences. Her academic research interests encompass urban and regional geography and tourism. She is an active researcher: an author/co-author of around 30 scientific peer-reviewed publications, a participant in national and international projects, attendant of conferences, seminars, training, and, workshops around the globe.

Salma N. Talhouk Professor of landscape Horticulture in the Department of Landscape Design and Ecosystem Management (LDEM), in the Faculty of Agricultural and Food Sciences (FAFS), at the American University of Beirut (AUB) in Lebanon.

Alix Varnajot is Postdoctoral Researcher at the Geography Research Unit at the University of Oulu and is affiliated with the Arctic Interactions (ArcI) interdisciplinary program. His research interests include tourism and climate change adaptation, place-making and place representations, and conceptual developments on Arctic tourism geographies.

Sayyeda Zonah comes from Pakistan. Sayyeda Zonah's interests lie in exploring diverse cultures and digitalization in tourism and hospitality for a sustainable future. In 2020, Zonah completed her Bachelor's in Business Administration from NUST, Pakistan where her final project was based on promoting e-tourism companies for Northern Pakistan.

Zonah graduated with her Master's in Tourism Management from Yaşar University, Turkey; having successfully defended her thesis "Digitalization of Heritage Tourism: A Perspective of European Museums" in 2022. Zonah was also an Erasmus exchange student at the University of Milano-Bicocca, Italy.

Abbreviations and Acronyms

ADB	African Development Bank
BEST	Beyond Sustainable Tourism
CBC	Competency-Based Curriculum
CC	Carrying Capacity
CMNP	Cumbres de Majalca National Park
CONANP	Comision Nacional de Areas Naturales Protegidas
CREST	Centre for Responsible Tourism
DM	Demand Management
DMO	Destination Management Organization
EPI	Environmental Performance Index
GDP	Gross Domestic Product
GHG	Greenhouse Gas
GSTC	Global Sustainable Tourism Council
ICT	Information Communication Technology
IDAL	Investment Development Authority of Lebanon
ILO	International Labour Organization
IMF	International Monetary Fund
INE	National Statistics Institute of Spain
KAP	Knowledge-Attitudes-Practices model
KWCA	Kenya Wildlife Conservancies Association
LC	Local Community
LMTLT	Research Council of Lithuania
NAM	Norm-Activation Model

OECD	Organization for Economic Cooperation and Development
PATA	Pacific Asia Travel Association
PETB	Pro-Environmental Travel Behavior
PNAs	Protected Natural Areas
RTS	Resident Travel Survey
SADC	South African Development Council
SARSCOv2	Severe Acute Respiratory Syndrome Coronavirus 2
SD	Smart Decisions
SDG	Sustainable Development Goals
SM	Seasonality Management
TBL	Tripple Bottom Line
TE	Tourist Experience
TIC	Tourism Information Centre
TIES	The International Ecotourism Society
TOURQUAL	Tourism Quality
TPZ	Tourism Promotion Zone
UNED	National University of Distance Education
UNEP	United Nations Environmental Programme
UN-OCHA	United Nations Office for the Coordination of Humanitarian Affairs
UNWTO	United Nations World Tourism Organisation
WHO	World Health Organization
WTC	World Trade Centre
WTES	Willingness to Make Economic Sacrifices
WTO	World Trade Organization
WTTC	World Travel and Tourism Council
WUR	Wageningen University & Research

List of Figures

List of Tables

1

Introduction: The Interplay Between Tourism Resilience and Sustainability in the New Normal

Vanessa GB Gowreesunkar, Shem Wambugu Maingi, and Maximiliano E Korstanje

While Volume I of our book dealt with the changing behavior of tourists in the new normal and its impact on industry resilience, Volume II offers readership an interesting account of its impacts on sustainability. Tourism resilience and sustainability are in fact inexorably linked. This is well documented in the work of Lew et al. (2020) who clearly demonstrate how New Zealand creatively responded to the COVID-19 pandemic and reopened its domestic tourism market earlier than most countries. The country launched a campaign to encourage locals to travel within their

V. GB Gowreesunkar (✉)
Indian Institute of Management, Sirmaur, Himachal Pradesh, India

S. W. Maingi
Kenyatta University, Nairobi, Kenya
e-mail: maingi.shem@ku.ac.ke

M. E Korstanje
University of Palermo, Buenos Aires, Argentina

S. W. Maingi et al. (eds.), *Tourist Behaviour and the New Normal, Volume II*,
https://doi.org/10.1007/978-3-031-45866-8_1

own country and support local businesses. This example shows how tourism resilience, that is, the ability to withstand and recover from shocks and stresses, can enhance tourism sustainability, that is the ability to balance the economic, social, and environmental impacts of tourism activities. The new normal is a business landscape that has fundamentally changed (Prayag, 2023). As a result, tourism in the post-pandemic era requires a profound adaptation and innovation of tourism products, services, and policies to meet the changing needs and demands of tourists, as well as to ensure the sustainability of tourism development in economic, social, and environmental terms (Gowreesunkar et al., 2021; Hall et al., 2020). There is indeed a major restructuring in the world economy driven by technological innovation on the supply side. From the demand side, the new normal involves a preference for flexible and small-group travel, special interest tourism, and new health and hygiene protocols among others (Prayag, 2023). This changing trend in demand and supply of tourism products has also affected the sustainability of the global tourism industry, a point echoed in the study of Buhalis (2000): "…following the COVID-19 Pandemic, tourism suppliers at destinations need to understand that they should not compete with each other at the destination level. Instead, they should join forces and pool resources…" (Buhalis & Cooper, 1998, p. 21). This is also well-detailed in the work of Gowreesunkar et al. (2022)—Management and Restoration of the Tourism Ecosystem Services Post Pandemic (2022) and Korstanje et al. (2023)—"Tourism in Troubled Time."

As the world is moving toward the *next normal* of an ongoing pandemic, it is observed that tourism resilience can have positive or negative effects on tourism sustainability depending on how tourists behave and how destinations manage their tourism systems. For instance, tourism resilience can enhance tourist confidence and trust in traveling to destinations that have demonstrated effective responses to shocks and stresses (Prayag, 2023). This can increase tourist loyalty and satisfaction, which are important factors for tourism sustainability. This point is also well illustrated by the Jamaican Tourism Minister who emphasizes the collective responsibility of tourism stakeholders to protect the environment and renewed his call for the establishment of a global tourism resilience fund. The Ministry of Tourism developed several initiatives to support small and medium tourism enterprises while holding them accountable to sustainable business practices aligned with industry standards. Technology was a major tool utilized to build resilience and substantiality in tourism and connect the smaller players in the industry to the larger enterprises. The Jamaican Ministry of Tourism also established a platform (ALEX) that connected the small farmers of deep-rural regions to procurement officers in hotels all across Jamaica. On their smartphones, they work out prices and delivery systems that allow for products to be brought directly from the farm gate to the hotel (source: eTurbo News, 2023). Likewise, tourism resilience can foster tourist adaptation and innovation in the new normal. Tourists may seek new experiences, destinations, modes of travel, or ways of engaging with local communities that are more aligned with their values and preferences in the new normal. This can create opportunities for tourism sustainability by diversifying tourism products and markets, reducing seasonality and overcrowding, and increasing social and cultural benefits. Tourism resilience can also challenge tourist behavior in the new normal. For instance, tourists may face trade-offs between their personal interests and the collective interests of destinations and host communities. They have to comply with strict health and safety measures, respect local norms and customs, or contribute to environmental conservation and social development. This can require tourists to be more responsible, ethical, and aware of their impacts on tourism sustainability.

The COVID-19 crisis has indeed exposed several vulnerabilities and weaknesses in the tourism industry. In the next normal, addressing these challenges will be paramount. Researchers argue that rethinking tourism models, emphasizing local engagement and sustainable practices, and enhancing collaboration between public and private sectors are critical steps toward building resilience (see Gössling et al., 2021; Prayag, 2023; Gowreesunkar et al., 2022). This inclusive approach enables tourism destinations to better withstand future shocks while benefiting local communities and preserving natural and cultural resources. Innovation and technology will play instrumental roles in shaping the tourism industry's future. This is also well illustrated in the Jamaican tourism example above. Moreover, Xiang et al. highlight how digital technologies, such as artificial intelligence, data analytics, and contactless services, can enhance efficiency, safety, and personalized experiences. Adopting and leveraging these innovations will not only drive competitiveness but also contribute to the resilience of tourism businesses and the sustainability of tourism destinations in the face of changing circumstances.

In conclusion, to achieve both resilience and sustainability in the new normal, tourists need to be informed, educated, motivated, and empowered to make sustainable choices. Destinations need to be proactive, adaptive, collaborative, and inclusive to create a conducive environment for sustainable tourism development. These ideas are captured in this volume. With contributions from Finland, Spain, Palestine, India, Argentina, Hungary, Lebanon, Mexico, Turkey, Kenya, Ethiopia, the Netherlands, Portugal, Lithuania, Brazil, Mauritius, UAE, and Pakistan, Volume II seeks to offer a rich insight on the impacts of consumer behavior on the sustainability of the tourism industry.

The book is structured in five parts: Part A focuses on the relationship between tourism behavior and sustainable tourism. Authors are Shem aingi, Maximiliano Korstanje, Vanessa Gowreesunkar, Mahendar Reddy Gavinolla, Maraísa da Silva Soares Costa, Cláudia Aparecida Avelar Ferreira, Sweety Mishra, and Nimit Chowdhary. Part B delves into environmental sustainability and its influence on tourist behavior, addressing SDGs related to clean energy, responsible consumption, climate action, and marine and terrestrial conservation. The authors are Isaac Rami, Manuel Gonzalez Herrera, Julián Alberto Álvarez Hernández, Silvia Giralt Escobar, Deborah Kangai, Eliyas Ebrahim Aman, Árpád Ferenc Papp-Váry, Darius Liutikas, Edis Kriaučiūnas, Viktorija Baranauskienė, Gintarė Pociūtė-Sereikienė, Joseph K Muriithi, and Philip Ireri. Part C explores economic sustainability and governance trends, highlighting the importance of SDGs related to poverty eradication, peace, justice, and strong institutions. The contributing authors are Silvia Fernandes and Francis Boadu, Aapo Lundén and Alix Varnajot, Adrián Mendieta-Aragón and Raquel Arguedas-Sanz, Isaac Kimunio, Martin W. Nandelenga, and Steve Makambi, Riza Mae Mirandilla, Chloe Marie Novo, and Maria Criselda Badilla. Part D addresses the socio-cultural dimension of tourist behavior and visitor behavior from the host perspective, encompassing SDGs related to poverty and health. Contributing authors are Sneha Bhattacharyya, Alain Daou, Leila El Zeenni, Anna

Hourani, and Salma Talhouk. The diversity of the authorship in the book brings a richness to the readership, as this provides a comprehensive analysis of the subject matter, encompassing various cultural, social, and economic contexts.

Finally, the COVID-19 pandemic has been a wake-up call for the tourism industry. It has taught the world how to enhance tourism resilience while driving sustainable tourism development and it has taught tourism stakeholders how to develop coping capacity, adaptive strategy, and resiliency in order to win the race against sustainability. A good crisis is after all never waste (Winston Churchill)!

References

Buhalis, D. (2000). Marketing the competitive destination of the future. *Tourism management, 21*(1), 97–116.e in tourism, 324–46.

Buhalis, D., & Cooper, C. (1998). Competition or co-operation? Small and medium sized tourism enterprises at the destination. *Embracing and managing chang.*

Gössling, S., Humpe, A., Fichert, F., & Creutzig, F. (2021). COVID-19 and pathways to low-carbon air transport until 2050. *Environmental Research Letters, 16*(3), 034063.

Gowreesunkar, V., Maingi, S., & Mingate, F. (2022). *Tourism management and restoration of ecosystem service in a post pandemic context: Global issues and challenges*. Routledge. ISBN 9781032248080. http://www.routledge.com/9781032248080/

Gowreesunkar, V., Maingi, S., Roy, H., & Micera, R. (2021). *Tourism destination management in a post pandemic context: Global issues and destination management solutions* (1st ed.). Emerald Publishing Limited.

Grzywacz, R., & Zeglen, P. (2016). Typology of tourists and their satisfaction level. *Scientific Review of Physical Culture, 6*(1), 5–16.

Hall, C. M., Scott, D., & Gössling, S. (2020). Pandemics, transformations and tourism: Be careful what you wish for. *Tourism Geographies, 22*(3), 577–598. https://doi.org/10.1080/14616688.2020.1759131

Hohnholz, L. (2023, May). Need to boost global tourism resilience and sustainability. *ETurbo News*. https://eturbonews.com/need-to-boost-global-tourism-resilience-and-sustainability/

Korstanje, M., Gowreesunkar, V., & Maingi, S. (2023). *Tourism in crisis*. Nova Science Publishers. https://novapublishers.com/shop/tourism-in-crisis/

Lew, A., et al. (2020). COVID-19 and the recovery of tourism. *Tourism Geographies, 22*(3), 455–464.

Pocinho, M., et al. (2021). Wellbeing and resilience in tourism: A systematic literature review. *Frontiers in Psychology, 12*, 748947.

Prayag, G. (2023). Tourism resilience in the 'new normal': Beyond jingle and jangle fallacies? *Journal of Hospitality & Tourism Management, 54*, 513–520.

Tourism New Zealand. (2020). Do something new, New Zealand. https://www.tourismnewzealand.com/news/do-something-new-new-zealand/

UNESCO. (2020). Sustainable & resilient tourism: Questions & answers. https://en.unesco.org/sites/default/files/info_sheet_sustainable_tourism.pdf

UNWTO. (2021). International tourism down 74% in worst year on record. Retrieved January 10, 2023, from https://www.unwto.org/news/2020-international-tourism-down-74-in-worst-year-on-record

Wachyuni, S. S., & Kusumaningrum, D. (2020). The impact of Covid-19 pandemic on tourist behavior and its implication on tourism marketing strategy. *Journal of Business and Management, 22*(12), 1–8. https://doi.org/10.9790/487X-2212010108

Part I

Tourism Behaviour and Sustainable Tourism

2

Tourism Decolonization, Geopolitics, and *Degrowth:* A Theoretical Case for Tourism Sustainability in East Africa

Shem Wambugu Maingi
and Vanessa GB Gowreesunkar

Introduction

Degrowth and Sustainable Tourism in East Africa: Opportunities and Challenges in the Global South

Tourism growth has had a number of paradigms and implications for destinations in the Global North and South. The monolithic models of tourism growth that have traditionally dominated worldview in the past have failed in addressing the changing tourism landscapes in the Global South. Over the years, there have been numerous problems associated

S. W. Maingi (✉)
Kenyatta University, Nairobi, Kenya
e-mail: maingi.shem@ku.ac.ke

V. GB Gowreesunkar
Indian Institute of Management, Sirmaur, Himachal Pradesh, India

S. W. Maingi et al. (eds.), *Tourist Behaviour and the New Normal, Volume II*,
https://doi.org/10.1007/978-3-031-45866-8_2

with the traditional tourism growth models, such as the unsustainable levels of tourism, tourism gentrification, and excessive pollution coupled with social and economic inequalities, therefore, outranking long-term development prospects (Fletcher et al., 2021; Mowforth & Munt, 2016). This traditional tourism growth model has had serious social and ecological consequences in destinations such as a degraded environment, disruption of the local cultures as well as income inequalities in developing countries (Fang et al., 2021), degradation of the environment, and generation of wastage and tourism gentrification (Blazquez et al., 2019; Creaco & Querini, 2003; Lakshmi & Shaji, 2016). To a certain extent, the "Southern thought" has dominated most developing nations in the Global South driven by the need to develop territorial alternatives to growth-based tourism development. This Southern thought as defined by Franco Cassano's thesis relates to the prioritization of community values such as slowness, moderation, and conviviality (Kallis et al., 2022). Shaban and Datta (2019) describe slowness as the alternative and slow form of tourism development that leads to harmonious eurhythmia and more sustainable cities. In the recent past, there has been wide criticism of the traditional growth notion of development, with the arguments that the natural limits to growth have so far been surpassed and that the current unsustainable and inequitable growth trajectory cannot be sustained (Andriotis, 2014). Victor (2012) notes that the notion of slow growth or degrowth is suitable for addressing the physical and social limits of economic growth. Similarly, the concept of moderation in tourism entails consideration of social and environmental justice issues in tourism growth (Demaria et al., 2019).

The form and extent of the moderation are based on the roles and impacts of tourism development on the existing stakeholders. According to Kallis et al. (2022), the degrowth discourse entails a planned slowdown to respect planetary limits, a social well-being as well as a convivial reclamation of the commons. It opposes and challenges extractivist and unilateral growth of the economy without due regard to the people and the planetary impacts. In this decade where environmental and social justice discourse has dominated global action, the need to focus on human rights, sustainability, as well as climate justice, has been paramount in bringing meaningful prosperity to the people. The global need to restate the basics of life, freedom, morality, and community globally has resulted in the promulgation, global recognition, and Universal Declaration of Human Rights (UNHR) (Witte & Green, 2012; Veal, 2015). In a sense, the tourism sector has gone through a decolonization process that has strongly been influenced by the geopolitical discourses within countries. Tourism and decolonization in the African context have been centred on addressing the dual ideological tensions that promote inclusive and people-centred growth paradigms as vis-à-vis historical structural inequalities and exclusion of local communities in decision-making (Akama et al., 2011). Chambers and Buzinde (2015) from a post-colonization perspective noted that tourism scholars need to move away from a Eurocentric approach to a more value-led humanistic approach. This value-led humanistic approach towards tourism focuses on developing tourism towards a more cultured behaviour and observing that critical balance between people, planet, and profits.

In the African context, the focus of tourism growth has been based on its economic contribution as well as its ability to improve livelihoods as well as stimulate production and service sectors (World Bank, 2013). However, tourism growth in Africa has, by no means, been linear as the sector has been limited by environmental degradation (Ehigiamusoe, 2019), financial crisis (Ekeocha et al., 2021), regional conflicts (Karimi et al., 2022), and political risk (Muzindutsi & Manaliyo, 2016; Brown, 2000). Developing countries in the region face an existential developmental challenge of sustaining tourism livelihoods (Maingi, 2021), tourism infrastructure pressures (Rogerson, 2016), the effects of climate change (Hoogendoom & Flitchett, 2018) as well as sustaining debt-fuelled tourism economies (Edo et al., 2020). As an export sector, tourism is much affected by the social and economic conditions in the African countries. The COVID-19 pandemic triggered an unprecedented economic crisis in the African region. The IMF reported that tourism-dependent countries in Africa in the post-pandemic era have been facing increasing inflation and global interest rates as well as exchange rate pressures and higher borrowing costs (IMF, 2022). In East Africa, significant tourism policy priorities in the new normal have focused on enforcing "Tourism regulations" to curb the effects of unregulated tourism on the environment, communities, and local economies. Despite the emphasis on tourism growth in most of the developing countries in the region, the rising effects of climate change, uncontrolled population growth, resource-use conflicts, and vanishing ecosystems have accentuated the physical limits to the growth and expansion of the tourism industry in the African region. Apart from the physical limits to growth, there have been human rights issues in tourism in the recent decades that have affected communities, tourists, and the tourism industry in developing countries which include issues such as insecurity, displacement, gentrification, discrimination, sex tourism, rights of children, and freedom of movement amongst other issues (Hashimoto et al., 2021; Maingi, 2019; Maingi & Gowreesunkar, 2022). This paradox has contributed to what Jim Butcher refers to as *"The moralization of tourism"* as well as rising activism concerning human rights, sustainability, and climate justice in tourism. This chapter seeks to address these issues as well as develop theoretical, management, and policy frameworks for tourism degrowth and promoting the New Moral Tourism in East Africa.

Decolonization in Tourism Discourse: *Ubuntu* Perspective

African societies need to shift paradigms from the colonial *Gilles Deuleze's* philosophy of control societies to the *Yuk Hui's* concept of modulated societies (Hui, 2015). These guiding philosophies of societies are followed as they transition from a form-imposing mode to a self-activating society. Rutternberg (2022) noted that colonized tourism narratives in the Global North were form-imposing and have changed local cultures, ecologies, and economies of the Global South in a myriad of ways. To a particular extent, there have been cases where there is a blend of two cultures, that is, the indigenous and the imposed Western cultures. According to Vong and Ung (2012), this blend has provided unique opportunities for destinations to reposition themselves as a balance of both East and West cultures as in the case of Macau Heritage. Decolonization in tourism discourse has been a basis for affirming the glocal narratives, economies, cultures, and images that reflect the indigenous states of the local people, places, and positions in a more politicized and socialized way (Boer, 2015; Hall & Tucker, 2004). Decolonization in this context involves the realization of economic, psychological, and cultural freedom to achieve indigenous sovereignty. The right towards indigenous decolonization had been guaranteed with the achievement of the sovereignty of African states. However, in an article on the political economy of imperialism, Gartzke and Rohner (2011) note that real economic freedom has not been guaranteed to the former colonized countries since decolonization took more of a territorial and political route than an economic route.

There are destinations that have developed tourism products around their colonial heritage. These are colonies that have not been liberated from their imperial regimes. In these societies, colonization and ultimately decolonization have affected their economic, social, and political processes resulting in the colonization of tourist places, and cultural gentrification of communities. Colonial assets have become objects of heritage and tourist attractions, therefore confirming the dominance of the colonial cultures as compared to the indigenous cultures (Simmons, 2004; Craik, 1994). The urgency of the post-colonial transformation of the tourism sector has become the subject of discussion in most countries (Wong et al., 2016; Palmer, 1994). The key priorities of post-colonial and decolonization discourse have been centred on addressing the current and future challenges facing the tourism communities in the Global South. From an African perspective, there has been an urgency for tourism transformation towards a sustainable development perspective as well as towards the normative moral theory of Ubuntu. The *Ubuntu* philosophy represents a paradigm shift towards the normalization of society based on Africa's humanness and values. The notion relies on the basic principles of community care, well-being, and prosperity for all which is rooted in Africa (Francis, 2010). As popularized by Desmond Tutu, ubuntu relates to the fact that *"a person is a person through other persons"*. In his short essay, "Ubuntu: On the Nature of Human Community", tutu wrote:

> Ubuntu is the essence of being human. It speaks of how my humanity is caught up and bound up inextricably with yours. It says, not as Descartes

did, "I think, therefore I am" but rather, "I am because I belong." I need other human beings to be human. Tutu (2011)

African cultures have a lot to contribute to the global tourism sectors in terms of social, philosophical, humanitarian, and economic principles (Nussbaum, 2003). Indigenous knowledge of the African people plays an important role in shaping the nature of the tourism experience. However, relational actors such as tour operators, hotels, and airlines play an important role in co-creating the overall experience that is the subject of tourism production and consumption. Westernized imaginaries of African tourism have more often ignored the local needs, priorities, and interests in tourism for a long time (Sindiga, 1999). For a long time now, images of Africa remain misconstrued with an elevated feature of African nature and minimal features of African knowledge base and diverse cultures, therefore leading to misunderstandings about the African continent. In a post-colonial context, the need for African culture to form a central facet of tourism growth and development cannot be underscored. This is quite evidenced in studies such as Mkono (2019) and Angula et al. (2018) that African cultures and opinions are important and cannot be overrun by Western cultures and opinions which eventually guide tourism and wildlife conservation policies. The developmental approach needs to ensure congruence between the local cultures, conditions, and the specific context and needs of the local people. Decolonization in tourism needs to be values-led, humane, and based on indigenous knowledge (Chambers & Buzinde, 2015; Pritchard et al., 2011). For decolonization to take place in tourism, there needs to be an urgent focus on local needs, priorities, and interests as opposed to market and industry needs and interests in tourism. Benjamin and Dillete (2021) show that decolonial thinking in tourism focusing on local representation and awareness can contribute to local transformation in tourism as a result of cultural power.

Geopolitics and Tourism Behaviour in the Global South

Globally, tourism destinations face diverse crises precipitated by social, political, economic, and environmental havoc. In Africa, tourism destinations have been adversely affected by global changes in the struggle for political, geographical, and market dominance. Internationally, tourism is deeply embedded in global politics and global politics is also deeply embedded in tourism (Hall, 2017; Burns & Novelli, 2007). Todd Cleveland in his book *A History of Tourism in Africa*, wondered, *"How do we do tourism in Africa?"* by focusing on the weighty considerations within the context of exoticization, exploitation, and enrichment in post-colonial Africa (Cleveland, 2021). In his work, Cleveland stated,

> Tourism also negatively affects Africa. Perhaps most troubling from an economic perspective is the volatility of the industry which is dependent on the whims of individuals, most of whom reside in the Global North. This relationship is at its most problematic when developing nations become overdependent on tourism, whose clientele is largely out of their control. (Cleveland, 2021, p. 13)

Tourism behaviour from the African perspective has been largely subject to the uncontrollable global trends. Despite the deleterious ecological, economic, and social impacts of tourism on indigenous communities in the African context, there have been geopolitical factors that have contributed to *"symbolic colonialism"* in the African context. From a geopolitical context, multinationals' development aggression (Baleva, 2019), global diplomatic conflicts (Neacsu et al., 2018), and geopolitical competition in Africa among global powers such as China, the US, Russia, and the EU have largely contributed to disruptions in the African tourism system. These political disruptions have negatively affected tourism-dependent countries in the Global South leading to impoverishment, social exclusion, discrimination, and destruction of indigenous peoples' individual and collective rights.

Concluding Note: Sustainable Tourism and Tourism Politics in the New Normal

Studies such as Yasarata et al. (2010) assert that in reality, political systems and power structures in society are key to understanding sustainable tourism policy development and implementation. Butler (1999) noted that the rapid global political changes have mirrored the global changes in tourism. According to Odemba and Maingi (2023), established notions and policies of sustainable tourism can only be achieved through government regulations, policies, and political support. It is critical to determine the political dimensions of sustainability to regulate sustainable tourism in destinations as evidenced by studies such as Scheyvens (2011) and Gkoumas (2019). Government support has been featured as a critical factor in engaging sustainable tourism. Tourism policy priorities in the new normal enforce "Tourism regulations" to curb the effects of unregulated tourism on the environment, communities, and local economies. However, despite national policies, there is a need to transition towards self-activating African tourism societies through integrating tourism decolonization narratives and promoting inclusive tourism in the African context. In the wake of the developmental challenges facing the tourism sectors in Africa, there is a need for a local development approach that produces integrated and indigenous solutions for addressing the needs of local people, local economies, and local environments.

References

Akama, J. S., Maingi, S. W., & Camargo, B. A. (2011). Wildlife conservation, safari tourism and the role of tourism certification in Kenya: A postcolonial critique. *Tourism Recreation Research, 36*(3), 281–291.

Andriotis, K. (2014). Tourism development and the degrowth paradigm. *Turističko poslovanje, 13*, 37–45.

Angula, H. N., Stuart-Hill, G., Ward, D., Matongo, G., Diggle, R. W., & Naidoo, R. (2018). Local perceptions of trophy hunting on communal lands in Namibia. *Biological Conservation, 218*, 26–31.

Baleva, M. K. A. (2019). *Regaining paradise lost: Indigenous land rights and tourism*. Brill Nijhoff.

Benjamin, S., & Dillete, A. K. (2021). Black travel movement: Systematic racism informing tourism. *Annals of Tourism Research, 88*. https://doi.org/10.1016/j.annals.2021.103169

Blazquez, M., Blanco-Romero, A., Gual-Carbonell, J., & Murray, I. (2019). Tourist gentrification of retail shops in Palma (Majorca). In C. Milano, J. Cheer, & M. Novelli (Eds.), *Overtourism: Excesses, discontents and measures in travel and tourism*. CABI.

Boer, D. V. (2015). Toward decolonization in tourism: Engaged tourism and the Jerusalem tourism cluster. *Jerusalem Quarterly, 65*(1), 9–21.

Brown, D. O. (2000). Political risk and other barriers to tourism promotion in Africa: Perceptions of US-based travel intermediaries. *Journal of Vacation Marketing, 6*(3), 197–210.

Burns, P. M., & Novelli, M. (2007). *Tourism and politics: Global frameworks and local realities*. Elsevier.

Butler, R. W. (1999). Sustainable tourism: A state-of-the-art review. *Tourism Geographies, 1*(1), 7–25.

Chambers, D., & Buzinde, C. (2015). Tourism and decolonization: Locating research and self. *Annals of Tourism Research, 51*, 1–16.

Cleveland, T. (2021). *A history of tourism in Africa: Exoticization, exploitation and enrichment*. Ohio University Press.

Craik, F. I. (1994). Memory changes in normal aging. *Current directions in psychological science, 3*(5), 155–158.

Creaco, S., & Querini, G. (2003, August 27th–30th). *The role of tourism in sustainable economic development*. Paper presented at the 43rd Congress of the European Regional Science Association: "Peripheries, Centres, and Spatial Development in the New Europe". https://www.econstor.eu/bitstream/10419/115956/1/ERSA2003_084.pdf

Demaria, F., Kallis, G., & Bakker, K. (2019). Geographies of degrowth: Nowtopias, resurgences and the decolonization of imaginaries and places. *Environment and Planning E: Nature and Space, 2*(3), 431–450.

Edo, S., Osadolor, N. E., & Daning, I. F. (2020). Growing external debt and declining export: The concurrent impediments in the economic growth of sub-Saharan African countries. *International Economics, 161*(1), 173–187.

Ehigiamusoe, K. U. (2019). Tourism, growth and environment: Analysis of non-linear and moderating effects. *Journal of Sustainable Tourism, 8*(1), 1174–1192.

Ekeocha, D. O., Ogbuabor, J. E., Orji, A., & Kalu, U. I. (2021). International tourism and economic growth in Africa: A post-global financial crisis analysis. *Tourism Management Perspectives, 40*, 100896.
Fang, J., Gozgor, G., Paramati, S. R., & Wu, W. (2021). The impact of tourism growth on income inequality: Evidence from developing and developed economies. *Tourism Economics, 27*(8), 1669–1691.
Fletcher, R., Blanco-Romero, A., Blanquez-Salom, M., Canada, E., Mas, M., & Sekulova, F. (2021). Pathways to post-capitalist tourism. *Tourism Geographies*. https://doi.org/10.1080/14616688.2021.1965202
Francis, C. (2010). Ubuntu and skills development in the Overberg region of the Western cape. *Skills at Work: Theory and Practice Journal, 3*(1), 27–43.
Gartzke, E., & Rohner, D. (2011). The political economy of imperialism, decolonization, and development. *British Journal of Political Science, 41*(3), 525–556.
Gkoumas, A. (2019). Evaluating a standard for sustainable tourism through the lenses of local industry. *Heliyon, 5*(11), e02707.
Hall, D. (2017). Bringing geopolitics in tourism. In D. Hall (Ed.), *Tourism and geopolitics: Issues and concepts from central and Eastern Europe*. CABI.
Hall, M. C., & Tucker, H. (2004). Tourism and postcolonialism: An introduction. In M. C. Hall & H. Tucker (Eds.), *Tourism and postcolonialism: Contested discourses, identities, and representations*. Routledge.
Hashimoto, A., Harkonen, E., & Nkyi, E. (2021). *Human rights issues in tourism*. Routledge.
Hoogendoom, G., & Flitchett, J. M. (2018). Tourism and climate change: A review of threats and adaptation strategies for Africa. *Current Issues in Tourism, 21*(7), 742–759.
Hui, Y. (2015). Modulation after control. *New Formations, 84*(84–85), 74–91.
IMF. (2022). *Regional economic outlook: Sub-Saharan Africa*. International Monetary Fund.
Kallis, G., Varvarousis, A., & Petridis, P. (2022). Southern thought, islands and real-existing degrowth in the Mediterranean. *World Development, 157*. https://doi.org/10.1016/j.worlddev.2022.105957
Karimi, M. S., Kherzri, M., & Razzaghi, S. (2022). Impacts of regional conflicts on tourism in Africa and the Middle East: A spatial panel data approach. *Current Issues in Tourism, 25*(10), 1649–1665.
Lakshmi, S. R., & Shaji, T. L. (2016). Transformation of coastal settlements due to tourism. *Procedia Technology, 24*, 1668–1680.

Maingi, S. W. (2019). Sustainable tourism certification, local governance and management in dealing with Overtourism in East Africa. *Worldwide Hospitality and Tourism Themes, 11*(5), 532–551.

Maingi, S. W. (2021). Safari tourism and its role in sustainable poverty eradication in East Africa: The case of Kenya. *Worldwide Hospitality and Tourism Themes, 13*(1), 81–94.

Maingi, S. W., & Gowreesunkar, V. G. B. (2022). Child rights and inclusive sustainable tourism development in East Africa: Case of Kenya. In H. Seraphin (Ed.), *Children in sustainable and responsible tourism.* Emerald Publishing.

Mkono, M. (2019). Neocolonialism and greed: Africans' views on trophy hunting in social media. *Journal of Sustainable Tourism, 27*(5), 689–704. https://doi.org/10.1080/09669582.2019.1604719

Mowforth, M., & Munt, I. (2016). *Tourism and sustainability: Development, globalisation, and new tourism in the third world* (4th ed.). Routledge.

Muzindutsi, P. F., & Manaliyo, J. C. (2016). Effect of political risk shocks on tourism revenue in South Africa: Time series analysis. *International Journal of Business and Management Studies, 8*(2), 169–186.

Neacsu, M. C., Negut, S., & Vlasceanu, G. (2018). The impact of geopolitical risks on tourism. *Amfiteatru Economic, 20*(12), 870–884.

Nussbaum, B. (2003). African culture and Ubuntu: Reflections of a south African in America. *Perspectives, 17*(1), 1–12.

Odemba, K., & Maingi, S. W. (2023). East African tourism policies review and crisis preparedness in pandemic situations. In M. E. Korstanje, V. G. B. Gowreesunkar, & S. W. Maingi (Eds.), *Tourism in crisis.* NOVA Science Publishing.

Palmer, C. A. (1994). Tourism and colonialism, the experience of The Bahamas. *Annals of Tourism Research, 21*(4), 792–811.

Pritchard, A., Morgan, N., & Ateljevic, I. (2011). Hopeful tourism: A new transformative perspective. *Annals of Tourism Research, 38*(3), 941–963.

Rogerson, C. M. (2016). Public-private partnerships for tourism infrastructure development: Evidence from the cradle of humankind, South Africa. *Tourism: An International Interdisciplinary Journal, 64*(4), 419–430.

Rutternberg, T. (2022). Alternatives to development in surfing tourism: A diverse economies approach. *Tourism Planning, and Development.* https://doi.org/10.1080/21568316.2022.2077420

Scheyvens, R. (2011). The challenge of sustainable tourism development in the Maldives: Understanding the social and political dimensions of sustainability. *Asia Pacific Viewpoint, 52*(2), 148–164.

Shaban, A., & Datta, A. (2019). Towards 'slow' and 'moderated' urbanism. *Economic & Political Weekly, 54*(48), 36–42. https://discovery.ucl.ac.uk/id/eprint/10092641/2/Datta_EPW_2019.pdf
Simmons, B. A. (2004). Saying the same old things: A contemporary travel discourse and the popular magazine text. In M. C. Hall & H. Tucker (Eds.), *Tourism and postcolonialism: Contested discourses, identities, and representations*. Routledge.
Sindiga, I. (1999). *Tourism and African development: Change and challenge of tourism in Kenya*. Routledge.
Tutu, D. (2011). Ubuntu: On the nature of human community. In Tutu (Ed.), *God is not a Christian: And other provocations*. Harper Collins.
Veal, A. J. (2015). Human rights, leisure and leisure studies. *World Leisure Journal., 57*(4), 249–272.
Victor, P. A. (2012). Growth, degrowth, and climate change: A scenario analysis. *Ecological Economics, 84*, 206–212.
Vong, L. T., & Ung, A. (2012). Exploring critical factors of Macau's heritage tourism: What heritage tourists are looking for when visiting the City's iconic heritage sites? *Asia Pacific Journal of Tourism Research, 17*(3), 231–245.
Witte, J., & Green, M. C. (2012). Introduction. In J. Witte & Green (Eds.), *Religion and human rights*. Oxford University Press.
Wong, I. A., McKercher, B., & Li, X. (2016). East meet west: Tourist interest in hybrid culture at postcolonial destinations. *Journal of Travel Research, 55*(5), 628–642.
World Bank. (2013). *Tourism in Africa: Harnessing tourism for growth and improved livelihoods*. The World Bank.
Yasarata, M., Altinay, L., Burns, P., & Okumus, F. (2010). Politics and sustainable tourism development—Can they co-exist? *Voices from North Cyprus, Tourism Management, 31*(3), 345–356. https://doi.org/10.1016/j.tourman.2009.03.016

3

The Shadows of Hospitality: The Figure of Tourist Gaze Reconsidered

Maximiliano E Korstanje

Introduction

Let us start this book chapter by clarifying that John Urry was an Emeritus sociologist who notably contributed to the sociology of tourism and mobilities. Based on Foucauldian legacy, he coined the term tourist gaze to connote the empescence of lay citizens who connect to a much deeper cultural matrix indicating what can be gazed at and what cannot be gazed at. This tourist gaze is not strictly associated to the Foucault medical gaze, which is used as a mechanism of control. Per Urry's viewpoint, the concept of tourist gaze [far from being an instrument of discipline—as Foucault adheres] corresponds with a quest for something outstanding and unique while shaping the tourist experience. To put the same in other terms, the tourist gaze was mainly marked by the rise of a new process known as the globalization of culture as well as the capitalist expansion which commoditized and exchanged merchandise, cultures, and people as interchangeable products. In this respect, visual

M. E Korstanje (✉)
University of Palermo, Buenos Aires, Argentina

S. W. Maingi et al. (eds.), *Tourist Behaviour and the New Normal, Volume II*,
https://doi.org/10.1007/978-3-031-45866-8_3

consumption—associated with the quest for authenticity—occupies a central position in the postmodern social imaginary (Urry, 1992, Urry & Larsen, 2011; Lash & Urry, 1993). Doubtless, this seems to be a world of high mobility where the tourist gaze dispossesses, re-appropriates, and interprets the geographical space according to a hyper-globalized and abstract economy of signs (Sheller, 2016, 2017; Tzanelli & Korstanje, 2016; Tzanelli, 2004). Having said this, the mobilities theory not only has advanced considerable and firm steps in recent years but has also consolidated as an academic leading paradigm worldwide [of course with some discrepancies and conceptual contradictions]. What is more than important to discuss, the tourist gaze and globalization are inextricably intertwined (Korstanje, 2018a). Over the recent decades, some critical voices have alerted that the explosion of tourism, which was accompanied by the multiplication of international flights and off-the-beaten-track-destinations, evinces the decomposition of social ties. Beyond the quest for novelty, for these scholars, "the Other" remains an object of fear and neglect in the Western social imagination. The world ultimately divides into two clear-cut spheres: rich tourists who can pay for their travels and asylum-seekers or labor migrants who are carefully scrutinized at airports, ports, and transport hubs (Bauman, 1996; Franklin, 2003). Quite aside from this point, the socio-economic crisis precipitated by the nineteenth-century global trade grounded not only global trade but also the tourism industry into an unparalleled halt. It is tempting to say that the post-COVID-19 context opened a valuable opportunity for slow tourism and more sustainable forms of consumption while reversing the effects of the current ecological crisis (Higgins-Desbiolles, 2020; Ratten, 2020; Hall & Seyfi, 2020).

As the previous argument is given, the present book chapter interrogates further the real and long-lasting consequences of the nineteenth century in travel behavior globally. At the same time, the chapter discusses critically the future of hospitality in a new normal where the "other" is considered an undesired guest.

In the first section, we analyze the geopolitical consequences of the Pandemic in the configuration of global and national identities. Beyond the socio-economic effects, COVID-19 ignites a new climate of re-feudalization [if not fracture] of the world that terminates with the globalization. Complementarily, the second section discusses to what extent the postmodern ethos is mutating toward a new stage. Centered on John Urry's contributions, we place the concept of tourist gaze under the critical lens of scrutiny. In former decades, tourists were estimated and valorized as ambassadors of democracy and economic prosperity. It is noteworthy that developed nations delivered their citizens who had an accommodated position to visit the four corners of the world. This resulted from two combined factors, just World War II (WWII). To the technological breakthrough deriving from innovation we need to add the excess of leisure time that was a product of the working-time reduction. In the first world citizens enjoyed from further leisure time and better salaries while foreign investors launched to invest funds in tourist enterprises geographically located in former colonies. As a result of this tourism development greased the rails of a long dormant dependency between the center and its periphery (Britton, 1982; Cohen, 1984; Korstanje, 2017). The tourist gaze reaffirms an old state of dependency mainly given by the expansion of European economies globally. This process seemed to be radically shifted just after the turn of the twentieth century when the industry was facing a set of different global risks which comprised terrorist attacks, natural disasters, and of course virus outbreaks (Korstanje et al., 2022). In these dark days, tourists are suspected or treated as potential carriers of a lethal virus while the high mobilities temporarily ceased. Last but not least, the third section is limited to focus on the end of hospitality just after the nineteenth century.

Tourism and COVID-19: A Conflicted Marriage

Many studies have focused their attention on the socio-economic effects of COVID-19 in service sectors (including tourism) (Zhang et al., 2021; Gössling et al., 2020). Those developing economies geographically situated in the global South or those economies with high dependency on international tourist demand have faced devastating consequences because of COVID-19 (Rogerson & Rogerson, 2020, 2021). The original outbreak was reported in Wuhan, China, by the end of 2019. In the question of weeks [if not months], the virus (baptized as SARSCOv2) disseminated rapidly to Europe, Latin America, Africa, and the US. COVID-19 became the global pandemic that jolted the Western nations from the lingering complacency (Khalid et al., 2021). Those economies which recovered after the stock and market crisis in 2007 witnessed a new downturn resulting from several restrictions imposed to stop the virus. Governments desperately imposed restrictive measures which comprised the closure of borders and airspaces, the cancellation of all international flights as well as strictest lockdowns that suspended the right to free transit in the national territory (Korstanje & George, 2021). For more than two years, lay citizens lived confined or simply changed their travel behavior. As stated, some studies enthusiastically applauded the idea that COVID-19 was the cornerstone toward a more sustainable form of consumption (Higgins-Desbiolles, 2020; Romagosa, 2020). To the socio-economic effects that wreaked havoc on global trade in general and tourism in particular, one might add the geopolitical tensions or disputes (among states) (Korstanje & George, 2022), the rise of chauvinist expression or separatist movements (Mostafanezhad et al., 2020; Cheer et al., 2021), the imposition of heavier travel bans or restrictions, and the increase of academic dropout rates in higher tourism education because of the higher levels of uncertainty. Many students left their careers behind, given the higher uncertainty and mistrust of the future of the tourism industry (Ye & Law, 2021; Reichenberger & Raymond, 2021). Amid this mayhem, the notion of social [ecological] justice has come to stay. This academic paradigm is rife with the belief that the expansion of tourism was mainly based on the technological revolution that greased the rails toward global capitalism. However, this expansion was achieved according to some material asymmetries [inequalities] which included the acceleration of climate change, internal wars and ethnic cleansings, political violence, and a state of economic exclusion over a whole portion of the citizenry of the under-developed nations. In part culturally rooted in the colonial past, or engulfed in the inter-class inequalities resulting from the monopoly of a privileged class of the production means. Whatever the case may be, COVID-19 ignites a type of justice [reversion] of the economic and social calamities derived from global capitalism. Per these scholars, the new normal would give a fertile climate not only for degrowth tourism but for a fairer wealth distribution worldwide (Prayag, 2020; Rastegar et al., 2021). Given the problem in these terms, some relegated voices and cosmologies are tentatively adopted as valid solutions to the current ecological crisis (Prayag, 2020). Some interesting essay reviews mentioned the urgency to introduce interdisciplinary research to expand the current understanding of proximity tourism in a world precisely without tourists (Wen et al., 2021; Liu et al., 2021). Without any doubt, the nineteenth-century pandemic has shown two important implications for the industry. On one hand, the lack of reliable instruments to predict and mitigate global risks in tourism research. On another, the failure of risk perception theory [as well as the precautionary doctrine] to design safer destinations, without mentioning the urgency of adopting new paradigms as post-disaster management to placate the aftermaths of the pandemic. The notion of risk perception has been replaced by adaptancy in academic circles (2020/05/ 2020). Part of the specialized literature overlooks the fact that probably we have to be accustomed to live with COVID-19 in the years to come. Hence, post-COVID-19 strategies in tourism destination management are at least necessary.

What Has Changed After COVID-19?

In the introductory section, we have mentioned the figure of John Urry, a senior emeritus sociologist fully dedicated to understanding the complex interplay between geography, globalization, and tourism. Urry argues convincingly that modernity has passed to a new stage [dubbed postmodernity] which mainly marks by the hegemony of the sign. Put the problem simply, the hegemony of the sign entails two important aspects of modernity. On one hand, there is an aesthetic reflexibility where the subject confronts decentralized forms of production and consumption. On another, the expansion of capitalism is centered on the re-elaboration of cultures and landscapes through a tourist gaze. In Urry, the tourist gaze engages directly with a cultural matrix that says what can be seen as a tourist attraction or plausibly avoided. Of course, the multiplication of international flights and arrivals coincides with what he knows as "the mobilities theory." Global societies are moving in a type of aesthetic saturation where the geographical borders are blurred. The act of traveling is essentially motivated by the quest for novelty, which is a cultural consequence of the rise of postmodernity. In this new ethos, as Urry adheres, nation-states leave the monopoly of identity in the market while delineating new frontiers which are subordinated to the co-dominance of spectacle and multiculturalism. Anyway, this does not mean the subject is unilaterally dominated by ideological instruments; rather, each person negotiates, re-appropriates, and rejects the consumed cultural heritage according to their own constituted biography. In this way, the net of experts divides the world into two: the safe and wild places. While the former signals international tourist destinations, the latter refers to spaces of exclusion and immobility. Urry ultimately understands that mobilities do not engage with all citizens indistinctively, rather it successfully structures a class hierarchy where some actors are doomed to a state of immobility and political oppression (Sheller & Urry, 2016; Urry, 1992, 2002, 2009). Although Urry illustrated a lot of scholars and academic circles (Franklin, 2001; Tzanelli, 2007; Larsen, 2010; Adey & Bissell, 2010; Korstanje, 2017), what is more than important to discuss, Urry's diagnosis is represented by a moment of capitalism in ongoing expansion, in which case, it is not otiose to say, he writes in a scenario pretty different than the post-COVID-19 world. This point puts a more than interesting point on the tapestry: What is changing with the COVID-19 pandemic?

The above-mentioned question is very hard to answer, at least with accuracy. The real effects of the pandemic are not only long-lasting but also far from ending (Romagosa, 2020). Dominic Lapointe (2020) alerts that the restriction measures adopted by governments led the industry into a gridlock [standstill]. The new normal seems to evince a paradoxical situation. The micro-mobilities have been expanded in view of the constraints of macro-mobilities. Proximity tourism is still an option but not the only one. The figure of the alterity has been re-signified according to new circuits of mobilities as well as the re-embedding of local economies into enclavic tourism. The transformations of the tourism industry just after COVID-19 include the creation of a new sense of alterity. The classic double alterity means those visitors who coexist with locals in specific consumerist enclaves. Nowadays these enclaves have been liberated and locals engage directly with their neighbors. At the same time, local accommodations (hospitality) have been recycled in order for health workers to be lodged. In consonance with this, Rodanthi Tzanelli (2021) proffers a more than interesting explanation. In her seminal book entitled *Cultural Immobilities and the Virocene*, she holds the thesis that "the concept of grand mobilities" originally introduced by Urry should be reconsidered in view of the COVID-19 virus outbreak. Today's capitalism is mainly characterized by countless frictions, discrepancies, and a climate of slow mobility as never before. Not surprisingly, the pandemic not only affected negatively to the tourism industry but ignited a new climate where the bio-medical discourse prevailed. Henceforth, the virocene speaks to us of a set of plagues determined by the circulation of different viruses that escape to the control of nation-states. Of course, in the virocene, the capitalist society enters in a crisis, which far from being isolated, is engulfed in a dense net of other spoken crises (such as environmental pollution, political violence, racism and xenophobia, and illegal migration). The old material inequalities that resulted in the colonial age are reaffirmed according to what Tzanelli dubbed "the colonial archive of the virocene." It is difficult to resist the impression that the virus ignites a climate of unpredictability where all political structures are amply questioned, but beyond this horizon, the medical reason [fabricated in the global North] is never confronted. To wit, the bio-discipline not only instrumentalizes the "Otherness" but also domesticates the "uncertainness." Paradoxically, the sense of immobility sparks a sentiment of recrimination [if not conspiracy] against local authorities and governments worldwide. Historically, racism was instilled to enhance internal cohesion. Tzanelli toys with the belief that COVID-19 engenders a paradoxical situation. While it allows the rise of xenophobia and hostility against foreign tourists, there are a set of unmet claims that internally ruin the legitimacy of democracies. In Tzanelli's argumentation, tourism appears to be far from disappearing, rather it is gradually mutating into a virtual form.

Last but not least, Korstanje and George describe the post-COVID-19 community as a continuation of the aftermath of 9/11. The terrorist attacks perpetrated against the US and Europe opened a climate of intolerance to bear uncertainty where "the foreigner" became an undesired guest. Terrorism not only weaponized four commercial airplanes but also instilled a sentiment of mistrust where people felt not safe anytime and any longer. To put the same bluntly, terrorism kicked off a new climate where anybody would be suspected to be a potential terrorist, a friend, a neighbor, or a colleague. COVID-19 is not a foundational event, but the reconfirmation [foreclosure] of this tendency. In a post-COVID-19 scenario, tourists are potential carriers [killers] who should be carefully identified and quarantined. Having said this, the Western sense of hospitality has been seriously affected [harmed] by terrorism sooner and COVID-19 later (Korstanje & George, 2021, 2022). The crisis of sense accelerated by COVID-19 eloquently probes the decline of hospitality, at least as we know it (Ritzer, 2007; Barbosa et al., 2021). This aspect, which is the first point of entry in this debate, will be deciphered in the next section.

The End of Hospitality in the West

It is important not to lose sight that many signs show hospitality is finally dying in the West. Tom Selwyn (2019) calls attention to the complex interplay between the Windrush Scandal and Brexit, as two significant events that mark a serious crisis in the UK. In this direction, the hostility against foreigners is still taking the lead in the British social imaginary. As Jacques Derrida (2000) puts it, hospitality and hostility share the same etymological root. Hostility activates whenever the borders of the host are vulnerated by the guest. Hostility and hospitality are inevitably entwined. Hospitality is a sacred rite that can be given or not, but in any case, the host asks "who are you and what do you want?" Hospitality is culturally enrooted in the language which unites those who are native speakers while excluding others who are not. The conditioned hospitality is given to those who can retribute for the received shelter, while unconditioned hospitality never asks for anything in return. For Derrida, conditioned hospitality is the cornerstone of modern politics and the origin of the nation-state. In this token, Daniel Innerarity (2017) equates the lack of hospitality to those persons obsessed to achieve a risk-zero society. Those nations that erect barriers against foreigners are close to being culturally exhausted. Like the guest, who suddenly knocks on the doors in the quest for shelter or food, risks are part of our daily life. Negating risks, like negating guests, lead to standardized forms of relation mainly marked by the obsession to domesticate uncertainness. This raises two significant questions: What are the clear signs hospitality is in decline? And of course, why?

Amitai Banerjee (2021) speaks to us of a "fractured society" as a neologism that explains the impacts of COVID-19 on social life. The restrictive measures not only interrupted but affected those rites celebrated in daily life to enhance the social contract. Psychologically speaking, there was an inevitable over-valorization of scientists who struggled against a common enemy [a virus] but at the same time, popular opinion was subject to contradictory actions or confusion given by errors [even from the side of experts] or media misunderstanding. COVID-19 witnessed the end of all certainness. The concept of a fractured society connotes a double dynamic. At a closer look, the rise of social pathologies derived from the higher levels of anxiety but secondly the re-drawing of the symbolic borders and identities in a grim landscape traversed by countless restrictions and confinement. As Graham Scambler (2020) notably observes, the lockdown and other restrictive steps are far from confronting the neoliberal agenda. Rather, it is a breaching experiment oriented to calibrate neoliberal governance through the employment of medical reasoning. The fractured society characterizes not only by an economic precarity of some groups but also by the increasing income inequality that enthralled populist movements. These movements create a post-national "othering" where some minorities are labeled as enemies of the state. In a nutshell, this leads citizens to too much deep sentiment of abandonment or a chasm that is filled by the media. Racism was classic in some radical voices but now it affects not only migrants and refugees but to traveling businessmen and tourists. COVID-19 as a natural experiment successfully exposes the fractures of liberal society.

In earlier works, we have discussed critically the connection between colonialism and Western hospitality. During the colonial epoch, the expansion of European powers was previously determined by the need to index new economies and cultures to the great imperial matrix. In the process, the natives were objects of fear but at the same time, of curiosity (Korstanje, 2018a). The notion of Western hospitality was certainly given by a strange paternalism where the "stranger" should be re-educated according to European cultural values. In the first wave of imperialism, European powers colonized over-seas territories putting the figure of the native as a key player in the literary genre (Korstanje, 2018b). The "non-Western Other" has a significant position not only in the literature [cultivating a new genre known as travel or captivity writing] but also in science. In the nineteenth century, ethnography and anthropology started from the premise that aboriginal cultures—as distant others—should be collected and documented earlier than their disappearance. Without any doubt, the "Other" as a co-creator of novel experiences was a fertile ground for the modern tourism industry. For the European imaginary, the outside world was an unknown territory that should be conquered, documented, and indexed to the authority of the state. Of course, the end of WWII accelerated a political decomposition in a whole portion of the European Empires while forcing mass migration of workers coming from former colonies to Europe. Meanwhile, the successful expansion of globalization not only blurred borders but the distinction between being safe here or being unsafe there. The second wave of Imperialism began with the attacks on the World Trade Center in September 2001. Plausibly, the native who had inspired different literary genres and travel in the past is now a neighbor who potentially can act as a potential terrorist. The "Other" living here not only looks to humiliate but also hates the Western lifestyle. In the days after 9/11, anybody would be suspected to be a potential terrorist, even a friend, a neighbor, or a colleague. The precautionary doctrine that emphasized the urgency to detect the potential enemies of democracy set the pace for a climate of paranoia and fear that ultimately affected negatively the sense of Western hospitality (Korstanje, 2019; Scribano, 2020). What is equally true, 9/11 caused radical transformations in Western democratic institutions prompting the end of hospitality. Needless to say, long-lasting psychological panic left by terrorism activated new supremacist discourses, protocols of surveillance, travel bans as well as xenophobia and tourist-phobia. To put the same bluntly, the stranger living here [like us] was an object of fear, an undesired guest who should be controlled (Korstanje, 2018c). The third wave of imperialism starts with the COVID-19 outbreak, an event that reaffirms the post-9/11 context. Nowadays, foreign tourists are potential disseminators of a lethal virus that places the health system in jeopardy. Having said this, the COVID-19 pandemic is a foundational event that deepened a tendency to self-cannibalize. Like the terrorist who lives within, the virus is hosted by our body exhibiting in the figure of the lockdown a vivid expression for the "Other's negation." This face of imperialism is mainly marked by low mobilities and an atmosphere of political atomization. The new normal is seen as a new opportunity for tourism rebirth but travel behavior is notably changed. Passengers are often subject to more bureaucratic travel bans and regulations whereas the classic passport as a document issued by the state set the pace for the health passport as a triumph of biotechnology which is never falsified (Korstanje, 2021a). Medicine and biotechnology played a leading role in the configuration of emerging [hybridized] identities in the new normal. In the new normal, passengers' identities are not validated by nation-state but by the grades of immunity before COVID-19. Those who are not vaccinated or inoculated are simply discriminated or—so to speak—forced to live in immobility. If in modernity capital marked the difference between rich nomad tourists and the rest, now immunity takes the lead in drawing the borders of a new geopolitical space. The decline of hospitality, as well as the neglect of the "Non-Western Other," underpins a profound crisis of sense in the West (Korstanje, 2018c, 2021b; Barbosa et al., 2021). To what extent it remains an open door toward a more just or sustainable society is a question of time.

Conclusion

Recently, the tourism industry has faced some major risks which included natural disasters, terrorist attacks, and political instability as well as the latest virus outbreak COVID-19 which paralyzed the global trade. The real effects of the COVID-19 pandemic are still uncertain and plausibly alerting part of policy-makers worldwide. Put things in these terms, experts estimate that the industry will recover in almost two years at the least. This new normal, which is based on geopolitical tensions, travel bans, and migratory crisis, exhibits not only the end of globalization but a crisis of sense in the West. John Urry coined the term tourist gaze to denote the intersection of global geography and a cultural matrix, but this position should be reconsidered. Doubtless, global tourism has been replaced by proximity tourism opening the doors to a new sustainable behavior. In this book chapter, we interrogate further the conception of sustainability and the shifts in travel behavior. In the next years, the figure of the tourist would experience considerable transformations. We have enumerated the three stages of European imperialism from the colonial epoch to date. Far from being a foundational event, the COVID-19 pandemic aligns with a tendency originated in the turn of the twentieth century.

References

Adey, P., & Bissell, D. (2010). Mobilities, meetings, and futures: An interview with John Urry. *Environment and Planning D: Society and Space, 28*(1), 1–16.

Banerjee, A. (2021). COVID-19: Fractured society and future challenges. *Industrial Psychiatry Journal, 30*(Suppl 1), S3.

Barbosa, R. B., Costa, J. H., Handayani, B., & Korstanje, M. E. (2021). The effects of COVID-19 in the tourist society: An anthropological insight of the trivialisation of death and life. *International Journal of Tourism Anthropology, 8*(2), 179–192.

Bauman, Z. (1996). From pilgrim to tourist–Or a short history of identity. In P. du Gay & H. Stuart (Eds.), *Questions of cultural identity, 1* (pp. 18–36). Sage Publications.

Britton, S. G. (1982). The political economy of tourism in the Third World. *Annals of Tourism Research, 9*(3), 331–358.

Cheer, J. M., Lapointe, D., Mostafanezhad, M., & Jamal, T. (2021). Global tourism in crisis: Conceptual frameworks for research and practice. *Journal of Tourism Futures, 7*(3), 278–294.

Cohen, E. (1984). The sociology of tourism: Approaches, issues, and findings. *Annual Review of Sociology, 10*, 373–392.

Derrida, J. (2000). *Of hospitality*. Stanford University Press.

Franklin, A. (2001). The tourist gaze and beyond: An interview with John Urry. *Tourist Studies, 1*(2), 115–131.

Franklin, A. (2003). The tourist syndrome: An interview with Zygmunt Bauman. *Tourist studies, 3*(2), 205–217.

Gössling, S., Scott, D., & Hall, C. M. (2020). Pandemics, tourism and global change: A rapid assessment of COVID-19. *Journal of Sustainable Tourism, 29*(1), 1–20.
Hall, C. M., & Seyfi, S. (2020). COVID-19 pandemic, tourism and degrowth. In M. Hall, L. Lundmark, & J. J. Zhang (Eds.), *Degrowth and tourism* (pp. 220–238). Abingdon.
Higgins-Desbiolles, F. (2020). The "war over tourism": Challenges to sustainable tourism in the tourism academy after COVID-19. *Journal of Sustainable Tourism, 29*(4), 551–569.
Innerarity, D. (2017). *Ethics of hospitality*. Routledge.
Khalid, U., Okafor, L. E., & Burzynska, K. (2021). Does the size of the tourism sector influence the economic policy response to the COVID-19 pandemic? *Current Issues in Tourism, 24*(19), 2801–2820.
Korstanje, M. E. (2017). *The mobilities paradox: A critical analysis*. Edward Elgar.
Korstanje, M. E. (2018a). *Terrorism, tourism and the end of hospitality in the 'West'*. Palgrave Macmillan.
Korstanje, M. E. (2018b). *Tracing spikes in fear and narcissism in Western democracies since 9/11*. Peter Lang.
Korstanje, M. E. (2018c). *The challenges of democracy in the war on terror: The liberal state before the advance of terrorism*. Routledge.
Korstanje, M. E. (2019). The war on terror. In *Terrorism, technology and apocalyptic futures* (pp. 63–79). Palgrave Macmillan.
Korstanje, M. E. (2021a). The COVID-19 and the figure of the undesired guest: The right to travel in scrutiny. *International Journal of Human Rights and Constitutional Studies, 8*(4), 336–349.
Korstanje, M. E. (2021b). Mining the sacred law of hospitality: Terrorism and what being native means today. In Y. Amrouni & B. Majoul (Eds.), *Poetics of the native, 66*. Cambridge Scholar Publishing.
Korstanje, M. E., & George, B. (2021). *Mobility and globalization in the aftermath of COVID-19: Emerging new geographies in a locked world.* Palgrave Macmillan.
Korstanje, M. E., & George, B. (2022). *The nature and future of tourism: A post-COVID-19 context.* CRC Press.
Korstanje, M. E., Seraphin, H., & Maingi, S. W. (2022). Tourism through troubled times: Emerging issues and new pathways for the 21st century tourism. In M. E. Korstanje, H. Seraphin, & S. W. Maingi (Eds.), *Tourism through troubled times* (pp. 1–5). Emerald Publishing Limited.

Lapointe, D. (2020). Reconnecting tourism after COVID-19: The paradox of alterity in tourism areas. *Tourism Geographies, 22*(3), 633–638.

Larsen, J. (2010). Goffman and the tourist gaze: A performative perspective on tourism mobilities. In M. H. Jacobsen (Ed.), *The contemporary Goffman* (pp. 327–346). Routledge.

Lash, S. M., & Urry, J. (1993). *Economies of signs and space* (Vol. 26). Sage.

Liu, X., Wen, J., Kozak, M., Jiang, Y., & Li, Z. (2021). Negotiating interdisciplinary practice under the COVID-19 crisis: Opportunities and challenges for tourism research. *Tourism Review., 77*(2), 484–502.

Mostafanezhad, M., Cheer, J. M., & Sin, H. L. (2020). Geopolitical anxieties of tourism: (Im)mobilities of the COVID-19 pandemic. *Dialogues in Human Geography, 10*(2), 182–186.

Prayag, G. (2020). Time for reset? COVID-19 and tourism resilience. *Tourism Review International, 24*(2-3), 179–184.

Rastegar, R., Higgins-Desbiolles, F., & Ruhanen, L. (2021). COVID-19 and a justice framework to guide tourism recovery. *Annals of Tourism Research, 91*, 103161.

Ratten, V. (2020). Coronavirus (covid-19) and entrepreneurship: Changing life and work landscape. *Journal of Small Business & Entrepreneurship, 32*(5), 503–516.

Reichenberger, I., & Raymond, E. M. (2021). Students' perceptions of tertiary tourism education and careers during the Covid-19 pandemic. *Journal of Teaching in Travel & Tourism, 21*(4), 380–402.

Ritzer, G. (2007). Inhospitable hospitality? In P. Lynch, A. Morrison, & C. Lashley (Eds.), *Hospitality* (pp. 129–140). Routledge.

Rogerson, C. M., & Rogerson, J. M. (2020). COVID-19 tourism impacts in South Africa: Government and industry responses. *Geo Journal of Tourism and Geosites, 31*(3), 1083–1091.

Rogerson, C. M., & Rogerson, J. M. (2021). African tourism in uncertain times: COVID-19 research progress. *Geo Journal of Tourism and Geosites, 38*(4), 1026–1032.

Romagosa, F. (2020). The COVID-19 crisis: Opportunities for sustainable and proximity tourism. *Tourism Geographies, 22*(3), 690–694.

Scambler, G. (2020). Covid-19 as a 'breaching experiment': Exposing the fractured society. *Health Sociology Review, 29*(2), 140–148.

Scribano, A. (2020). The other as a threat: Consequences of 9/11 for politics of sensibilities in the Global South. In M. Korstanje (Ed.), *Allegories of a never-*

ending war: A sociological debate revolving around the war on terror and 9/11 (pp. 149–166). Nova Science.
Selwyn, T. (2019). Hostility and hospitality: Connecting Brexit, Grenfell and Windrush. In B. Rowson & C. Lashley (Eds.), *Experiencing hospitality* (pp. 35–45). Nova Science Pubs.
Sheller, M. (2016). Moving with John Urry. *Theory, Culture & Society, 33*(7-8), 317–322.
Sheller, M. (2017). From spatial turn to mobilities turn. *Current Sociology, 65*(4), 623–639.
Sheller, M., & Urry, J. (2016). Mobilizing the new mobilities paradigm. *Applied Mobilities, 1*(1), 10–25.
Tzanelli, R. (2004). Constructing the 'cinematic tourist': The 'sign industry' of The Lord of the Rings. *Tourist Studies, 4*(1), 21–42.
Tzanelli, R. (2007). *The cinematic tourist: Explorations in globalization, culture and resistance*. Routledge.
Tzanelli, R. (2021). *Cultural (im)mobilities and the Virocene: Mutating the crisis.* Edward Elgar Publishing.
Tzanelli, R., & Korstanje, M. E. (2016). Tourism in the European economic crisis: Mediatised worldmaking and new tourist imaginaries in Greece. *Tourist Studies, 16*(3), 296–314.
Urry, J. (1992). The tourist gaze 'revisited'. *American Behavioral Scientist, 36*(2), 172–186.
Urry, J. (2002). *The tourist gaze*. Sage.
Urry, J. (2009). Mobilities and social theory. In B. Turner (Ed.), *The new Blackwell companion to social theory, 18* (pp. 477–495). Sage.
Urry, J., & Larsen, J. (2011). *The tourist gaze 3.0*. Sage.
Wen, J., Wang, C. C., & Kozak, M. (2021). Post-COVID-19 Chinese domestic tourism market recovery: Potential influence of traditional Chinese medicine on tourist behaviour. *Anatolia, 32*(1), 121–125.
Ye, H., & Law, R. (2021). Impact of COVID-19 on hospitality and tourism education: A case study of Hong Kong. *Journal of Teaching in Travel & Tourism, 21*(4), 428–436.
Zhang, H., Song, H., Wen, L., & Liu, C. (2021). Forecasting tourism recovery amid COVID-19. *Annals of Tourism Research, 87*, 103149.

4

Accessible Tourism: A Review of Recent Research Trends and Future Agenda

Maraísa da Silva Soares Costa,
Claudia Aparecida Avelar Ferreira,
and Mahendar Reddy Gavinolla

Introduction

As per World Health Organization (WHO), there are about 1 billion people or over 15% of the world's population with different disabilities across the globe (World Health Organization, 2020), and their exclusion can lead to a loss of 3 to 7% of countries' Gross Domestic Product (The World Bank, 2021). Persons with disabilities exhibit similar needs and desires to experience tourism as others (Yau et al., 2004). However, these individuals find it difficult to participate in tourism and leisure activities, due to the lack of accessibility and adaptation of assistance resources.

M. da Silva Soares Costa • C. A. A. Ferreira
Program Post Graduation in Administration, Pontificial Catholic University of Minas Gerais, Belo Horizonte, Brazil

M. R. Gavinolla (✉)
Department of Tourism, National Institute of Tourism and Hospitality Management, Hyderabad, India

HESPI, the Vidzeme University of Applied Sciences, Valmiera, Latvia

S. W. Maingi et al. (eds.), *Tourist Behaviour and the New Normal, Volume II*,
https://doi.org/10.1007/978-3-031-45866-8_4

Accessibility is a fundamental right for all citizens, and governments should ensure this at different levels for the sustainable development of society (Julienti et al., 2016).

Accessibility must include all people with different types of disabilities, whether physical, mental, visual, intellectual, or hearing, and not just people with reduced mobility. Accessibility is understood as the "possibility and condition of reach for the safe and autonomous use of spaces, furniture, urban equipment, buildings, transportation, information, and communication, including their systems and technologies" according to the technical standard (Brazilian Association of Technical Standards [ABNT], 2015, p. 2).

In tourism, one of the barriers for tourists who have one or more types of disabilities, such as physical, visual, or reduced mobility, refers to the architectural barrier that hinders the passage and prevents the socialization of these people. This impediment to circulation leads to the emergence and action of social movements that fight in favor of the visibility of the cause and the fulfillment of the rights and guarantees of these people, and to prevent risks and accidents. Reducing physical barriers depends on public policies and/or political interest to adapt to public and private spaces, favoring greater social participation by people with disabilities or reduced mobility (Costa et al., 2022).

Communication barriers are obstacles that make it impossible for people with sensory (visual and/or auditory) disabilities to access tourism services, making it impossible for them to equally enjoy physical spaces and social interactions (Condessa et al., 2020). Thus, the inclusion practices that involve the concern with accessibility in tourism is a challenge to overcome so that they can have the autonomy and ease of enjoying tourist sites. Providing access to people with disabilities and reduced mobility favors the inclusion of these people in tourism (environment, products, and services) on equal terms for all (Rucci & Porto, 2022), this contributes to the global goal of sustainable development.

Tourism was one of the sectors most impacted by COVID-19 (Livina et al., 2021). The pandemic has created an unprecedented challenge for accessible tourism (Streimikiene et al., 2021). There are studies carried out on various aspects of tourism for disability people in the post-COVID-19 pandemic (Moura et al., 2018). However, a detailed review of the perspectives of access for these people in tourism in the new standard is a field that has not yet been exhausted in the literature, which is an important contribution to tourism. With this background, this review text provides a synthesis of stakeholder perspectives on access to tourism for disability people, which is an important feature of quality accessible tourism. Further, it presents issues, challenges, benefits, advances, and emerging trends that broaden the future research agenda in accessible tourism. In doing so, the authors reviewed the articles published in the journals indexed by the Scopus database from 2020 to 2022, a series of suggestions are suggested for future research agendas for greater accessible tourism for disability people in aspects related to planning, development, and management.

Tourism and Accessibility: Barriers and Challenges and Opportunities

Universal accessibility in tourism is a significant opportunity to participate in tourism not only as a tourist or visitor but also enables enhanced, equal economic and employment opportunities for the disabled. The United Nations World Tourism Organization (UNWTO) is one of the important organizations in tourism, committed to the promotion and practice of accessible tourism, a concept which stemmed from the "Global Code of Ethics for Tourism, a fundamental frame of reference for responsible and sustainable tourism development." UNWTO suggests that developing accessible tourism should be a priority for all countries. UNWTO defines accessible tourism as

> A form of tourism that enables people with access requirements, including mobility, vision, hearing, and cognitive dimensions of access, to function independently and with equity and dignity. It enables this through the delivery of universally designed tourism products, services, and environments, all of which require a collaborative process among stakeholders. (World Tourism Organization, 2014)

Tourism accessibility is considered a promising solution for people with disabilities, as it supports the rebalancing of personal and social resources, collaborating positively with health and well-being. This stimulates the development of new products for tourism focused on a population that depends on certain conditions and specific policies (Moura et al., 2018).

Accessible and inclusive tourism has been much touted, but it does not only refer to architectural barriers, but it is also important to have qualified personnel to serve this public. Valente (2019) surveyed graduates of short-term tourism courses in the Azores. And among the 534 training schools, only five contemplated the theme of accessibility in tourism and inclusion of these tourists, and 60% did not have any module that addressed this public. Valente (2019) states that accessible, inclusive, and universal tourism is not enough, but universal accessibility in tourism is understood as "the creation, construction and/or adaptation of structures, infrastructures, and superstructures, as well as services, products, and goods that allow enjoyment by the greatest possible number of users, considering universality and human diversity" (p. 65). Information and communication technology (ICT) has transformed the behavior of most people in several countries over the last 30 years, and tourists with disabilities and reduced mobility have started to use these resources [internet, smartphones] for decision-making (Ferst et al., 2020) and online word-of-mouth advertising through social networks favors obtaining information about accessibility before purchasing a tour package (Ferst et al., 2020).

Methodological Framework

The study aims to understand the tourism stakeholder's perspective on underlying issues, challenges, and advantages in the tourism and hospitality industry while emphasizing the broadening of the future research agenda in accessible tourism. In doing so, the authors reviewed the articles from two scientific databases Web of Science and Scopus, but prioritized Scopus in the results part.

We first searched the literature, touching on the aspects of accessibility for tourism to the disabled and assistive technology for people with disabilities in tourism. Accordingly, a review of the scientific production on accessibility and on assistive technology for people with disabilities in tourism was carried out by reviewing the articles published from the year 2020 to 2023. One of the important reasons to select this period was to understand the progress of research toward the mentioned objective of the study in the post-pandemic era of COVID-19 or in the new normal. Through an exploratory search in the Scopus (2023) database was carried out in two phases. The keywords [tourism and accessibility and "disabled people"] were used for the research from 1960 to 2023. However, when filtering the same keywords for the period from 2020 to 2023, only 12 articles were found. Also, in the search using the keywords [tourism and "assistive technology" and "disabled people"] only three articles were found in addition and were used as Boolean operators. Scopus is considered as a standard database for indexing scientific journals that are having quality check through peer review process. Another reason is that compared to Web of Science the number of indexed sources and articles is high (Fig. 4.1 and Table 4.1). A detailed overview of the search for articles on accessibility and assistive technology in the context of tourism is provided in Fig. 4.1.

Result Discussion, Conclusion, and Implications

The analysis of the studies examined the perspectives on access to tourism for disabilty people, in terms of issues, challenges, benefits, advances, and emerging trends that broaden the future research agenda in accessible tourism. Mainly, three themes emerged. Firstly, part of the study discusses issues and challenges for people with disabilities in tourism. The second part emphasizes alternatives to improve accessible and inclusive tourism and the last one deals with tours behavior and accessible tourism. The result of the survey on accessibility, people with disabilities, and tourism is provided in Table 4.1.

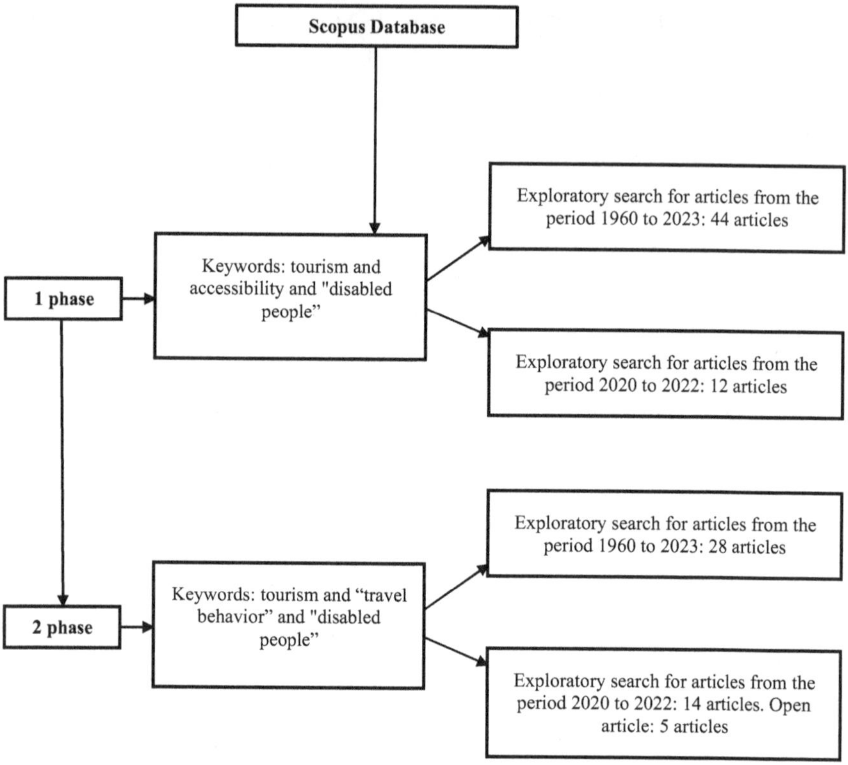

Fig. 4.1 Search for articles on accessibility and assistive technology in the context of tourism

Issues and Challenges for the People with Disabilities in Tourism

Tourists with disabilities and/or reduced mobility [elderly people, pregnant women, people with chronic diseases, among others] (Ricci & Porto, 2022), around 66.41%, can change their choice of route if there is accessibility for a better one (Ferst et al., 2020). Also, around 65.63% use social networks to choose their travel destination; 46.09% choose the trip through social networks; and 22.66% pay attention to online word-of-mouth marketing (Ferst et al., 2020).

Table 4.1 Result of the survey on accessibility, people with disabilities, and tourism

#	Author(s) and year	Article title	Kind of study	Study contribution
1	Altinay et al. (2020)	Evaluation of the Barrier-Free Tourism and Sustainability of the Barrier-Free Society in Cyprus Open access	Qualitative	It encourages education in universities, secondary and primary schools regarding barrier-free tourism and the provision of health services for people with disabilities. Also, raise public awareness of barrier-free tourism.
2	Borbála and Gábor (2020)	The Role of Nádasdy Castle in Tourism of Sárvár—The Appearance of Disabled People in Cultural Tourism	Qualitative and quantitative	Ensure accessibility for people with disabilities in tourism.
3	Cerutti et al. (2020)	Europe Without Barriers: Accessible Tourism Between Places and Projects	Qualitative	It highlights the need to consider the priorities of people with disabilities in all their diversity and to respond to the multiplicity of access needs in innovative ways. It is considered that the strategies permeate good practices that inspire public and private organizations and civil society in advancing the cause of accessible tourism for all.

(*continued*)

Table 4.1 (continued)

#	Author(s) and year	Article title	Kind of study	Study contribution
4	Somnuxpong and Wiwatwongwana (2020)	The Ability to Support Accessible Tourism in Chiang Mai, Thailand	Qualitative	The agenda of accessibility for people with disabilities must be prioritized in the political agenda, thus establishing marketing strategies for the accommodation of tourists with disabilities. The cause of accessibility will favor a greater number of tourists, but for that, information about the tourist spot must be available through mobile applications to promote equitable accessibility.

(*continued*)

Table 4.1 (continued)

#	Author(s) and year	Article title	Kind of study	Study contribution
5	Singh et al. (2021)	Compliance of Accessibility in Tourism Websites: A Pledge Towards Disability	Quantitative	It is necessary to create social awareness and understanding among tourism organizations and policymakers about the importance of web accessibility, its compliance to act in the prevention of discrimination, and its interconnection with the rehabilitation of people with disabilities. Thus, it is possible to encourage the creation of accessible web content in tourism.
6	Vignaroli et al. (2021)	The Touristic Sector in the 5G Technology Era: The 5G-TOURS Project Approach Open access	Qualitative	Use of technology for tourists with disabilities to improve remote tourism based on augmented/ virtual reality experiences.

(*continued*)

Table 4.1 (continued)

#	Author(s) and year	Article title	Kind of study	Study contribution
7	Özcan et al. (2021)	Determinants of Travel Participation and Experiences of Wheelchair Users Traveling to the Bodrum Region: A Qualitative Study Open access	Qualitative	Need for managers and employees to be attentive, socially aware, and educated with people in wheelchairs, and hotels adapted to meet this potential public.
8	Sakarneh and Katanani (2021)	Obstacles Facing Disabled People in Accessing the Historical and Archaeological Sites in Jordan Open access	Qualitative	It emphasizes the role of the government and the private sector in Jordan in improving and implementing resources so that people with disabilities are more aware of the resources offered to make destinations more accessible to them.
9	Singh and Sibi (2021)	Accessibility and Readability of Website: An Analysis of Online Travel Aggregators (OTAs) of India	Quantitative	Improve accessibility and readability for people with disabilities.

(*continued*)

Table 4.1 (continued)

#	Author(s) and year	Article title	Kind of study	Study contribution
10	Sisto et al. (2022)	Sustainable and Accessible Tourism in Natural Areas: A Participatory Approach	Qualitative	Participatory approach from design to implementation of a brand that enables people with disabilities in tourism accessibility to protected areas.
11	Tarhuni (2022)	Accessibility for Disabled People at Peshmerga Park, Erbil, Iraq	Qualitative	The need to create accessibility in Erbil parks for people with disabilities.
12	Zahari et al. (2022)	Profiling Disabled Facilities and Accessibility Provided in National Heritage Buildings in Malaysia Open access	Qualitative	Accessible facilities in historic buildings to ensure the participation of people with disabilities in social activities.

Ricci and Porto (2022) made a counterpoint that accessibility is not the only main issue for tourist destinations in Spain, a country with extensive experience in accessible tourism, but solar destinations present positive performances. Therefore, cultural destinations need to improve accessibility and access information channels before destinations are chosen. The difficulty of accessibility refers to the various barriers when the tourist prepares and plans a trip to a destination, therefore, abolishing these barriers gives these people chances to exercise their right to travel and contribute to the competitive advantages of destinations, guaranteeing universality and growth of the tourism industry.

Several constraints can limit leisure and tourism and are classified into three types: intrapersonal [related to people's skills, characteristics, and level of functioning]; interpersonal [referring to communication and social interaction]; and structural [environmental context where the tourist experiences tourism] (Ricci & Porto, 2022; Crawford & Godbey, 1987; Smith, 1987). Therefore, planners should be aware of these three types when managing the best destinations.

Interactivity has been intensified with the internet, but search engines need to be improved daily to reach all audiences. The authors Singh et al. (2021) investigated the websites of the United States of 57 states, detected serious and significant access difficulties for people with disabilities, and to remedy such problems the site must be less accessible in the WCAG 2.0 guidelines (Web Content Accessibility) and Section 508. This would make it compatible, navigable, have text options, be distinguishable, and be more adapted, so the design becomes universal and makes the rehabilitation of these individuals more accessible (Singh et al., 2021).

Accessible tourism in Slovakia is incipient, for people with disabilities who want to travel, because of the lack of attractions, structural barriers, and tourist facilities (Marčeková et al., 2021). The offer of access to tourism is low due to the lack of adaptations, with only 20% of the total or partially accessible. As the population with physical disabilities is more numerous as compared to other areas, the adjustment of supply makes the destination more attractive for these visitors, in addition to achieving better service provision; this justifies investing in Slovakia.

It is the right of tourists with disabilities and/or reduced mobility to enjoy accessible tourism, with a full meaningful experience of the trip, under equal conditions to people without disabilities. For this, it is necessary to remove physical, architectural, transport, sensory, communication, technological, and attitudinal barriers (Ricci & Porto, 2022; Ferst et al., 2020). Government websites should be audited for accessibility for all and involve external experts and the disability community to ensure they meet essential requirements and equal access (Singh et al., 2021).

Alternatives to Have and/or Improve Accessible and Inclusive Tourism

Accessible tourism after the COVID-19 pandemic is the path to sustainability because many people have morbidities after the disease (Moura et al., 2018), increasing the population that needs attention and accessible services and products. In addition, the elderly population is increasing, which tends to have greater participation in the tourism industry.

Table 4.2 presents some alternatives to have and/or improve accessible and inclusive tourism with planning and suggestions.

Table 4.2 Alternatives to have and/or improve accessible and inclusive tourism with planning and suggestions

Authors	Planning	Suggestions
Ferst et al. (2020)	Tourism and marketing professionals in planning to publicize the tourist destination.	The tourism professional must know more precisely the profile of this traveler and understand the impact of accessibility and use of the internet in the field of tourism.
Hutter et al. (2020)	Digital platforms, as alternative assistive technology resources, can provide detailed information for planning and booking hotel stays for people with disabilities.	Guests can be offered an accessible application so that they can access digital information about their stay, such as information about reservation data, accommodation, and reserved services, among others.
Singh et al. (2021)	Websites must comply with national and international accessibility standards, due to non-compliance with regulations and being audited.	DMO web developers must be trained in an effective program and aware of accessibility for all.
Marčeková et al. (2021)	Focus on the development of sustainable tourism and increase the competitiveness of the destination.	Better the offer of destinations and thus provide better services.

(*continued*)

Table 4.2 (continued)

Authors	Planning	Suggestions
Rucci and Porto (2022)	Tour agencies and operators and management of destination organizers in all countries should investigate the specific access needs of tourists with disabilities and reduced mobility, and provide reliable, specific, personalized data that is easily accessible and readily available as an inventory on infrastructure, services, and possibilities for improvement.	All tourist destinations need to work to increase access and visibility and improve ever new situations. It is important to demand public policies to enable accessibility and ensure the sustainability of cities. Ask tourists to evaluate the package or route taken after the experience.
Zabłocki et al. (2022)	It is necessary to consider accessibility in means of transport	Thinking about innovative devices and equipment that support people with this type of disability means socially including these people. The development of specialized products that facilitate different leisure activities aims to increase the participation of people with disabilities in activities in the urban space.

These are some possibilities presented for a research agenda for accessible and inclusive tourism. Reflecting on and studying these study approaches are ways to build societies focused on the accessibility and inclusion of people with disabilities.

Travel Behavior and Accessibility Affect Tourism

The regulations directed to tourists with disabilities in tours are impacted by discrimination (barriers and practices) for access to national parks (Bianchi et al., 2020). These authors question the low participation of entrepreneurial stakeholders in the dissemination of information, leaving it solely up to the public administration to take responsibility for the flow of information and exercise the essential role of implementing public policies that favor the accessibility of parks. The consequence is the limitation of the number of actors who are involved in granting accessibility for tourists with disabilities (Bianchi et al., 2020).

People with disabilities seek travel above the average of the population without disabilities, therefore, an important niche for tourism businesses, however, this public with disabilities does not like to be segregated during the trips and wants to feel included. For this, it is necessary to understand their expectations and their specific consumption habits to promote an adequate tourism offer and provide equal and equitable access to services (Gonda, 2021). Özcan et al. (2021) add to the position of Gonda (2021) that every day has been growing the number of people with disabilities and the types of disabilities, and this public still has low participation in travel, because tourism is not accessible, either in the accommodation or travel sectors (e.g., Wheelchair users). For people with physical disabilities who depend on wheelchairs, some tourist tours encounter nature barriers, sand, beach, and water, so these types of tours are not suitable. It is essential to assess the expectations and conditions of people with disabilities before offering a package tour. Because tourism is one of the most profitable sectors in expansion, accessibility to everything (hospitality, products, and services) is fundamental.

Farkas et al. (2022) argue that it is essential to research the diverse relationship between accessibility and tourism today. The travel habits of people with disabilities point out that the lack of accessibility leads to the phenomenon of crisis in tourism, because "the spirit of accessibility is simply missing from both the professional and everyday practical thinking" (p. 1), which directs to an experience of partial access and concomitant develop a negative outcome, so it is important to grant the paradigm of accessibility.

People with visual impairment who use guide dogs can have greater mobility, but suffer constraints to move around (Rickly et al., 2022). These authors evidenced that there is confidence to work the dog in the external campo of their daily life and has a positive effect in relation to the number of nocturnal trips made with the dog, and the affective qualities of the relationship influence the behavioral change of travel for adaptation to the limitations and the comfort of the guide dog.

The review raises a discussion on tourist accessibility, a core to moving the tourism industry, but it is necessary to invest and adapt the routes so that the tourist with disabilities can feel included and not segregated and have a positive experience on the trip. The lack of accessibility affects tourism for disability people, which justifies the low frequency of traveling, so it is necessary to plan focused on diversified tourism.

References

Altinay, F., Altinay, Z., Altinay, M., & Dagli, G. (2020). Evaluation of the barrier-free tourism and sustainability of the barrier-free society in Cyprus. *European Journal of Sustainable Development, 9*(4), 137–146. https://doi.org/10.14207/ejsd.2020.v9n4p137

Bianchi, P., Cappelletti, G. M., Mafrolla, E., Sica, E., & Sisto, R. (2020). Accessible tourism in Natural Park areas: A social network analysis to discard barriers and provide information for people with disabilities. *Sustainability, 12*(23), 9915, 1–14. https://doi.org/10.3390/su12239915

Borbála, G., & Gábor, W. (2020). The role of Nádasdy castle in tourism of Sárvár—The appearance of disabled people in cultural tourism. *Annales—Series Historia et Sociologia, 30*(2), 177–342. https://doi.org/10.19233/ASHS.2020.18

Brazilian Association of Standards and Techniques. (2015). ABNT NBR 9050—*Accessibility to buildings, furniture, spaces, and urban equipment.* https://docs.wixstatic.com/ugd/a107db_3708a40ba3f24192969db6500c6a3084.pdf

Cerutti, S., Falco, S., & Trono, A. (2020). Europe without barriers: Accessible tourism between places and projects. *Journal of Research and Didactics in Geography, 1*(9), 115–127. http://www.j-reading.org/index.php/geography/article/view/246

Condessa, A. M., Giordani, A. J. M., Neves, M., Hugo, F. N., & Hilgert, J. B. (2020). Barriers and facilitators to communication in the care of people with sensory impairment in primary health care: A multilevel study. *Brazilian Journal of Epidemiology, 23*, 1–14. https://doi.org/10.1590/1980-549720200074

Costa, M. S. S., Santos, N. K. F. S., & Ferreira, C. A. A. (2022). Accessibility for people with disabilities: Analysis of the cartoon genre, *Revista Expectativa, 21*(2), 73–90. https://www.researchgate.net/publication/362594051_REVISTA_EXPECTATIVA_Acessibilidade_de_pessoas_com_deficiencia_analise_do_genre_charge_Accessibility_for_people_with_disabilities_analysis_of_the_cartoon_genre

Crawford, D., & Godbey, G. (1987). Reconceptualizing barriers to family leisure. *Leisure Sciences, 9*(2), 119–127. https://doi.org/10.1080/01490408709512151

Farkas, J., Raffay, Z., & Dávid, L. D. (2022). Fundamental accessibility and technical accessibility in travels—The encounter of two worlds which leads to a paradigm shift. *Sustainability, 14*(7), 3765, 1–24. https://doi.org/10.3390/su14073765

Ferst, M. C., Anjos, S. J. G., Coutinho, H. R. M., & Flores, L. C. S. (2020). Electronic Word of Mouth (e-WOM) and choice of tourist destination for people with disabilities or reduced mobility (PcDMr). *PODIUM Sport, Leisure and Tourism Review, 9*(3), 435–461. https://doi.org/10.5585/podium.v9i3.16015

Gonda, T. (2021). Travelling habits of people with disabilities. *Geojournal of Tourism and Geosites, 37*(3), 844–850. https://doi.org/10.30892/gtg.37315-717

Hutter, H. P., Darvishy, A., Roth, S., Gaumann, S., Kaspar, H., Thimm, T., Gaiduk, M., Evans, S., & Rosenberg, M. (2020). Service design for accessible tourism. *Conference on Universal Access in Human-Computer Interaction, UAHCI 2020, held as part of the 22nd International Conference on Human-Computer Interaction, Copenhagen*, 12189 LNCS, 409–419. https://doi.org/10.1007/978-3-030-49108-6_29

Julienti, L. A. B., Mohamada, A., Abdullaha, S., Intan, N. S. S., & Hanafiaha, N. M. A. (2016). *Entrepreneurial Attitude of Disabled People Towards Edu–Tourism Industry*, Proceedings of the 2nd UUM International Qualitative researcch Conference, UM International QUALITATIVE RESEARCH CONFERENCE, 24–26 May 2016. Penang, Malaysia.

Livina, A., Bukovska, G., Abols, I., & Reddy, G. M. (2021). The recovery tactics of the tourism industry amid COVID-19 pandemic conditions in the Baltic states. In V. G. Gowreesunkar, S. W. Maingi, H. Roy, & R. Micera (Eds.), *Tourism destination management in a post-pandemic context (Tourism security-safety and post conflict destinations)* (pp. 11–23). Emerald Publishing Limited. https://doi.org/10.1108/978-1-80071-511-020211001

Marčeková, R., Šebová, Ľ., Pompurová, K., & Šimočková, I. (2021). Accessible tourism—Current state in Slovakia. *Entrepreneurship and Sustainability Issues, 9*(1), 66–86. https://doi.org/10.9770/jesi.2021.9.1(5)

Moura, A. F. A., Kastenholz, E., & Pereira, A. M. S. (2018). Accessible tourism and its benefits for coping with stress. *Journal of Policy Research in Tourism, Leisure and Events, 10*(3), 241–264. https://doi.org/10.1080/19407963.2017.1409750

Özcan, E., Güçhan Topcu, Z. G., & Arasli, H. (2021). Determinants of travel participation and experiences of wheelchair users traveling to the Bodrum region: A qualitative study. *International Journal of Environmental Research and Public Health, 18*(5), 2218. https://doi.org/10.3390/ijerph18052218

Rickly, J. M., Halpern, N., Hansen, M., & Welsman, J. (2022). Travelling with a guide dog: Trust, constraints and affective qualities of the human-guide dog relationship. *Tourism Management, 93*, 104617. https://doi.org/10.1016/j.tourman.2022.104617

Ricci, A. C., & Porto, N. (2022). Accessibility in tourist sites in Spain: Does it really matter when choosing a destination? *European Journal of Tourism Research, 31*(3108), 1–30. https://doi.org/10.54055/ejtr.v31i.2165

Sakarneh, M. A., & Katanani, H. J. (2021). Obstacles facing disabled people in accessing the historical and archaeological sites in Jordan. *Journal of Educational and Social Research, 11*(2), 267–275. https://doi.org/10.36941/jesr-2021-0045

Scopus. (2023). https://www.scopus.com/search/form.uri?display=basic#basic

Singh, R., & Sibi, P. S. (2021). Accessibility and readability of website: An analysis of Online Travel Aggregators (OTAs) of India. *E-Review of Tourism Research, 18*(5), 692–716.

Singh, R., Ismail, A., Sibi, P. S., & Singh, D. (2021). Compliance of accessibility in tourism websites: A pledge towards disability. *Journal of Hospitality and Tourism Insights, 4*(3), 263–281. https://doi.org/10.1108/JHTI-05-2020-0092

Sisto, R., Cappelletti, G. M., Bianchi, P., & Sica, E. (2022). Sustainable and accessible tourism in natural areas: A participatory approach. *Current Issues in Tourism, 25*(8), 1307–1324. https://doi.org/10.1080/13683500.2021.1920002

Smith, R. W. (1987). Leisure of disabled tourists: Barriers to participation. *Annals of Tourism Research, 14*(3), 376–389. https://doi.org/10.1016/0160-7383(87)90109-5

Somnuxpong, S., & Wiwatwongwana, R. (2020). The ability to support accessible tourism in Chiang Mai, Thailand. *African Journal of Hospitality, Tourism and Leisure, 9*(1), 1–16. https://www.ajhtl.com/uploads/7/1/6/3/7163688/article_69_vol_9_1__2020_thailand.pdf

Streimikiene, D., Svagzdiene, B., Jasinskas, E., & Simanavicius, A. (2021). Sustainable tourism development and competitiveness: The systematic literature review. *Sustainable Development, 29*(1), 259–271.

Tarhuni, S. M. (2022). Accessibility for disabled people at Peshmerga Park, Erbil, Iraq. *2022 8th International Engineering Conference on Sustainable Technology and Development* (IEC), Erbil, 199–204. https://doi.org/10.1109/IEC54822.2022.9807483

The World Bank. (2021). The inclusion of persons with disabilities is crucial for the sustainable development of Latin America and the Caribbean. https://www.worldbank.org/pt/news/press-release/2021/12/02/la-inclusion-de-las-personas-con-discapacidad-clave-para-el-desarrollo-sostenible-de-america-latin-y-el-caribe

Valente, I. T. (2019). Training and education in accessible tourism: The case of Faial, Pico and São Jorge professional schools. *Geography Notebooks, 39*, 67–74. https://www.researchgate.net/publication/333864928_Training_and_education_in_accessible_tourism_the_case_of_Faial_Pico_and_Sao_Jorge_professional_schools

Vignaroli, L., Gramaglia, M., Fuentes, M., Casella, O. R., Natale, L., & Altman, B. (2021). *The touristic sector in the 5G technology era: The 5GTOURS project approach.* In 2020 IEEE Globecom Workshop, Virtual Workshop. IEEE, 7–11. https://doi.org/10.1109/GCWkshps50303.2020.9367418

World Health Organization [WHO]. (2020). *Disability and health.* https://www.who.int/news-room/fact-sheets/detail/disability-and-healt

World Tourism Organization. (2014). In *Highlights of the 1st UNWTO Conference on Accessible Tourism in Europe* (San Marino, 19–20 November 2014). https://doi.org/10.18111/9789284417902

Yau, M. K. S., McKercher, B., & Packer, T. L. (2004). Traveling with a disability: More than an access issue. *Annals of Tourism Research, 31*(4), 946–960.

Zabłocki, M., Branowski, B., Kurczewski, P., Gabryelski, J., & Sydor, M. (2022). Designing innovative assistive technology devices for tourism. *International Journal of Environmental Research and Public Health, 19*, 14186. https://doi.org/10.3390/ijerph192114186

Zahari, N. F., Ani, A. I. C., & Rashid, R. A. (2022). Profiling disabled facilities and accessibility provided in national heritage buildings in Malaysia Open Access. *Journal of Critical Reviews, 7*(20), 109–115.

5

"Are We There Yet?": Mindful Consumption and Tourist Behavior in the Post-COVID World

Sweety Mishra and Nimit Chowdhary

Introduction

Tourism studies have always strived to explore alternative pathways to develop a tourist destination wherein the pressing concerns for environmental preservation and socioeconomic development go hand-in-hand. However, even after decades of efforts made by humans to promote sustainable consumption, we are on a path to encountering the worst-case climate change scenario (Thiermann & Sheate, 2020). Tourism is generally considered less harmful to the environment than other industries. Nonetheless, its widespread presence and sheer size have already created negative social, physical, and environmental consequences (Amara, 2010). The irresponsible exploitation and excessive development of tourist destinations can lead to their deterioration, resulting in adverse effects on the national economy and threatening the livelihoods of those reliant

S. Mishra (✉) • N. Chowdhary
Department of Tourism and Hospitality Management, Jamia Millia Islamia, New Delhi, India
e-mail: nchowdhary@jmi.ac.in; sweety2008716@st.jmi.ac.in

S. W. Maingi et al. (eds.), *Tourist Behaviour and the New Normal, Volume II*,
https://doi.org/10.1007/978-3-031-45866-8_5

on tourism. However, the world coming to a standstill during the COVID-19 pandemic provided the needed attention and opportunity to change the consumption-oriented worldview (Stankov, Filimonau, & Vujičić, 2020a). This unprecedented event made us rethink the revival of the tourism industry meaningfully. The needs and desires of people shape the current nature, character, and dynamics of the tourism industry, and a shift in the demand side of the tourism market can only bring about any significant change. Thus, the global pandemic ironically brings about that change in the tourism ecosystem and gives many consumers a place and time to consider their past and future travel behavior.

Mindfulness is a concept of engaging in moment-by-moment experiences and, at the same time, addresses the pressing issue of sustainability and sustainable consumption. The resurgence of mindfulness in tourism research requires the implementation of a sustainability agenda for tourism and coming up with new ways to incorporate the mindfulness concept into various types of tourist behavior. According to Booking.com (2021), the COVID-19 pandemic is said to have influenced people's commitment to take a more mindful approach to future tourism activities. Post-COVID-19, consumers have become more aware and careful about their insentient behavior, purchasing patterns, and improved ability to resist the promise of fictitious happiness. Although some alternative tourism markets currently offer mindfulness-driven travel, the expanding market of consumers motivated by mindfulness could spur the growth of responsible or mindful tourism development in the future (Stankov & Filimonau, 2021). In light of this, the primary objective of this study is to understand the viewpoint of urban Indian tourists and their understanding of mindfulness and mindfulness-driven tourism consumption post-COVID-19. In addition, the researchers aim to know whether the new-age travelers in India are driven by compassion and meaningful experiences at a tourist destination.

Literature Review

The concept of mindfulness has existed since ancient times and is understood in various ways. It is omnipresent. Eastern philosophical approach toward mindfulness is based on spiritual and therapeutic context, and the construct mainly focuses on peace of mind, meditation, transcendent experiences, attention, and awareness (Amel et al., 2009). The modern world, too, approaches mindfulness in the same way. As per Bishop et al. (2004), mindfulness is an approach for enhancing awareness and activating consciousness without worrying about the past and future (Kabat-Zinn & Kabat-Zinn, 2013) to effectively deal with mental problems that cause emotional suffering and maladaptive behavior (Stankov, Filimonau, Gretzel, & Vujičić, 2020b; Stankov & Filimonau, 2021).

In 2020 the sudden outbreak of the COVID-19 pandemic locked people at home. People experienced an entirely new way of life due to unexpected changes in everyday routines and freedom from social and business responsibilities, which made them anxious and stressed (Stankov & Filimonau, 2021). Due to the prevalence of poor mental health, many individuals are searching for self-help techniques, such as meditation, Pilates, and healthy living, to cope with their conditions. Several studies have been conducted during and after the COVID-19 pandemic to understand mindfulness behavior as a coping strategy to deal with stress, anxiety, fear, loss, uncertainty, or simple boredom (Weis et al., 2020; Dillard & Meier, 2021). Globally people are adopting mindfulness practices for relaxation and enjoyment.

In tourism, mindfulness is the concept of engaging in moment-by-moment experiences. At the same time, it addresses the pressing issue of sustainability and sustainable consumption of tourism products and services (Chen et al., 2017; Kirwin et al., 2019; Loureiro et al., 2019). There is widespread recognition of the positive effects of mindfulness on tourists' well-being and happiness, as well as its capacity to enrich their travel experiences (Jang et al., 2020). Mindful tourists can be defined as those who escape from their past and future, current emotions and thoughts, and live to the present moment by being attentive to the actual somatic sensations at the destination (Loureiro et al., 2019). Moreover, as per Vujičić et al. (2022), incorporating mindfulness practices into the travel industry would have far-reaching consequences. The transformational potential of mindfulness has been emphasized by an increasing number of voices stating that focusing on the present moment may help address the world's most pressing social, economic, and environmental problems (Stankov, Filimonau, & Vujičić, 2020a).

Recently, there has been a surge in the adoption of the mindfulness concept in tourism-related contexts. Numerous researchers have investigated the effects of mindfulness on sustainability and responsible tourism (Barber & Deale, 2013; Moscardo, 2017; Taylor & Norman, 2018; Stankov, Filimonau, Gretzel, & Vujičić, 2020b). In recent years, "responsible tourism" has gained widespread acceptance and endorsement from tourism destinations, industry professionals, government officials, and academic institutions everywhere (Del Chiappa et al., 2016). In their view, the idea of "responsible tourism" is better equipped to deal with the adverse effects of travel (Mondal & Samaddar, 2021). As per François-Lecompte and Prim-Allaz (2009), tourist responsible behavior refers to actions taken by visitors that benefit the local community and environment, such as supporting locally owned businesses or volunteering. Tourists' levels of sensitivity and interest in their surroundings have been used by many scholars in various fields to categorize them into several groups (Zgolli & Zaiem, 2018; Gonzalez et al., 2009). François-Lecompte and Prim-Allaz (2009) outlined five factors that define responsible tourist behavior within the tourism domain. These factors include a readiness to give up personal comfort, a preference for traveling with a responsible travel manager, an intention to preserve the local environment, a limitation on travel distance, and a desire to protect the cultural and natural heritage of visited areas. People with an interest in the environment exhibit a proclivity toward abstaining from the use of products that are deemed to be less ecologically sound (Yu et al., 2011), as well as engaging in boycotts of companies that are perceived to be acting irresponsibly (Chalamon & Nabec, 2013). These days, the destinations also focus on responsible tourism by encouraging tourists to mindful practices. Several contemporary destinations with a nature-based or new-age theme have implemented techniques to cultivate a sense of place attachment and foster more genuine lifestyles through mindfulness (Stankov & Filimonau, 2021). However, despite the positive effects of mindfulness, the tourism industry has yet to fully embrace its transformative potential (Filep et al., 2022).

Methodology

In qualitative research, purposeful sampling is used to pick information-rich instances and make the most of available resources (Patton, 2015). Since deep insight into the reasoning of travelers was desirable, the data for the study was collected by purposive sampling using semi-structured interviews with open-ended questions. Twenty respondents were chosen to respond to the questionnaire to reach the theoretical saturation point. Identifying and selecting individuals or groups with relevant experience or knowledge about a topic of interest is essential (Creswell & Clark, 2017). Apart from having experience and knowledge, the importance of being present and eager to take part and the ability to convey one's experiences and viewpoints in a well-spoken, vivid, and thoughtful manner is also essential (Palinkas et al., 2013). Therefore, the tourists-respondents are screened based on their understanding of sustainability and sustainable consumption, and also their willingness to participate in the interview for data collection.

Discussion and Analysis

Tourism and mindfulness are connected in that both can offer opportunities for spiritual growth, tranquility, relaxation, and self-awareness. During their journey, people participate in activities like yoga, meditation, and nature walks, promoting mindfulness. These activities may help people disconnect from their daily routines, reduce tension, and gain a fresh perspective on their lives (Dillard & Meier, 2021). Also, getting exposed to different cultures, customs, and ways of life through tourism may help people develop a broader perspective and foster empathy and understanding. As more and more travelers learn about the value of protecting the environment and cultural heritage, sustainable destinations and mindful practices have been growing popular (Geiger et al., 2019).

In 2015, the United Nations World Tourism Organization took significant steps toward promoting sustainable development by setting its sights on 17 sustainable goals aimed at achieving global sustainability.

The twelfth goal of sustainable development pertains to the promotion of responsible consumption and production practices. This initiative aims to underscore the imperative of advancing sustainable consumption and production practices to ensure the well-being of both present and future generations (Unwto, 2015). The sustainable development of tourist sites is largely dependent on the responsible behavior of the visitors. Visitors who act responsibly help keep natural areas, historic sites, and local cultures intact while also benefiting the local economy (Mondal & Samaddar, 2021). Tourist responsible behavior helps in achieving sustainable goals and a better future for all (Musavengane, 2019). Sustainable destinations balance the requirements of visitors and those of the local community, focusing on all aspects of visitor and local satisfaction (Saleem et al., 2020). Sustainable destinations strike this delicate balance through conscious practices consisting of ecotourism, cultural tourism, community-based tourism, and responsible tourism. These methods aim to reduce waste and pollution by giving preference to eco-friendly products, renewable energy, and low-impact means of transportation (Mondal & Samaddar, 2021; Vujičić et al., 2022). Sustainable tourism destinations can help protect native traditions and ecosystems for future generations while providing visitors with a more unique and memorable experience using these measures (Xu et al., 2018; Vujičić et al., 2022).

On the other hand, the negative way of consuming tourism products and services by tourists can cause severe damage to the destination. Negative behaviors undertaken by tourists to a destination, such as littering, vandalizing, or damaging natural or cultural landmarks, as well as neglecting local customs and traditions, are detrimental to both the destination and the host community (Matiza & Slabbert, 2021). Additionally, destination overcrowding strains the destination's infrastructure and resources (Tiwari et al., 2021), which can lead to the destruction of cultural heritage assets and a loss of authenticity (Namberger et al., 2021). Furthermore, COVID-19 has brought attention to health and safety issues. Now in the post-COVID-19 scenario, it is necessary to understand the viewpoint of urban Indian tourists and their understanding of mindfulness and mindfulness-driven tourism consumption, compassion, and meaningful experiences at a tourist destination. To achieve these objectives, the researcher framed five questions asking respondents to share their viewpoints.

S. No.	Questions
Q. 1	Do you think the way tourists consume services/products have a negative impact on the destination?
Q. 2	What, according to you, can be called the negative impacts of tourist behavior on a destination, and how are these a matter of concern to you?
Q. 3	Elaborate on your thoughts on whether tourist awareness contributes to a change in behavior and ways of consuming tourism products/ services.
Q. 4	What are your thoughts on the role of communities, tourism providers/ government in creating awareness that contributes to promoting responsible consumption behavior among tourists?
Q. 5	What are your thoughts on tourists traveling responsibly and whether or not it results in a meaningful experience at a tourist destination?

Consumption Patterns of Tourists and Their Impact on Destination

Tourism is one of the world's largest industries, and it significantly impacts the economic growth of several countries. However, travelers' consumption patterns of tourism services and products can harm the destination (Xu et al., 2018). Consumption levels in early industrialized countries, which are at the center of these developments, continue to deplete the world's natural resources much faster than their capacity to regenerate (Geiger et al., 2019). Moreover, excessive consumption is frequently associated with negative socioeconomic and cultural factors, such as loss of cultural authenticity, intolerable working conditions, pollution, water shortage, and depletion of biodiversity, which cause environmental deterioration and prevent sustainable development (Patwary, 2022). Respondents' opinions were sought to understand consumption patterns. Almost every respondent acknowledged that travelers' consumption patterns have detrimental effects on the destination. Therefore, the second probing question was to understand if the tourist's behavior negatively impacted the destination. To which the respondent voiced their opinion. Some comments are word-for-word below.

R05, Female, Age 32: The most significant adverse impacts of tourist activity on a location are the exploitation and misuse of natural resources, which can lead to environmental degradation and a scarcity of resources for the natives.

R18, Male, Age 29: Tourists leave behind trash that harms the ecosystem and disturbs the local community. Items like plastic bags and bottles do not biodegrade and add to the pollution.

R12, Male, Age 36: Littering at a tourist destination is one of the negative impacts of tourists at a destination. Additionally, tourists generate a lot of waste, including plastic and food waste which impacts the destination's environment badly.

R09, Female, Age 26: News of tourists creating a ruckus by reaching in large numbers to a particular destination and polluting the atmosphere by littering. Consequently, it creates a negative image in the minds of potential tourists and they avoid visiting a particular place.

R01, Female, Age 21: One of the negative effects of tourist behavior is when visitors disregard the culture of that community. They also choose to buy imported things and do not help the local economy by purchasing locally produced products.

Based on the analysis of the data collected, it can be deduced that nearly all respondents acknowledge that tourist consumption patterns exert an adverse effect on the destinations. According to the response, environmental degradation, pollution, littering, and leaving trash behind can pose severe health risks for humans and wildlife. Few respondents raise concern over disrespectful manners toward the culture. Previous studies also mentioned that no destination could satisfy consumers' desires. Consumers' endless appetite is detrimental to the health of the planet, society, and individuals (Csikszentmihalyi, 2000).

Tourist Awareness and Behavioral Shift

The shifts in tourist behavior and consumption patterns of tourism goods and services can be primarily attributed to the increased awareness among tourists (Lee et al., 2021). Tourist awareness improves the tourism sector, the environment, and local communities. There are several ways to increase tourist awareness, such as through education programs and communication efforts through different channels (Luo et al., 2020). Furthermore, when visitors become more environmentally concerned, they actively look for sustainable travel approaches. To cater to the needs of conscious tourists, tourism services providers are implementing sustainable practices to accommodate the growing interest in eco-friendly vacationing (Zgolli & Zaiem, 2018). Moreover, tourists who are more mindful of the consequences of their activities are more willing to take culturally and environmentally appropriate actions (Ozdemir & Kizilirmak, 2022).

R12, Male, Age 36:	Tourist awareness can make a difference, but I believe awareness programs should be promoted vigorously and reach the ground level as the habits of people need to change, who can become potential tourists in the future.
R07, Male, Age 23:	Tourist awareness plays a crucial role in shaping the way tourists consume tourism products/services. Tourists' choices can make a huge impact on the local economies and tourists can make sustainable choices to minimize negative impacts on the destination by respecting local customs and traditions and supporting local communities and the environment.
R01, Female, Age 21:	As we all know tourism contributes to the country's GDP (Gross domestic product) and helps in creating employment opportunities for many local communities as they are dependent on the tourist for survival. If we pay more attention to

this and respect each other's decisions and values on how to deal with them, represent our culture instead of just selling in front of them for earning, and know the importance of consuming tourism services help to create awareness among them.

R17, Male, Age 28: Definitely, tourists should learn about the place, the local traditions, customs & rituals before visiting the place. By doing this their acceptance in the place will increase.

R09, Female, Age 26: Yes, tourism awareness certainly contributes to changing the behavior of tourists to the stimuli of the environment and local culture. It is effective in educating people about practices that could negatively affect the environment.

Upon examining the responses, it can be inferred that tourist awareness can change the behavior of tourists. Research has indicated that tourists with a deeper awareness of sustainable tourism practices demonstrate a higher propensity to engage in environmentally conscious behaviors, such as minimizing waste, public transportation, participating in cultural activities, and encouraging local businesses (Ozdemir & Kizilirmak, 2022). The earlier studies also linked mindfulness, responsible behavior, and sustainable consumption of tourism products and services (Geiger et al., 2019).

Stakeholders' Awareness-Raising Role

Awareness of responsible consumption behavior among tourists necessitates a collective endeavor from different stakeholders such as communities, government, and tourism service providers. As part of their role in spreading knowledge, the tourism industry may adopt green policies, create responsible itineraries, and share guidelines for more eco-friendly behavior (D'Alessandro, 2016). Similarly, communities have a significant role in encouraging visitors to participate in responsible tourism by offering them chances to learn about and appreciate local culture, extend support to local businesses, and lessen their negative ecological footprint (Mathew & Sreejesh, 2017). By enforcing environmental and cultural protection rules and offering rewards for environmentally friendly business practices, governments may help spread a culture of responsibility (Hardy & Pearson, 2018). Awareness among tourists requires collaborative efforts from all parties involved. The Sustainable Development Goals (SDGs) and "Tourism for SDGs" emphasize the need to partner to promote responsible tourism and travel (Unwto, 2015).

R19, Male, Age 32:	The Government & tourism providers play an important role in creating awareness. They can develop an app which will tell you about the information of the place or timings to visit. Apart from this, the tourism providers can also distribute a kind of brochure about the place to the tourists in their language before they start the journey.
R09, Female, Age 26:	Government should try to make laws that are sustainable in nature without hampering and disturbing the regional species and landscape. The laws ought to promote environmentally and socially beneficial responsible tourism practices.
R12, Male, Age 36:	As you all know it's a major source of income and especially for the natives as the money earned from a tourist is usually spent on the development of infrastructure which helps to improve the standard of living of people and it also gives a sense of understanding about different cultures and traditions of that particular community. That's why the stakeholders make their best effort to create awareness to save tourism.
R01, Female, Age 21:	Education and awareness campaigns can be launched to inform tourists to promote responsible behavior among tourists' is also essential for

	the government to engage local communities in implementing schemes related to tourism promotion at a destination. Local communities are the best saviors of the tourist destination.
R05, Female, Age 32:	I believe they all are at the receiving end and are the first ones whom all tourists encounter. They can make a huge difference if they make tourists aware of their roles and responsibility.
R03, Male, Age 25:	Critical to examine any role anyone can play to combat this negative effect. Stakeholders majorly focus on the economy generated by this industry.

The respondent agreed that stakeholders could create awareness for tourism sustainability. The responses show that as a stakeholder, significant emphasis has been given to the government. Governments may encourage people to behave responsibly as visitors by setting up information centers at tourist destinations, running public awareness campaigns via different channels, outlining specific guidelines, and implementing severe laws (Hardy & Pearson, 2018). Moreover, in their annual reports, tourism businesses often discuss the company's efforts to be socially and ecologically responsible (Roxas et al., 2020). Prior research has shown that tourists will buy these businesses' tourism services. On the other hand, the natives can be a game changer in creating awareness when interacting with tourists (Camilleri, 2015). Few respondents, however, also mentioned that responsible tourism will lessen the tourist industry's potential to increase revenue. Therefore, the negative impact on the destination cannot be mitigated.

Responsible Traveling and Meaningful Experiences

It has been found that engaging in responsible travel practices can enhance the overall experience of tourists during and post-trip (Filep et al., 2022). The exercise of responsible tourism practices, such as waste reduction and utilization of public transportation, has the potential to strengthen the bond between tourists and the local culture and environment (Luo et al., 2020). Execution of responsible tourism practices can facilitate prospects for cultural exchange and knowledge acquisition. Tourists who show responsible behavior tend to develop a deeper understanding and appreciation for local customs and traditions by engaging in meaningful interactions, such as supporting locally owned businesses and participating in cultural events (Mondal & Samaddar, 2021). Contrarily, careless behavior on the side of visitors might result in disappointing experiences and damage the ecology and culture of the destination.

R17, Male, Age 28:	Tourists travelling responsibly & respecting the native traditions will always have a positive impact on the tourist destination. The local people & the tourists exchange their knowledge, learn about each other's traditions & culture and much more. This gives tourists immense satisfaction and a memorable experience.
R05, Female, Age 32:	Personally, I believe that travelling responsibly would lead to a much more enriching and meaningful experience as I would make meaningful connections with the locals and the destination.
R19, Male, Age 32:	Yes, it can result in a meaningful experience at a tourist destination. Also, it will encourage tourists to visit the destination again in the future.
R03, Female, Age 24:	Today's tourists are irresponsible and have destroyed natural tourism by bringing artificial fast food, vehicular movements, littering etc. This creates a negative experience for the hosts and they would not like to welcome the tourists wholeheartedly.
R12, Male, Age 36:	Sometimes some tourists are responsible for their behaviour but they get affected by the other irresponsible tourist's behaviour which gives them a negative experience.

Previous research shows that mindfulness helps people to make better choices by eliminating mindless consumption and gaining a better experience of the places visiting (Bahl et al., 2016; Loureiro et al., 2019). Responsible tourists who exercise discipline seek an appropriate balance between taking advantage of a place's unique attractions and somewhat contributing to the local economy (Mondal & Samaddar, 2021). Whereas the irresponsible behavior of tourists can generate an unfavorable ambience, leading to insecurity or discomfort among other tourists. It can lessen travelers' overall satisfaction and make other travelers tense and uneasy (Su et al., 2022). Moreover, post-pandemic people's perspectives on their own mental and physical well-being and their travel behavior have changed more positively (Matiza, 2020).

Conclusion

The tourism industry is crucial to the global economy because of the money it brings in. However, with the emphasis mostly placed on boosting the economy through tourism, many social and environmental aspects have suffered badly. This is a significant issue, and increased emphasis has been placed on sustainable and responsible tourism. Tourists are expected to be careful about how their actions affect natives and the environment. Making more thoughtful decisions about what to buy and consume can lessen environmental impact and encourage eco-friendly tourism. Tourists should exhibit mindfulness and responsible behavior when participating in any tourism-related activities. As traveling and tourism activities resume post-pandemic, tourists' inclination toward mindfulness also increases. The researcher tried to get the opinion of urban new-age travelers about the consumption pattern of tourism services and products and their impact on the destination. It also examined how mindfulness contributes to satisfaction and meaningful experiences for travelers. The responses of the study show that tourists' consumption patterns are creating a negative impact on the destination. Tourists are destroying the authenticity and culture of the destination. Littering, overcrowding, pollution, and excess consumption of natural resources are the issues at a tourist destination. The respondents believe that increasing awareness among tourists can work effectively to conserve and preserve the environment. Post-COVID tourists have grown more conscious of their impact on the destination and are taking steps to reduce their environmental footprint. Tourists are more likely to participate in environmentally responsible behavior and engage in sustainable product and service choices when they know that doing so would improve their own lives and that of others and the planet. The adoption of responsible behavior by tourists is a crucial factor in attaining sustainable tourism and enhancing the overall quality of the tourist experience. The collaborative effort of the other stakeholders is recommended for creating awareness among tourists. Thus, this chapter is an effort to bring the viewpoints of urban new-age travelers to the limelight and assist the stakeholders and policymakers in bringing in the required changes in tourism products and services. This study attempted to understand how travelers can practice mindfulness while traveling. Future research can explore visitors' responsible behavior and conscientious practices during the trip, from planning to ending.

References

Amara, D. F. (2010). Tourism as a tool of development: The case study of Siwa oasis—Egypt western desert. *WIT Transactions on Ecology and the Environment.* https://doi.org/10.2495/st100461

Amel, E. L., Manning, C. M., & Scott, B. A. (2009). Mindfulness and sustainable behavior: Pondering attention and awareness as means for increasing green behavior. *Ecopsychology, 1*(1), 14–25. https://doi.org/10.1089/eco.2008.0005

Bahl, S., Milne, G. R., Ross, S. M., Mick, D. G., Grier, S. A., Chugani, S. K., Chan, S. S., Gould, S., Cho, Y.-N., Dorsey, J. D., Schindler, R. M., Murdock, M. R., & Boesen-Mariani, S. (2016). Mindfulness: Its transformative potential for consumer, societal, and environmental well-being. *Journal of Public Policy & Marketing, 35*(2), 198–210. https://doi.org/10.1509/jppm.15.139

Barber, N. A., & Deale, C. (2013). Tapping mindfulness to Shape Hotel Guests' sustainable behavior. *Cornell Hospitality Quarterly, 55*(1), 100–114. https://doi.org/10.1177/1938965513496315

Bishop, S. R., Lau, M., Shapiro, S., Carlson, L., Anderson, N. D., Carmody, J., Segal, Z. V., Abbey, S., Speca, M., Velting, D., & Devins, G. (2004). Mindfulness: A proposed operational definition. *Clinical Psychology: Science and Practice, 11*(3), 230–241. https://doi.org/10.1093/clipsy.bph077

Booking global. (2021). *Sustainable travel report 2021*. Booking.com's 2021 Sustainable Travel Report affirms potential watershed moments for industry and consumers. https://globalnews.booking.com

Camilleri, M. A. (2015). Responsible tourism that creates shared value among stakeholders. *Tourism Planning & Development, 13*(2), 219–235. https://doi.org/10.1080/21568316.2015.1074100

Chalamon, I., & Nabec, L. (2013). Les pratiques de lecture des étiquettes Nutritionnelles: Une analyse sémiotique des représentations des règles de nutrition. *Décisions Marketing, 70*, 59–74. https://doi.org/10.7193/dm.070.59.74

Chen, L. I.-L., Scott, N., & Benckendorff, P. (2017). Mindful tourist experiences: A Buddhist perspective. *Annals of Tourism Research, 64*, 1–12. https://doi.org/10.1016/j.annals.2017.01.013

Creswell, J. W., & Clark, V. L. P. (2017). *Designing and conducting mixed method research*. Sage Publication.

Csikszentmihalyi, M. (2000). The costs and benefits of consuming. *Journal of Consumer Research, 27*(2), 267–272. https://doi.org/10.1086/314324

D'Alessandro, F. (2016). Green building for a green tourism. A new model of eco-friendly agritourism. *Agriculture and Agricultural Science Procedia, 8*, 201–210. https://doi.org/10.1016/j.aaspro.2016.02.094

Del Chiappa, G., Grappi, S., & Romani, S. (2016). Attitudes toward responsible tourism and behavioral change to practice it: A demand-side perspective in the context of Italy. *Journal of Quality Assurance in Hospitality & Tourism, 17*(2), 191–208. https://doi.org/10.1080/1528008x.2015.1115254

Dillard, A. J., & Meier, B. P. (2021). Trait mindfulness is negatively associated with distress related to covid-19. *Personality and Individual Differences, 179*, 110955. https://doi.org/10.1016/j.paid.2021.110955

Filep, S., King, B., & McKercher, B. (2022). Reflecting on tourism and Covid-19 research. *Tourism Recreation Research*, 1–5. https://doi.org/10.1080/02508281.2021.2023839

François-Lecompte, A., & Prim-Allaz, I. (2009). Les Français et le Tourisme durable: Proposition d'une typologie. *Management & Avenir, 29*(9), 308–326. https://doi.org/10.3917/mav.029.0308

Geiger, S. M., Grossman, P., & Schrader, U. (2019). Mindfulness and sustainability: Correlation or causation? *Current Opinion in Psychology, 28*, 23–27. https://doi.org/10.1016/j.copsyc.2018.09.010

Gonzalez, C., Korchia, M., Menuet, L., & Urbain, C. (2009). Comment les consommateurs socialement responsables se représentent-ils la consommation? Une approche par les associations libres. *Recherche et Applications En Marketing (French Edition), 24*(3), 25–41. https://doi.org/10.1177/076737010902400302
Hardy, A., & Pearson, L. J. (2018). Examining stakeholder group specificity: An innovative sustainable tourism approach. *Journal of Destination Marketing & Management, 8*, 247–258. https://doi.org/10.1016/j.jdmm.2017.05.001
Jang, J., Jo, W. M., & Kim, J. S. (2020). Can employee workplace mindfulness counteract the indirect effects of customer incivility on proactive service performance through work engagement? A moderated mediation model. *Journal of Hospitality Marketing & Management, 29*(7), 812–829. https://doi.org/10.1080/19368623.2020.1725954
Kabat-Zinn, J., & Kabat-Zinn, J. (2013). *Full catastrophe living: Using the wisdom of your body and mind to face stress, pain, and illness*. Bantam Books.
Kirwin, M., Harper, N. J., Young, T., & Itzvan, I. (2019). Mindful adventures: A pilot study of the outward-bound mindfulness program. *Journal of Outdoor and Environmental Education, 22*(1), 75–90. https://doi.org/10.1007/s42322-019-00031-9
Lee, C.-K., Olya, H., Ahmad, M. S., Kim, K. H., & Oh, M.-J. (2021). Sustainable Intelligence, Destination social responsibility, and pro-environmental behaviour of visitors: Evidence from an ecotourism site. *Journal of Hospitality and Tourism Management, 47*, 365–376. https://doi.org/10.1016/j.jhtm.2021.04.010
Loureiro, S. M., Stylos, N., & Miranda, F. J. (2019). Exploring how mindfulness may enhance perceived value of travel experience. *The Service Industries Journal, 40*(11–12), 800–824. https://doi.org/10.1080/02642069.2019.1600672
Luo, W., Tang, P., Jiang, L., & Su, M. M. (2020). Influencing mechanism of tourist social responsibility awareness on environmentally responsible behavior. *Journal of Cleaner Production, 271*, 122565. https://doi.org/10.1016/j.jclepro.2020.122565
Mathew, P. V., & Sreejesh, S. (2017). Impact of responsible tourism on destination sustainability and quality of life of community in tourism destinations. *Journal of Hospitality and Tourism Management, 31*, 83–89. https://doi.org/10.1016/j.jhtm.2016.10.001
Matiza, T. (2020). Post-Covid-19 crisis travel behaviour: Towards mitigating the effects of perceived risk. *Journal of Tourism Futures, 8*(1), 99–108. https://doi.org/10.1108/jtf-04-2020-0063

Matiza, T., & Slabbert, E. (2021). Tourism is too dangerous! perceived risk and the subjective safety of tourism activity in the era of COVID-19. *GeoJournal of Tourism and Geosites, 36*(2spl), 580–588. https://doi.org/10.30892/gtg.362spl04-686

Mondal, S., & Samaddar, K. (2021). Responsible tourism towards sustainable development: Literature review and research agenda. *Trends in Asia Pacific Business and Management Research*, 85–122. https://doi.org/10.4324/9781003257325-5

Moscardo, G. (2017). Exploring mindfulness and stories in tourist experiences. *International Journal of Culture, Tourism and Hospitality Research, 11*(2), 111–124. https://doi.org/10.1108/ijcthr-11-2016-0108

Musavengane, R. (2019). Small hotels and responsible tourism practice: Hoteliers' perspectives. *Journal of Cleaner Production, 220*, 786–799. https://doi.org/10.1016/j.jclepro.2019.02.143

Namberger, P., Jackisch, S., Schmude, J., & Karl, M. (2021). Overcrowding, over-tourism and local level disturbance: How much can Munich handle? In *Travel and tourism in the age of overtourism* (pp. 100–120). https://doi.org/10.4324/9781003140610-7

Ozdemir, M. A., & Kizilirmak, I. (2022). Tourism activities' role on destination choice and post-visiting behavior: Evidence from Istanbul as a cultural destination. *Tourist Behavior*, 295–321. https://doi.org/10.1201/9781003282082-14

Palinkas, L. A., Horwitz, S. M., Green, C. A., Wisdom, J. P., Duan, N., & Hoagwood, K. (2013). Purposeful sampling for qualitative data collection and analysis in Mixed Method Implementation Research. *Administration and Policy in Mental Health and Mental Health Services Research, 42*(5), 533–544. https://doi.org/10.1007/s10488-013-0528-y

Patton, M. Q. (2015). *Qualitative research & evaluation methods: Integrating theory and practice*. SAGE Publications, Inc.

Patwary, A. K. (2022). Examining environmentally responsible behaviour, environmental beliefs and conservation commitment of tourists: A path towards responsible consumption and production in tourism. *Environmental Science and Pollution Research, 30*(3), 5815–5824. https://doi.org/10.1007/s11356-022-22577-w

Roxas, F. M., Rivera, J. P., & Gutierrez, E. L. (2020). Mapping stakeholders' roles in governing sustainable tourism destinations. *Journal of Hospitality and Tourism Management, 45*, 387–398. https://doi.org/10.1016/j.jhtm.2020.09.005

Saleem, M. A., Li, J., & Afzal, H. (2020). Protect for affinity? the role of destination brand engagement in forming environmentally responsible tourist behaviours. *Journal of Sustainable Tourism, 29*(8), 1344–1364. https://doi.org/10.1080/09669582.2020.1835932

Stankov, U., & Filimonau, V. (2021). Here and now—The role of mindfulness in post-pandemic tourism. *Tourism Geographies, 25*(1), 374–389. https://doi.org/10.1080/14616688.2021.2021978

Stankov, U., Filimonau, V., & Vujičić, M. D. (2020a). A mindful shift: An opportunity for mindfulness-driven tourism in a post-pandemic world. *Tourism Geographies, 22*(3), 703–712. https://doi.org/10.1080/14616688.2020.1768432

Stankov, U., Filimonau, V., Gretzel, U., & Vujičić, M. D. (2020b). E-mindfulness—The growing importance of facilitating tourists' connections to the present moment. *Journal of Tourism Futures, 6*(3), 239–245. https://doi.org/10.1108/jtf-11-2019-0135

Su, L., Jia, B., & Huang, Y. (2022). How do destination negative events trigger tourists' perceived betrayal and boycott? The moderating role of Relationship Quality. *Tourism Management, 92*, 104536. https://doi.org/10.1016/j.tourman.2022.104536

Taylor, L. L., & Norman, W. C. (2018). The influence of mindfulness during the Travel Anticipation Phase. *Tourism Recreation Research, 44*(1), 76–90. https://doi.org/10.1080/02508281.2018.1513627

Thiermann, U. B., & Sheate, W. R. (2020). The way forward in mindfulness and sustainability: A critical review and research agenda. *Journal of Cognitive Enhancement, 5*(1), 118–139. https://doi.org/10.1007/s41465-020-00180-6

Tiwari, P., Kainthola, S., & Chowdhary, N. R. (2021). Overtourism conflicts and their resolution. In *Overtourism as destination risk* (pp. 151–165). Emerald Publishing Limited.

UNWTO. (2015). *The 17 goals | sustainable development*. United Nations. https://sdgs.un.org/goals

Vujičić, M. D., Stankov, U., & Vasiljević, Đ. A. (2022). Tourism at a crossroads—Ignoring, adopting, or embracing alternative pathways for more sustainable post-pandemic tourism development. In *Crisis management, destination recovery and sustainability*, 11–21. https://doi.org/10.4324/9781003295839-3

Weis, R., Ray, S. D., & Cohen, T. A. (2020). Mindfulness as a way to cope with Covid-19-related stress and anxiety. *Counselling and Psychotherapy Research, 21*(1), 8–18. https://doi.org/10.1002/capr.12375

Xu, S., Kim, H. J., Liang, M., & Ryu, K. (2018). Interrelationships between tourist involvement, tourist experience, and environmentally responsible

behavior: A case study of Nansha Wetland Park, China. *Journal of Travel & Tourism Marketing, 35*(7), 856–868. https://doi.org/10.1080/10548408.2018.1439429
Yu, C. P., Chancellor, H. C., & Cole, S. T. (2011). Measuring residents' attitudes toward sustainable tourism: A reexamination of the sustainable tourism attitude scale. *Journal of Travel Research, 50*(1), 57–63.
Zgolli, S., & Zaiem, I. (2018). The responsible behavior of tourist: The role of personnel factors and public power and effect on the choice of Destination. *Arab Economic and Business Journal, 13*(2), 168–178. https://doi.org/10.1016/j.aebj.2018.09.004

Part II

Environmental Sustainability and Tourist Behavior in the New Normal (SDG 7, 12, 13, 14 and 15)

6

The Willingness of Dutch Travelers to Travel Pro-environmentally Post-COVID-19

Rami K. Isaac

Introduction

COVID-19 has increased environmental awareness globally (Severo et al., 2021) as more people have started to see the relationship between climate change, pandemics, and travel which could cause a shift in demand for pro-environmental travel (Galvani et al., 2020; O'Connor & Assaker, 2021). People have started to understand that there is a link between mental and physical health and climate change (Marazziti et al., 2021). The COVID-19 global travel restrictions gave the planet a rest from tourism and time to recover, resulting in clearer waterways, and decreased Carbon Dioxide (CO2) emissions. Thus, some have concluded that tourism is one of the direct causes of environmental problems (O'Connor & Assaker, 2021).

The acceptance of environmental problems has resulted in more people changing their consumption patterns and acting more environmentally

R. K. Isaac (✉)
Breda University of Applied Sciences, Breda, The Netherlands
e-mail: Isaac.R@buas.nl

S. W. Maingi et al. (eds.), *Tourist Behaviour and the New Normal, Volume II*,
https://doi.org/10.1007/978-3-031-45866-8_6

to prevent future pandemics (Hall et al., 2020). Bernstein and Salas (2020) state that our health entirely depends on the quality of our climate. This, combined with people realizing that they are not as invincible as they might think pre-pandemic, could result in a shift to Pro-Environmental Travel Behavior (PETB). However, the insight into future Pro-Environmental Travel Behavior (PETB) during and post the pandemic is an under-researched topic in tourism literature (Anwari et al., 2021; Bratić et al., 2021).

Considering that the willingness to change depends on the perception of the harm of COVID-19, in combination with the duration of the pandemic, could have massively changed the perspectives and, therefore, the behavior of travelers (Funke & Klenert, 2020; Hall et al., 2020).

Studies claim and predict that travel behavior will change because of the impacts of the pandemic since the world has become more environmentally conscious due to COVID-19 (Severo et al., 2021). However, no research has been done specifically on Dutch traveler behaviors predicting their future pro-environmental travel behavior in this (post-)pandemic stage by using the Norm-Activation Model (NAM). NAM stipulates that "awareness of environmental problems", "ascription of responsibility for environmental problems", and the "sense of obligation toward the environment" are key determinants of individuals' pro-environmental behaviors (De Groot & Steg, 2009; Onwezen et al., 2013). Meanwhile, the Economic Sacrifices Theory asserts that "Willingness to Make Economic Sacrifices (WTES) for environmental protection" is another determining factor of individuals' pro-environmental behaviors (Hedlund, 2011), with both theories widely endorsed in the literature to predict pro-environmental behavior in both the general and tourist contexts (e.g., Han et al., 2019). Risk perception of COVID-19 (see Dryhurst et al., 2020) will also be included in this study (O'Connor & Assaker, 2021; Shin et al., 2018). Therefore, this study aims to narrow down the literature gap by examining how the COVID-19 pandemic has affected the willingness of Dutch travelers to travel pro-environmental by using a theoretical model (NAM) of O'Connor and Assaker (2021). In general and in tourism situations,

NAM and economic sacrifices theory are utilized to predict pro-environmental behavior (O'Connor & Assaker, 2021). According to Wang (2009), it is critical to get knowledge of potential changes in demand during good crisis management, such as the COVID-19 outbreak. As a result, the findings will be useful for climate policymakers, tourist operations, and destinations, as they will need to know how future pro-environmental behavior may have changed.

Literature Review

The Impact of (Health) Crises on Traveler's Behavior and Demand

COVID-19 is a significant catastrophe that has affected not only health and travel behavior but also economics (Oqubay, 2020). Travel restrictions were put in place to slow the spread of the disease. The global health crisis caused worldwide economic recessions, increased travel expenses, and disrupted border access owing to travel restrictions around the world, resulting in a major socioeconomic impact (O'Connor & Assaker, 2021; Scott et al., 2012).

Several studies have found that a major or minor crisis at a place affects travel behavior (Isaac, 2021; Isaac & van den Bedem, 2020). Post-crisis travel behavior is continually changing. This health problem is causing trends and influencing consumer behavior. The COVID-19 crisis has undoubtedly influenced future travel plans, travel behavior, tastes, and attitudes. This also occurred following past health emergencies (Hajibaba et al., 2015). Overall, the COVID-19 pandemic altered people's travel habits (Neuburger & Egger, 2020) due to new safety laws, travel limits, and economic and psychological issues, as well as the way they can travel (Bratić et al., 2021). Furthermore, greater sustainability consciousness may play a crucial influence. Many travelers postponed or canceled their travel plans during the pandemic (Bratić et al., 2021; Neuburger & Egger, 2020; O'Connor & Assaker, 2021; Orîndaru et al., 2021).

Increased Global Environmental Consciousness

The rise in global environmental consciousness may have influenced willingness and demand for sustainable solutions, making it critical for tourism companies and policymakers to understand how to adjust their (marketing) strategy to current demand (Florén et al., 2017; Orîndaru et al., 2021). Despite the increased global sustainability consciousness, it does not mean people will start acting on it. Previous research has shown a difference between perception, willingness, and behavior depending on the context (Abdelkader, 2020; Wang et al., 2020).

The timing and duration of COVID-19 result in a sustainable consciousness globally. Galvani et al. (2020) state that humans are more likely to behave more aligned with sustainability goals since they must think differently compared to pre-COVID. Considering this, the pandemic forced humanity to be more conscious about global problems, leading to businesses, governments, and individuals thinking differently. Ever since COVID-19 became a global pandemic, it made the world think about and work together to get a solution and take responsibility; this helped to create a global mind (Galvani et al., 2020). Galvani et al. claim that the increased global environmental consciousness can reflect on tourism. The planet benefited from the absence of travelers caused by travel restrictions. Not only can an underlying health condition, combined with the novel virus, kill, but it also makes most aware that one factor (in this case, a virus) could change "normal day-to-day" life completely, even if you are from a wealthier country. Corona made people worry more about society and themselves (Bavel et al., 2020; Uehara & Sakurai, 2021).

Willingness to Make Economic Sacrifices

The economic sacrifices theory has been used to predict future sustainable behavior in tourism better. The individual's Willingness to Make Financial Sacrifices (WTES) or pay a higher price to protect the environment is another predictor of making sustainable choices or behaving pro-environmentally (Batool et al., 2021). Studies have shown a positive relationship between environmental concerns and the willingness to pay more for sustainable solutions (Wang et al., 2020). The willingness to pay for environmental solutions has been researched in recent years. Most people (61%) Dutch people are willing to pay extra for sustainable products. However, the willingness to pay more for environmental solutions depends on the context (DPG Media, 2019). According to The Press Group (DPG) Media's study (2019), such as electronics, clothing, transport, and energy, the Dutch population was least willing to pay more for sustainability when it comes to travel compared to the other segments. The research confirms that most people focus on sustainable products when it comes to food. ABN Amro's study has similar results (Swart, 2020). According to the research from More in Common (2021), when it comes to travel, 34% are willing to pay higher taxes on gas or own a car to protect the environment. However, research has yet to be conducted regarding the willingness to pay more for environmental travel solutions during and post-COVID-19.

The Norm-Activation Model

The Norm-Activation Model (NAM) is a popular and widely utilized theory in both general (e.g., De Groot & Steg, 2009; Onwezen et al., 2013) and tourism literature (e.g., Han et al., 2015, 2019; Shin et al., 2018). NAM claims that three basic components, generally called moral or personal norms, can predict pro-environmental behavior: awareness of consequences/problems, attribution of responsibility, and moral obligation. In this study, adapted from O'Connor and Assaker (2021), I refer to (a) a person's awareness of the negative consequences of his or her actions on the environment as "environmental concerns" (Han et al., 2015), (b) a person's ascription of responsibility for the environmental consequences of his or her actions as "environmental responsibility" (Han et al., 2015; De Groot & Steg, 2009), and (c) a person's sense of obligation on his and other people's part to take or refrain from taking actions to help preserve/protect the environment as an "environmental moral obligation" (Han et al., 2019). The primary goal of NAM is to establish a sequential relationship between its variables and pro-environmental behavior (i.e., awareness of consequences, the ascription of responsibility, moral obligation, and pro-environmental behavior), on the assumption that a person is more likely to be aware of negative environmental consequences before assuming/assigning responsibility for such consequences. This instills a sense of moral commitment to the environment, leading to pro-environmental behavior (De Groot & Steg, 2009; Han et al., 2019; Onwezen et al., 2013).

Environmental Concerns, Environmental Responsibility, and Environmental Moral Obligation

Pro-environmental behavior refers to the behavior that can minimize the negative impact of individuals' activity on the environment as well as can benefit the environment (Steg & Vlek, 2009; Wu et al., 2021). Likewise, the pro-environmental travel behavior (PETB) indicates tourists' likelihood to choose the frequency of travel, the destination, the mode of transportation, the tourism products, and the activities during the travel for environmental protection (O'Connor & Assaker, 2021; Ahmed et al., 2020). For instance, O'Connor and Assaker (2021) considered the Norm-Activation Model (NAM) and economic sacrifices theory to study the effect of COVID-19 on PETB.

One stream of research has considered pro-environmental behavior (PEB) as a rational decision-making process, while another has viewed it from a moral perspective. Several theories can serve as the basis for the study of environmentally sustainable tourist behavior, including the NAM (Schwartz, 1977), theory of reasoned action (Ajzen & Fishbein, 1980), model of responsible environmental behavior (Hines et al., 1987), theory of planned behavior (Ajzen, 1991), and value-belief-norm theory (Stern, 2000). These have been applied to determine the antecedents of an individual's PEBs (Kaiser et al., 2005; Moghimehfar & Halpenny, 2016) and have discovered various factors that influence the intention to perform PEBs in different contexts. There are also some concerns about the application of PEBs related to their measurement (Juvan & Dolnicar, 2016). The literature contains several methods and techniques for investigating people's PEBs, such as quantitative, experimental, and qualitative approaches.

A recent study by Booking.com (2021) has shown that over 61% of people globally state that the pandemic made them want to travel more sustainably. Not only are people willing to travel more sustainably, but it has also been translated into actions. Within a year, this represents a massive change compared to the outcomes of the study conducted by de Haas et al. (2020) and Gray and Jackson (2020). According to Booking.com, the responsibility to take more sustainable trips in the future shows the readiness to change. A shift can be recognized in considering travelers' ecological footprints (Booking.com, 2021). The norm changes will lead to normalizing a minimalistic impact on the planet (Groen, 2021). According to Marianne Gybels, Director of Sustainability for Booking.com, "*Over the six years, we have been conducting this research, it has been inspiring to see awareness of the importance of sustainable travel consistently grow, both with our customers and now with our partners, too*" (Booking.com, 2021, p. 1).

Risk Perception of COVID-19 and Travel Behavior

The pandemic increased awareness of climate change's risks and consequences (Dryhurst et al., 2020). Pandemics are a significant issue for society and tourism, which pre-COVID were often ignored or forgotten by travelers, the tourism industry, or politicians (Hall et al., 2020). Nevertheless, the perceived risk can highly affect the intention to change travel plans, choose destinations, or avoid traveling to a destination (Neuburger & Egger, 2020). Neuburger and Egger (2020) have shown that perceived risk highly affects if people will travel at all. Therefore, risk perception is applied in this study as it will impact people's travel behavior (Anwari et al., 2021).

This perceived future risk of climate change could change people's behavioral willingness to travel more eco-friendly during and after post-COVID-19 (Gibbons, 2020; Gray & Jackson, 2020). Climate change is already generating a sense of fear as COVID increases health concerns. A study by Bratić et al. (2021) confirms that the current change in behavior has to do with two reasons relating to this fear. Firstly, the imposed travel restrictions discourage people from traveling, even if they are willing to travel. Secondly, travel is the cause of the continuous spreading of the virus and its variants, bringing health concerns. These health concerns make traveling less appealing. Consequently, part of risk perception goes together with fear. Humans are generally more scared of something if it is new, and the more uncertain the situation is, the more afraid people are. Therefore, the survey for this study includes questions related to fear (Gray & Ropeik, 2002).

Behavioral Willingness

Behavioral willingness is mainly about the willingness to engage in perceived behavior. Behavioral willingness reflects an "individual's openness to opportunity, that is, his or her willingness to perform a certain behavior in situations" (Pomery et al., 2009, p. 896). For several reasons, behavioral willingness is the best way of measuring actual future behavior compared to behavioral intentions and expectations. The first reason is that behavioral willingness does not directly focus on the individuals' consequences of specific behavior. Moreover, it focuses less on the pre-contemplation phase, which is more relevant as it is already known that people did become more environmentally aware (O'Connor & Assaker, 2021; Pomery et al., 2009). Secondly, willingness has been shown as a better predictor of actual behavioral decisions than behavioral intentions (Hammer & Vogel, 2013). This is more relevant for the tourism industry as this gives the best insight into actual future behavior rather than predictions and shows the best-case version of humanity. Willingness takes the unintentional, reactive component into account, just as the social circumstances, other than rational intentions (Gibbons, 2020). Moreover, behavior willingness will test if people will engage in a particular behavior even if they have not intended to do so.

Methodology

The circumstances and timing of data collection in the Netherlands are critical in the COVID-19 literature. From 8 July to 29 July 2021, primary data has been collected online via a survey. Due to extraordinary and rapid changes surrounding the virus, the outcome of this study could have been slightly different if it was completed sooner or later; this could impact projected travel behavior (de Haas et al., 2020; Funke & Klenert, 2020).

The survey was promoted on social media platforms such as LinkedIn, YouTube, Facebook, Instagram, and WhatsApp. These social media channels were chosen because they provide the opportunity to contact people from various walks of life in the Netherlands. A request to participate has been issued in multiple Facebook groups and beneath several posts on Instagram, YouTube, and Facebook. Furthermore, those working and waiting for family or friends at various vaccination locations in the Netherlands have been asked to complete the survey. They received an email with a link or scanned a QR code. Convenience sampling is employed to obtain as many respondents as possible (Fink, 2021). However, one sample criterion was taken into account. The respondent must be Dutch and over the age of 18, as this is the research's target demographic. The total number of responses received is 329.

Data Analysis and Survey Instrument

The questionnaire has seven distinct sections: (1) risk perception of COVID, (2) environmental concerns, (3) environmental responsibility, (4) environmental moral obligation, (5) WTES for environmental protection, (6) PETB, and (7) socio-demographic characteristics. These dimensions are all used to determine if the COVID-19 pandemic has influenced future PETB. The questionnaire was used in a similar case study adopted by O'Connor and Assaker (2021). All items were measured using a 5-point Likert scale ranging from 1 (strongly disagree) to 5 (strongly agree). The scales used are based on the literature and adopted from O'Connor and Assaker's study (2021).

Questionnaires were utilized in multiple comparable case studies to answer research questions and predict PETB (Booking.com, 2021; Juvan & Dolnicar, 2016; Miller et al., 2014; Schneider et al., 2017). Other literature that includes the NAM (De Groot & Steg, 2009; Onwezen et al., 2013; Park et al., 2018; Shin et al., 2018), the economic sacrifices theory (Brieger, 2018; Hedlund, 2011) or risk perception (Bratić et al., 2021; Dryhurst et al., 2020; Isaac & van den Bedem, 2020; Neuburger & Egger, 2020) used a questionnaire as well.

Results

Socio-Demographics of the Respondents

In this study, 32% (N = 106) of the respondents were male, and 68% (N = 223) were female. One person is non-binary, and one prefers not to mention their gender. There is an over-representation of young people between 18 and 30; they make up 56% (N = 189) of the total participants. Overall, 62% (N = 207) of the respondents obtained a degree, either a bachelor's or a master's. Of all the respondents, 40% (N = 125) have an annual household income of more than 40.000 euros. Finally, 36% (N = 117) had at least one child or more.

Risk Perception

The results indicate that risk perception needs to be more generalizable as the results differ per statement. Respondents were asked to what extent do you agree or disagree with the following statements. The means vary from 4 (affect future travel plans) to 1.6 (worried I might die from COVID-19). For all statements, the lowest given score is 1, and the highest given score is 5, except for one statement. Most statements' standard deviation (SD) was higher than 1 (0.8 1x, 1,1 4x, and 1.2 3x), indicating that the participants had varying opinions. However, out of all given answers to these statements regarding risk perception, somewhat agree is mainly selected (875 times), followed by somewhat disagreed (611 times). It is shown in Fig. 6.1 that all answers per statement varied.

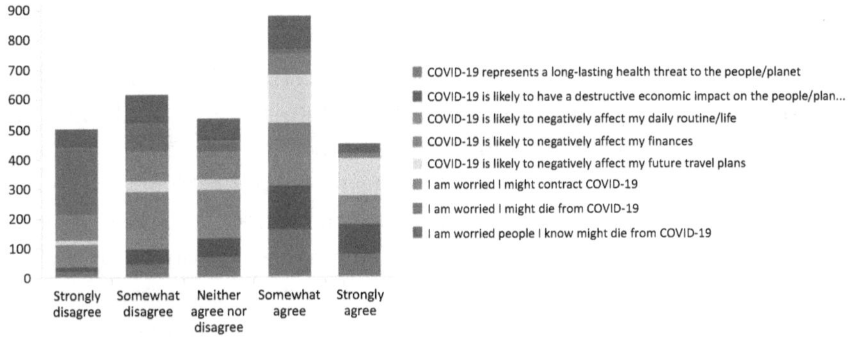

Fig. 6.1 Risk perception

Out of all the risk-perception-related statements, people mostly agree with the statement: 'COVID-19 is likely to affect my future travel plans (M 4) negatively'. Followed by 'COVID-19 is likely to have a destructive economic impact on the people/planet' (M 3.7). People are least worried that they will die from COVID-19 (M 1.6), just as they are not worried about getting infected (M 2.5).

Most respondents somewhat agree (43%, N = 159) or strongly agree (21%, N = 76) that 'COVID-19 represents a long-lasting health threat to the people and planet'. Two-thirds of the respondents find that 'COVID-19 is likely to have a destructive economic impact on the people/planet'. About 57% (N = 212) of the participants find that 'COVID-19 is likely to affect their daily routine/life negatively'. Most respondents disagreed (34%, N = 127) with the statement if 'Covid-19 would negatively affect their finances' followed by neither agree nor disagree (24%, N = 89).

Most people (N = 288), 77% in total, agree with the statement that 'COVID-19 will negatively affect their future travel plans'. On average, about 23% (N = 86) of people fear becoming infected with the virus. However, 23% (N = 87) are not afraid and strongly disagree with the statement. The mean for the answer is 2.5, and the SD is 1.2, meaning that the respondents had varying opinions regarding these statements.

Environmental Concerns

Four statements regarding environmental concerns have been given that could be answered. The results indicate that the environmental concerns are generalizable for these respondents in the research as the SD is equal to 1 or 0.9, which means that the participants did not have many varying opinions. For all statements, the lowest given score is 1, and the highest given score is 5. Most respondents somewhat agree or strongly agree with most statements, as shown in Fig. 6.2. The data suggests that most people are worried about the world. There is no significant difference in the answers per statement (M varying from 4 to 4.1).

For 79% (N = 275) of all participants, environmental protection is a critically important global issue. About 74% (N = 258) of all participants claim that they somewhat agree (41%, N = 143) or strongly agree (33%, N = 115) with the statement: I am concerned about the amount of environmental destruction. Secondly, 75% (N = 164) of respondents are concerned about environmental destruction. Finally, 72% (N = 255) (strongly) agree with the statement that if people continue interfering with nature, it will have disastrous consequences for the planet and people.

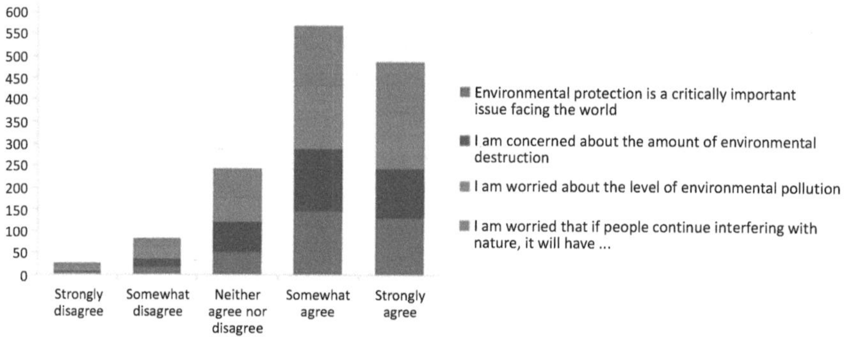

Fig. 6.2 Environmental concerns

Environmental Responsibility

In Fig. 6.3, two statements regarding responsibility have been given that could be answered. The results indicate that environmental responsibility is generalizable for the research sample. The findings clearly show that most Dutch travelers agree or strongly agree that environmental responsibility is a task for society. For all statements, the lowest given score is 1, and the highest given score is 5. The SD for the first statement is 0.8, and the second one is 1, indicating that the opinions within the sample did not differ much. Most of the sample (88%) somewhat agree (48%, N = 162) or strongly agree (40%, N = 137) with the statement that 'every person must take responsibility for the environment' (Fig. 6.3). Also, 84% (N = 230) of the respondents feel partly responsible for the environmental problems on the planet. Nevertheless, more people agree that every person must take responsibility (M 4.2) compared to feeling partly responsible for our planet's environmental problems (M 3.8).

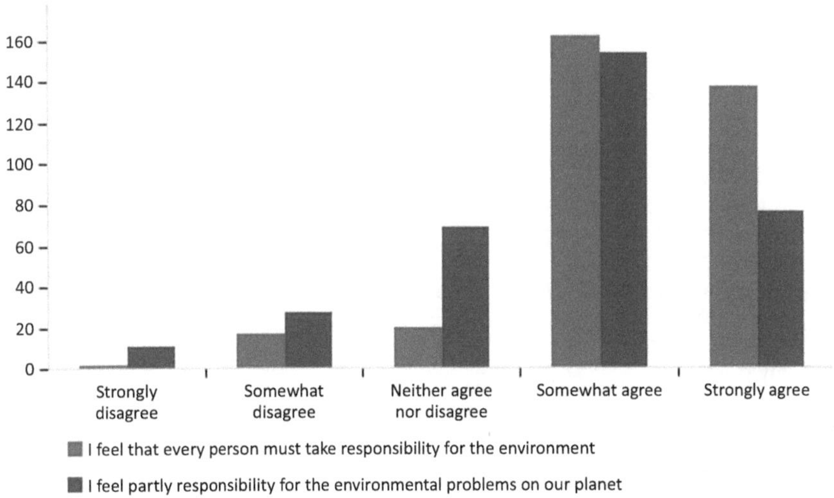

Fig. 6.3 Environmental responsibility

Environmental Moral Obligations

Four statements regarding responsibility have been given that could be answered from strongly disagree to agree (Table 6.1). The results regarding this topic are generalizable for the sample. All statements have an SD lesser or equal to one—indicating that the respondent's opinions did not differ much. Overall, most people somewhat agree with all statements that have to do with moral obligations. Only a tiny part of the respondents strongly disagreed with the statements regarding this topic. Even though the data suggests that people feel morally obligated to protect the environment (M 3.8), fewer participants also show that they imply

Table 6.1 Moral obligation toward the environment

#	Field	Minimum	Maximum	▲ Mean	Std Deviation	Variance	Count
4	Because of my own values/ principles, I believe people in general should act on a moral obligation to protect the environment	1.0	5.0	3.8	0.9	0.8	339
1	I have a moral obligation to protect the environment	1.0	5.0	3.8	0.9	0.8	340
2	I have a moral obligation to be environmentally responsible in my purchases and consumption	1.0	5.0	3.7	0.9	0.9	339
3	I have a moral obligation to be environmentally responsible in my travel end leisure activities	1.0	5.0	3.5	1.0	1.1	339

actions such as buying environmentally friendly and traveling (M 3.7 and 3.5). Compared to all statements, people tend to feel the least obligated to consider the environment during travel or leisure activities, as shown in Table 6.1. Overall, women tend to feel higher moral obligations to be environmentally responsible than men. This also shows in the moral obligation toward being environmentally responsible while traveling (female M 3.6 and male M 3.4).

Willingness to Make Economic Sacrifices (WTES)

Three statements have been given to the respondents. The results indicate that the willingness to pay is not generalizable as the answers to the statements are relatively spread out. However, somewhat agree has been selected chiefly (313 times, 33% of the time), followed by somewhat disagree (248 times, 26% of the time), and neither agree nor disagree (236 times, 25% of the time), then strongly disagree (98 times, 10%) and strongly agree (62 times, 6% of the time), which means that results per respondent can vary.

The data suggests that 32% (N = 103) of the respondents are not willing to pay much higher prices to protect the environment, 27% (N = 85) do not agree nor disagree, and 41% (N = 131) somewhat agreed to disagree with the statement (see Fig. 6.4). Secondly, 38% (N = 122) of the respondents are unwilling to accept a lower living standard to protect the environment. Finally, 38% (N = 121) of the respondents are not willing to pay higher taxes to protect the environment, 25% (N = 80) neither agree nor disagree with this statement, and 37% (N = 118) of the respondents are willing to do so. Nevertheless, strongly agree remains the least popular answer to all statements regarding the WTES.

Women are more willing to make economic sacrifices compared to men. Almost 50% of all men (strongly) disagreed with the statement to pay higher taxes for environmental protection. At the same time, 33% of the women (N = 70) tend to (strongly) disagree. Then 42% of all males (N = 101) (strongly) disagreed with the statement that they are willing to accept a lower living standard. In comparison, 37% (N = 79) of all females (strongly) disagreed with the same statement.

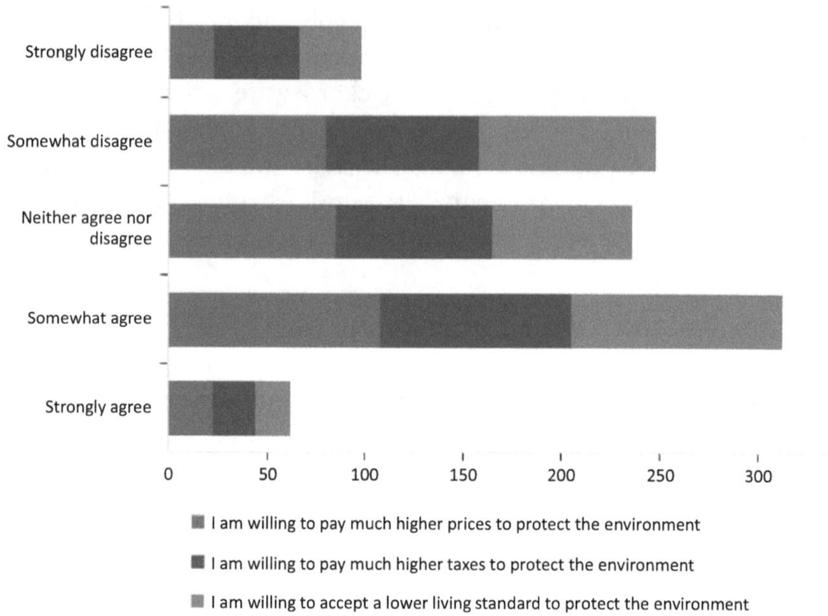

Fig. 6.4 Willingness to make economic sacrifices

Pro-Environmental Travel Behavior (PETB)

In Table 6.2, eight statements were given to respondents that can be answered from disagree to strongly agree. The outcome of these statements is not generalizable, as almost all statements have an SD of > 1. The mean to all statements varies from 3.4 to 2.6, indicating that it depends on the context if people are willing to change their behavior. However, the data suggests that out of all statements, people are most willing to purchase environmentally friendly tourism products, if possible, to avoid adverse environmental effects (M 3.4, N = 181); 56% tend to agree or strongly agree. However, compared to the other statements, the sample is the least likely to take fewer vacations to avoid causing adverse environmental effects (M 2.6), as 53% (N = 170) out of all respondents disagreed or strongly disagreed with this statement. Out of all respondents, 3% strongly disagreed with the statement regarding taking fewer holidays for environmental protection. Although most people are not planning on

Table 6.2 Pro-environmental travel behavior

#	Field	Minimum	Maximum	▲ Mean	Std Deviation	Variance	Count
8	I will purchase environmentally friendly tourism products if possible to avoid causing negative environmental effects	1.0	5.0	3.4	1.0	1.1	318
4	I will choose environmentally sustainable modes of transportation if possible to avoid causing air pollution	1.0	5.0	3.3	1.1	1.3	318
7	I will not participate in environmentally damaging vacation activities to avoid causing negative environmental effects	1.0	5.0	3.2	1.1	1.3	317
6	I will use environmentally certified tourism providers if possible to avoid causing negative environmental effects	1.0	5.0	3.2	1.1	1.1	318

(*continued*)

Table 6.2 (continued)

#	Field	Minimum	Maximum	▲ Mean	Std Deviation	Variance	Count
3	I will visit environmentally friendly tourism destinations to avoid causing negative environmental effects	1.0	5.0	3.1	1.1	1.2	319
2	I will vacation closer to home if possible to minimize transportation-related greenhouse gas emissions	1.0	5.0	2.9	1.2	1.6	319
5	I will purchase carbon offsets when I travel to a destination by plane to compensate carbon emissions from flying	1.0	5.0	2.8	1.2	1.4	318
1	I will take fewer vacations to avoid causing negative environmental effects	1.0	5.0	2.6	1.1	1.2	319

taking fewer holidays, 47% (N = 150) of the respondents will not participate in environmentally damaging vacation activities to avoid causing adverse environmental effects.

Although most people are not planning on taking fewer holidays, 47% (N = 150) of the respondents will not participate in environmentally damaging vacation activities to avoid causing negative environmental effects. In addition, 42% (N = 134) of the participants will use certified tourism providers, if possible, to avoid causing negative environmental effects. Besides, 42% (N-134) of respondents will visit environmentally friendly tourism destinations to avoid causing negative environmental effects.

While being on holiday (see Fig. 6.5), 51% (N = 165) of the sample would choose environmentally sustainable modes of transportation, if possible, to avoid causing air pollution (M 3.3). However, only 32% (N = 103) of all respondents are willing to purchase carbon offsets when traveling to a destination by plane to compensate for carbon emissions from flying. Additionally, 45% (N = 144) of the respondents (strongly) disagreed with the statement that I would vacation closer to home, if possible, to minimize transportation-related greenhouse gas emissions. The minority of the participants (9%, N = 27) strongly agreed with this statement.

Based on these answers (see Fig. 6.5), females are more likely to travel sustainably than males. Out of all respondents, only 1% strongly agreed with the statement: I will take a few vacations to avoid causing negative environmental effects. Whereas males are out of all options, they are most likely to purchase environmentally sustainable modes of transport. Females are more likely to buy environmentally friendly tourism products (Male M 3.1, female M 3.5). Out of all statements, females are less likely to take holidays closer to home besides taking fewer holidays. In contrast, men are less likely to purchase carbon offsets when they travel. Males are less likely to pick an environmentally friendly destination than women (Male M 3.0, female M 3.2). Although, it must be noted that all the SD is equal to one or higher, meaning that the respondents had varying opinions.

Overall, the willingness to travel pro-environmentally increases with age. Respondents 50 years and older tend to (strongly) agree more often than the younger participants as the means to their answers are higher. In contrast to the group aged 18–30, where 19% (N = 180) of the respondents are willing to vacation closer to home to minimize transportation-related greenhouse gas emissions, 37% of the group aged 61–70 (N = 16) are willing to do this. They are followed by 51–60 years old (N = 17). The younger travelers, aged 18–30, are the least likely to travel closer to home, as 60% (N = 108) strongly disagreed. They were followed by 31–40 years old (55%) who strongly disagreed. People aged 51–60 (M 3.3) and

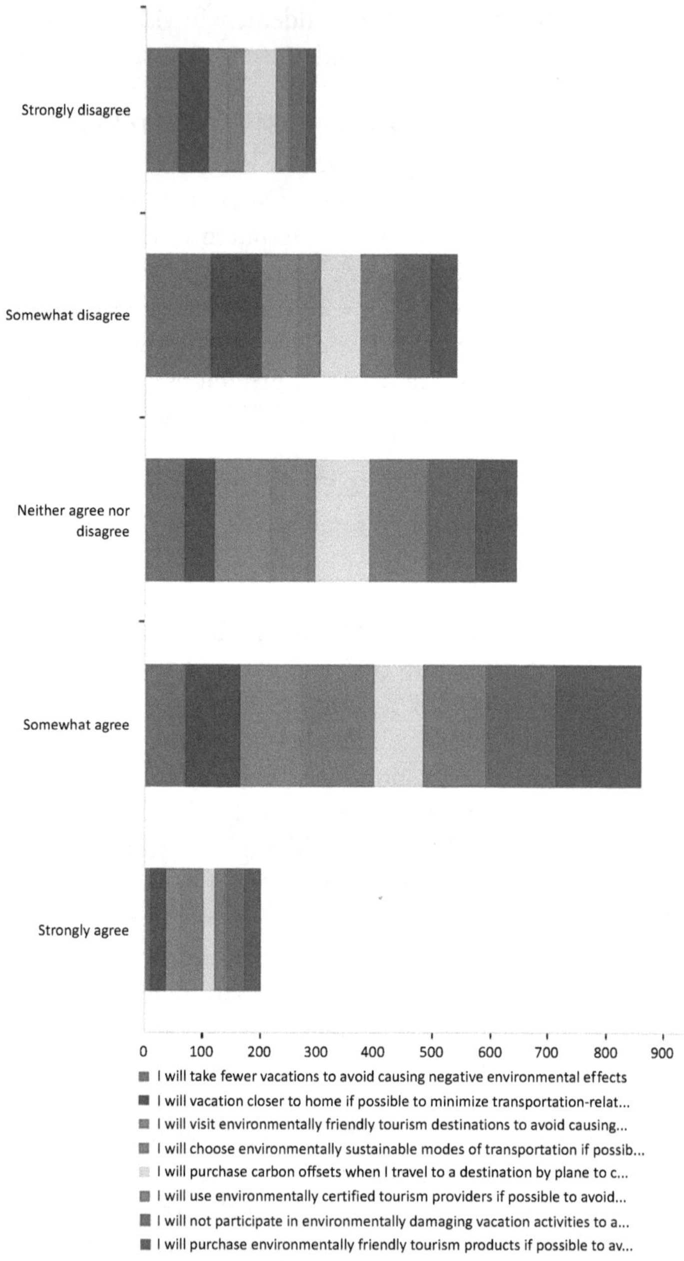

Fig. 6.5 Expected pro-environmental travel behavior

61–70 (M 3.5) are most likely to visit an environmentally friendly tourism location compared to the other groups (M<3.3).

Discussion

Of all respondents, 67% (N = 246) find that COVID-19 has a destructive economic impact. These findings build on the previous studies of Bavel et al. (2020) and Uehara and Sakurai (2021); 75% of the respondents are concerned about the environment. The results strengthen the claims of Gray and Jackson (2020), who found that 71% of the people globally consider COVID-19 as just as serious of a crisis as climate change. COVID-19 has shown that the willingness of citizens to get on board is crucial for a strategy to succeed. As Richards (2021) stated, to achieve a radical transformation, the tourist—rather than the industry—should be the focus of this transformation. The willingness to change depends on the perceived threat, making it likely that people will not fall back into their habits (Funke & Klenert, 2020; Wang et al., 2020).

Most of the samples find that everyone must take responsibility for the environmental problems on our planet. This is in line with the research conducted by More in Common (2021), which found that 78% of the Dutch population find that companies should reduce their carbon emissions to protect the environment. Furthermore, these results build on existing evidence that consumer behavior consists of complex patterns and is influenced daily. The data showed that most respondents felt morally obligated to protect the environment. However, the moral obligation is less likely to change into more significant actions such as less traveling or staying closer to home. Gibbons (2020) confirmed that people often predict their best-case version of humanity. The findings also align with DPG Media (2019) that the Dutch population was least willing to pay for environmental travel than other sectors. These results build on existing evidence that the pandemic increased awareness of the risks and consequences of climate change (Dryhurst et al., 2020).

Likewise, Balcetis and Dunning (2013) claim that the behavior is more complex to predict than other people's behavior. Considering that the responses are self-reported expected experiences, they differ slightly from actual future behavior. Furthermore, actual future behavior also depends on external factors such as finances and the supply side of tourism (de Haas et al., 2020). The literature suggests that paying higher prices for environmental protection could also predict PET (Batool et al., 2021). In this study, 32% of all the respondents claimed they are unwilling to pay extra for environmental protection, and strongly agree was the least selected answer. This contradicts the study of DPG Media (2019), which claims that 61% of Dutch people are willing to pay extra for sustainable products. However, only 2% of all respondents are willing to pay double the amount for sustainable alternatives.

Finally, the study by Gray and Jackson (2020) finds that 33% of respondents would prefer to replace flights with trains or buses to minimize pollution. The findings of this study are slightly different, and 23% of respondents would prefer to do so. Regarding certified products and accommodation, 42% of the respondents are open to this. Guterres (2020) claimed that rebuilding a more carbon-neutral sector is imperative. However, only 31% of this study's participants are willing to pay for this.

Conclusion

The purpose of this study was to investigate the pandemic's effect on Dutch travelers' future behavioral willingness to travel eco-friendly. This study has both managerial and empirical implications. From an empirical perspective, this study predicts the market readiness of Dutch travelers to travel pro-environmentally post-COVID-19. This confirms that the willingness to travel pro-environmentally increased with the perceived threat and duration of COVID-19. Furthermore, this study proved that the willingness to travel, pay, and fear depend on the context.

In contrast to most other studies, this study used three theories, NAM, economic sacrifices theory, behavioral willingness, and risk perception, all used to predict PETB. These theories have not been used to predict future travel for this target group—the Dutch market. The difference between this study compared to other studies that did aim to examine the openness to environmental travel. They all show slightly different outcomes, while they are all used to predict pro-environmental travel behavior. Indicating that studies only using one of these methods are missing a big part of the information to base their conclusions. In addition, the timing of the pandemic plays a vital role in how bad the impacts of COVID-19 are perceived. Studies conducted in the early stage of the pandemic show that the willingness to travel sustainably is lower than the willingness during the pandemic, which has been there for two years. Finally, many studies often had a global perspective or solely focused on the transport or accommodation sector, and they did not consider behavior in the whole sector.

From a practical perspective, this gives the travel and tourism industry a better understanding of the Dutch traveler's market in detail and their readiness to travel in environmentally friendly post-pandemic. The main insights indicate that the travel industry should transition into an affordable but more environmentally friendly industry as the gross of Dutch travelers is concerned; however, not all travelers seem willing to pay for sustainable solutions. Currently, the demand still needs to meet the supply side of tourism. However, people are unwilling to change their behavior drastically; the level of concerns and perceived risk indicates an openness to and understanding of why the change is needed.

A vital factor of this success is in the hands of the tourism industry and operators. Tourism providers should start by proactively changing the industry. They should design and adjust their product and service portfolio accordingly so that the tourist can behave more environmentally. This can be done by implementing greener solutions for transport, such as affordable and easy-access trains instead of planes. This research shows that the respondents are ready, understand the need, and are willing to make minor adjustments.

Policies and promotion programs should be implemented to enable the market to perceive walking and cycling as comfortable, flexible, attractive, joyful, relaxing, and pleasant. This could be done through built environment rehabilitation to make active modes of transport appear more comfortable, pleasant, and inviting. Additionally, tourism destinations or businesses could offer green tourist programs to induce and continuously reinforce tourists' identity as environmental travelers. Interventions introducing a feeling of moral obligations could be developed as direct reminders during travel planning. For example, pop-ups in booking tools could inform tourists about the greenhouse emissions caused by each available option.

Limitations

The study is not generalizable for the Dutch population. Demographics could have influenced the overall outcome of the study. However, in some cases, only a few respondents represented a group, such as those older than 70 (N = 2) or those with a doctoral degree (N = 4). The generalizability of this study can also be questioned since the final response count was 329. The sample used in this study is only partially representative of the Dutch population, and there is an over-representation of people younger than 30.

References

Abdelkader, O. A. (2020). Impact of perception on 'willingness and behavior' of individuals toward switching to sustainable energy practices in buildings. *Energy Reports, 6*, 2119–2125.

Ahmed, S., Adnan, M., Janssens, D., & Wets, G. (2020). A personalized mobility-based intervention to promote pro-environmental travel behavior. *Sustainable Cities and Society, 62*, 102397.

Ajzen, I. (1991). The theory of planned behavior. *Organizational Behavior and Human Decision Processes, 50*(2), 179–211.

Ajzen, I. M., & Fishbein, M. (1980). *Understanding attitudes and predicting social behavior*. Prentice-Hall.

Anwari, N., Tawkir Ahmed, M., Rakibul Islam, M., Hadiuzzaman, M., & Amin, S. (2021). Exploring the travel behavior changes caused by the COVID-19 crisis: A case study for a developing country. *Transportation Research Interdisciplinary Perspectives, 9*, 100334.

Balcetis, E., & Dunning, D. (2013). Considering the situation: Why people are better social psychologists than self-psychologists. *Self and Identity, 12*(1), 1–15.

Batool, N., Dada, Z. A., Shah, S. A., & Ganai, S. A. (2021). Interactive relationship between the environmental concern and the eco-tourism destination regions: A literature review. *Journal of Agriculture Sciences, 36*(1), 457–465.

Bavel, J. J. V., Baicker, K., Boggio, P. S., Capraro, V., Cichocka, A., Cikara, M., Crockett, M. J., Crum, A. J., Douglas, K. M., Druckman, J. N., Drury, J., Dube, O., Ellemers, N., Finkel, E. J., Fowler, J. H., Gelfand, M., Han, S., Haslam, S. A., Jetten, J., & Willer, R. (2020). Using social and behavioral science to support COVID-19 pandemic response. *Nature Human Behaviour, 4*(5), 460–471.

Bernstein, A., & R. N. Salas. (2020, July 6). *Coronavirus and climate change.* C-CHANGE | Harvard T. H. Chan School of Public Health. https://www.hsph.harvard.edu/c-change/subtopics/coronavirus-and-climate-change

Booking.com. (2021, June). *Booking.com's 2021 Sustainable Travel Report.* Retrieved May 1, 2021, from https://globalnews.booking.com/bookingcoms-2021-sustainable-travel-report-affirms-potential-watershed-moment-for-industry-and-consumers/

Bratić, M., Radivojević, A., Stojiljković, N., Simović, O., Juvan, E., Lesjak, M., & Podovšovnik, E. (2021). Should I stay or should I go? Tourists' COVID-19 risk perception and vacation behaviour shift. *Sustainability, 13*(6), 3573.

Brieger, S. A. (2018). Social identity and environmental concern: The importance of contextual effects. *Environment and Behaviour, 51*(7), 828–855.

De Groot, J., & Steg, L. (2009). Morality and prosocial behavior: The role of awareness, responsibility, and norms in the norm activation model. *The Journal of Social Psychology, 149*(4), 425–449.

de Haas, M., Faber, R., & Hamersma, M. (2020). How COVID-19 and the Dutch 'intelligent lockdown' change activities, work, and travel behavior: Evidence from longitudinal data in the Netherlands. *Transportation Research Interdisciplinary Perspectives, 6*, 100150.

DPG Media. (2019). *Duurzaamheid en de Nederlandse bevolking.* Retrieved May 1, 2021, from https://www.dpgmedia.nl/marketingadvies/thema/insights/artikel/duurzaamheid-en-de-nederlandse-bevolkingn

Dryhurst, S., Schneider, C. R., Kerr, J., Freeman, A. L. J., Recchia, G., van der Bles, A. M., Spiegelhalter, D., & van der Linden, S. (2020). Risk perceptions of COVID-19 around the world. *Journal of Risk Research, 23*(7–8), 994–1006.

Fink, A. (2021). *How to conduct surveys a step-by-step guide* (Vol. 6). Sage Publishing.

Florén, H., Frishammar, J., Parida, V., & Wincent, J. (2017). Critical success factors in early new product development: A review and a conceptual model. *International Entrepreneurship and Management Journal, 14*(2), 411–427.

Funke, F., & Klenert, D. (2020). Climate change after COVID-19. *VOX*, CEPR Policy Portal. Retrieved August 17, 2020, from https://voxeu.org/article/climate-change-after-COVID-19

Galvani, A., Lew, A. A., & Perez, M. S. (2020). COVID-19 is expanding global consciousness and the sustainability of travel and tourism. *Tourism Geographies, 22*(3), 567–576.

Gibbons, F. X. (2020). *Intention, expectation, and willingness* (pp. 1–17). Iowa State University.

Gray, E., & Jackson, C. (2020, April). *How does the world view climate change and COVID-19?* Ipsos. Retrieved June 1, 2021, from https://www.ipsos.com/sites/default/files/ct/news/documents/2020-04/earth-day-2020-ipsos.pdf

Gray, G. M., & Ropeik, D. P. (2002). Dealing with the dangers of fear: The role of risk communication. *Health Affairs, 21*(6), 106–116.

Groen, S. (2021, March 30). *Reizen na corona: kunnen we na de pandemie nog met goed fatsoen op vakantie?* de Volkskrant. Retrieved June 1, 2021, from https://www.volkskrant.nl/cultuur-media/reizen-na-corona-kunnen-we-na-de-pandemie-nog-met-goed-fatsoen-op-vakantie~bd553c2c/

Guterres, A. (2020). *It is imperative that we rebuild the tourism sector*. United Nations. Retrieved July 6, 2021, from https://www.un.org/en/coronavirus/it-imperative-we-rebuild-tourism-sector

Hajibaba, H., Gretzel, U., Leisch, F., & Dolnicar, S. (2015). Crisis-resistant tourists. *Annals of Tourism Research, 53*, 46–60.

Hall, C. M., Scott, D., & Gössling, S. (2020). Pandemics, transformations, and tourism: Be careful what you wish for. *Tourism Geographies, 22*(3), 577–598.

Hammer, J. H., & Vogel, D. L. (2013). Assessing the utility of the willingness/prototype model in predicting help-seeking decisions. *Journal of Counselling Psychology, 60*(1), 83–97.

Han, H., Hwang, J., Kim, J., & Jung, H. (2015). Guests' pro-environmental decision-making process: Broadening the norm activation framework in a lodging context. *International Journal of Hospitality Management, 47*, 96–107.

Han, H., Yu, J., Koo, B., & Kim, W. (2019). Vacationers' norm-based behavior in developing environmentally sustainable cruise tourism. *Journal of Quality Assurance in Hospitality & Tourism, 20*(1), 89–106.

Hedlund, T. (2011). The impact of values, environmental concern, and willingness to accept economic sacrifices to protect the environment on tourists' intentions to buy ecologically sustainable tourism alternatives. *Tourism and Hospitality Research, 11*(4), 278–288.

Hines, J. M., Hungerford, H. R., & Tomera, A. N. (1987). Analysis and synthesis of research on responsible environmental behavior: A meta-analysis. *The Journal of Environmental Education, 18*(2), 1–8.

Isaac, R. K. (2021). An exploratory study: The impact of terrorism on risk perceptions. An analysis of the German market behaviours and attitudes towards Egypt. *Tourism Planning & Development, 18*(1), 25–44.

Isaac, R. K., & van den Bedem, A. (2020). The impacts of terrorism on risk perception and travel behaviour of the Dutch market: Sri Lanka as a case study. *International Journal of Tourism Cities, 7*(1), 63–91.

Juvan, E., & Dolnicar, S. (2016). Measuring environmentally sustainable tourist behavior. *Annals of Tourism Research, 59*, 30–44.

Kaiser, F. G., Hübner, G., & Bogner, F. X. (2005). Contrasting the theory of planned behavior with the value-belief-norm model in explaining conservation behavior. *Journal of Applied Social Psychology, 35*(10), 2150–2170.

Marazziti, D., Cianconi, P., Mucci, F., Foresi, L., Chiarantini, I., & della Vecchia, A. (2021). Climate change, environment pollution, COVID-19 pandemic, and mental health. *Science of The Total Environment, 773*, 145182.

Miller, D., Merrilees, B., & Coghlan, A. (2014). Sustainable urban tourism: Understanding and developing visitor pro-environmental behaviours. *Journal of Sustainable Tourism, 23*(1), 26–46.

Moghimehfar, F., & Halpenny, E. A. (2016). How do people negotiate through their constraints to engage in pro-environmental behavior? A study of front-country campers in Alberta, Canada. *Tourism Management, 57*, 362–372.

More in Common. (2021). *The impact of COVID-19 on Dutch society*. Retrieved July 2, 2021, from https://www.moreincommon.com/media/4qefbsy5/more-in-common-the-new-normal-the-netherlands-en.pdf

Neuburger, L., & Egger, R. (2020). Travel risk perception and travel behavior during the COVID-19 pandemic 2020: A case study of the DACH region. *Current Issues in Tourism, 24*(7), 1003–1016.

O'Connor, P., & Assaker, G. (2021). COVID-19's effects on future pro-environmental traveller behavior: An empirical examination using norm activation, economic sacrifices, and risk perception theories. *Journal of Sustainable Tourism, 2021*(1), 1–19.

Onwezen, M. C., Antonides, G., & Bartels, J. (2013). The norm activation model: An exploration of the functions of anticipated pride and guilt in pro-environmental behaviour. *Journal of Economic Psychology, 39*, 141–153.

Oqubay, A. (2020). *COVID-19—An unprecedented global threat that deserves unprecedented leadership*. Development Matters. Retrieved November 12,

2021, from https://oecd-development-matters.org/2020/04/24/COVID-19-an-unprecedented-global-threat-that-deserves-unprecedented-leadership/

Orîndaru, A., Popescu, M. F., Alexoaei, A. P., Căescu, T. C., Florescu, M. S., & Orzan, A. O. (2021). Tourism in a post-COVID-19 era: Sustainable strategies for industry's recovery. *Sustainability, 13*(12), 6781.

Park, E., Lee, S., Lee, C. K., Kim, J. S., & Kim, N. J. (2018). An integrated model of travellers' pro-environmental decision-making process: The role of the New Environmental Paradigm. *Asia Pacific Journal of Tourism Research, 23*(10), 935–948.

Pomery, E. A., Gibbons, F. X., Reis-Bergan, M., & Gerrard, M. (2009). From willingness to intention: Experience moderates the shift from reactive to reasoned behaviour. *Personality and Social Psychology Bulletin, 35*(7), 894–908.

Richards, G. (2021). Transforming tourism in a post-Covid world. *Vita Magazine, 5*(May), 62–63.

Schneider, C. R., Zaval, L., Weber, E. U., & Markowitz, E. M. (2017). The influence of anticipated pride and guilt on pro-environmental decision making. *PLOS ONE, 12*(11), 0188781.

Schwartz, S. H. (1977). Normative influences on altruism. *Advances in Experimental Social Psychology, 10*, 221–279.

Scott, D., Hall, M. C., & Stefan, G. (2012). *Tourism and climate change: Impacts, adaptation, and mitigation* (1st ed.). Routledge.

Severo, E. A., de Guimarães, J. C. F., & Dellarmelin, M. L. (2021). Impact of the COVID-19 pandemic on environmental awareness, sustainable consumption, and social responsibility: Evidence from generations in Brazil and Portugal. *Journal of Cleaner Production, 286*, 124947.

Shin, Y. H., Im, J., Jung, S. E., & Severt, K. (2018). The theory of planned behavior and the norm activation model approach to consumer behavior regarding organic menus. *International Journal of Hospitality Management, 69*, 21–29.

Steg, L., & Vlek, C. (2009). Encouraging pro-environmental behaviour: An integrative review and research agenda. *Journal of Environmental Psychology, 29*(3), 309–317.

Stern, P. C. (2000). New environmental theories: Toward a coherent theory of environmentally significant behavior. *Journal of Social Issues, 56*(3), 407–424.

Swart, A. J. (2020, September). *Duurzaam verpakken*. ABN Amro. Retrieved June 1, 2021, from https://www.empack.nl/wp-content/uploads/sites/74/2020/12/Duurzaam-verpakken-de-haat-liefdeverhouding-tussen-consument-en-plasticverpakkingen.pdf

Uehara, T., & Sakurai, R. (2021). Have sustainable development goal depictions functioned as a nudge for the younger generation before and during the COVID-19 outbreak? *Sustainability, 13*(4), 1672.
Wang, Y.-S. (2009). The impact of crisis events and macroeconomic activity on Taiwan's international inbound tourism demand. *Tourism Management, 30*(1), 75–82.
Wang, L., Wong, P. P. W., & Narayanan Alagas, E. (2020). Antecedents of green purchase behaviour: An examination of altruism and environmental knowledge. *International Journal of Culture, Tourism and Hospitality Research, 14*(1), 63–82.
Wu, J., (Snow), Font, X., & Liu, J. (2021). Tourists' pro-environmental behaviours: Moral obligation or disengagement? *Journal of Travel Research, 60*(4), 735–748.

7

Tourist Behavior for Sustainable Development in the *Cumbres de Majalca* National Park, Mexico: Challenges in a Post-pandemic Context

Manuel Ramón González Herrera,
Silvia Giralt Escobar,
and Julián Alberto Álvarez Hernández

Introduction

The approaches to sustainable tourism in Protected Natural Areas (PNAs) worldwide have been characterized in recent years by a more careful orientation toward environmental protection and the sustainability of natural and cultural landscapes, as well as the study of tourist behavior and sustainability in these natural spaces, especially considering the behavior of tourists in the

The original version of the chapter has been revised. A correction to this chapter can be found at https://doi.org/10.1007/978-3-031-45866-8_20

M. R. González Herrera (✉) • J. A. Álvarez Hernández
Department of Tourism, Autonomous University of Ciudad Juárez,
Ciudad Juárez, Mexico
e-mail: manuel.gonzalez@uacj.mx; jualvare@uacj.mx

S. W. Maingi et al. (eds.), *Tourist Behaviour and the New Normal, Volume II*,
https://doi.org/10.1007/978-3-031-45866-8_7

post-pandemic context, due to the strengthening of new demands in favor of environmental protection and conservation in the current scenario (Becken & Job, 2014; Sharmin et al., 2020; Spenceley et al., 2021a). Thus, the study of environmental sustainability is increasingly oriented toward compliance with The United Nations 2030 Agenda, including 17 Sustainable Development Goals (SDGs) (European Commission, n.d.).

There is a very close relationship between the SDGs and the approaches to environmental sustainability, low environmental impact tourism, and experiential landscape, expressed in how tourism can contribute to achieving such goals. In this sense, sustainability for tourist destinations in PNAs means the care of water and sanitation (SDG 6); the use of affordable and clean energy (SDG 7); the promotion of sustained and inclusive economic growth, full and productive employment and decent work (SDG 8); responsible production and consumption (SDG 12); action on the climate crisis (SDG 13); the care of forests, desertification, and biological diversity (SDG 15); as well as peace and justice (SDG 16) (World Tourism Organization [UNWTO], n.d.).

To achieve the SDGs in PNAs, there is a Strategic Framework for Sustainable Tourism with a perspective of 2030 and management instruments such as the Management Program; Public Use Program; Tourism Carrying Capacity Study; and Acceptable Change Limit Studies. This management must receive special attention in national parks due to their strategic value, maximizing the social and economic benefits for the local community, visitors, and the natural and cultural heritage, through synergies among all stakeholders. In the context of sustainability, national parks are reserved areas for preserving the natural and cultural environment for different purposes, such as tourism and recreation or visits of historical or scientific interest; not all national parks focus on the same landscape elements. The Editors of Encyclopaedia Britannica (2023) note that in the US and Canada, national parks focus more on land and wildlife protection, while in the UK, they tend to focus more on land, and add that parks in Africa focus mainly on the conservation of fauna.

S. Giralt Escobar
Department of Economics and Business Management, University of Alcala, Alcalá de Henares, Spain
e-mail: silvia.giralt@uah.es

The source mentions that other countries such as Brazil, Japan, India, and Australia have large areas reserved for national parks.

According to The General Law of the Ecological Balance and Environmental Protection of Mexico, national parks are landscapes of scenic beauty and scientific, educational, recreational, historical, floristic, and faunal values. They constitute highly protected spaces in which only activities related to the conservation of ecosystems, research, recreation, tourism, and environmental education are allowed (Secretary of the Environment [SEMARNAT], 2017). This chapter aims to diagnose the current scenario and to project management strategies for tourism with low environmental impact based on sustainability and responsible environmental behavior of visitors in the *Cumbres de Majalca* National Park (CMNP), located in the State of *Chihuahua*, Mexico.

The research problem is based on the fact that the sources of environmental impact for this park are the presence of forest pests and diseases; forest fires; uncontrolled grazing; use of off-road vehicles; the opening of unestablished roads that destroy vegetation, cause soil loss, accelerate erosion, and generate dust pollution and harmful noise; as well as the overtourism associated with vacation periods. The massive flows of visitors increase the number of arrivals, occupation of houses located within the park, uncontrolled vehicular traffic, firewood extraction, solid waste generation, degradation of rock formations, and damage to the built infrastructure. Based on this problem, the following research questions were formulated: What are the characteristics of CMNP in the current scenario? What is the perception of the potential of recreational tourism use in CMNP? What are the challenges in the post-pandemic era to project the desired scenario for better sustainable ecotourism management and promoting responsible tourism behavior?

The research significance is associated with developing a proposal that contributes to the conservation of wild flora and fauna and the care and preservation of forest vegetation, as well as maintaining the recharge of aquifers and the functionality of the hydrological cycle, protecting the upper parts of the basins where the streams that feed the *Chuvíscar*, *Sacramento*, and *Santa Isabel* rivers are born. At the same time, the proposal suggests the reconditioning of the site, whose scenic beauty and geological formations satisfy the visitors' leisure needs in the context of the new normality of post-pandemic tourism.

Literature Review

Sustainable and Environmental Protection of Natural Landscapes

According to the UNWTO (2013), sustainable tourism is "tourism that takes full account of its current and future economic, social and environmental impacts addressing the needs of visitors, the industry, the environment, and host communities" (p.1). Taking into consideration this source, the five key pillars of sustainable tourism are tourism policy and governance; trade, investment, data, and competitiveness; employment, decent work, and capacity building; poverty reduction and social inclusion; and sustainability of the natural and social environment, the last of these being closely related to this research. These five pillars influence tourist behavior, satisfaction, and tourist experiences, which is of great interest for this study. The current approaches to sustainability assume the environmental protection of natural landscapes with an integrative approach based on the preventive management of environmental impacts. In this sense, environmental sustainability uses different management instruments such as environmental planning, impact assessment, environmental monitoring, and environmental interpretation and education. The latter is of great importance for the protection of natural and cultural landscapes since the behavior of visitors is strongly conditioned by the environmental culture of the visitors and their environmental commitments.

Responsible tourist behavior and landscape education for a memorable experience are essential for sustainable tourism development. The World Tourism Organization presents a helpful definition in which Sustainable Tourism is conceptualized as (UNEP—UNWTO, 2005) "Tourism that takes full account of its current and future economic, social and environmental impacts, addressing the needs of visitors, the industry, the environment, and host communities" (p. 11). It states that sustainable tourism development guidelines and management practices apply to all types of tourism in all destinations, including mass tourism and the various niche tourism segments. This source indicates that sustainability principles incorporate tourism's environmental, economic, and socio-cultural dimensions; they recommend promoting a balance between them to achieve the sustainability of tourism development in the long term. In order to guarantee protection and sustainability in the tourist use of natural landscapes, it is necessary to consider the chain of impacts established between the visitors and the receiving environment. This means identifying the anthropic actions caused by tourism development and the behavior of visitors, which leads to the manifestation of positive or negative environmental changes. Consequently, impacts are generated, which can be beneficial or adverse according to the consequences for the recipients. This complex process requires a preventive management system since corrective actions always have residual effects.

Under this focus, maximizing the positive impacts of tourist behavior and minimizing the negative ones is possible. According to the Samalayuca Dunes Management Program, Low Environmental Impact Tourism is defined as an (CONANP, 2013)

> environmentally responsible tourism modality that consists of traveling to or visiting relatively undisturbed natural spaces in order to enjoy, appreciate and study the natural attractions of these spaces; as well as any cultural manifestation of the present and the past that can be found there, through a process that promotes conservation, has low environmental and cultural impact and induces an active and socio-economically beneficial involvement of local populations. (p. 112)

This concept relates to soft tourism because it "encompasses environmental and social compatibility, optimum wealth creation, and a new culture of travel" (Federal Agency for Nature Conservation, 2020, p. 1).

Tourist Behavior and Sustainability in Protected Natural Areas

Tourist behavior theory has acquired great importance for research related to the interactions between visitors and the visiting site, especially for the sustainability in PNAs. This approach has contributed to increasing concerns about environmental protection and the sustainability of natural landscapes. According to the definition presented by Elliot (2014) in the Encyclopedia of Tourism, it is assumed that tourist behavior is

> the process a tourist undertakes, both observable and unobservable, when planning and participating in tourism. It is the dynamic of affect and cognition, as well as biological and cultural forces interacting with marketing and environmental stimuli. In marketing, consumer behavior is a relatively young subfield that seeks to understand the where, when, and why of consumers and how their behavior might be influenced… In tourism, the challenge is to understand the complexities of behavior with the added dimensions of space, time, and the consumption of experiences. (p. 1)

In the presented study by Zhang and Zhang (2020), related to Transport and Energy Research, the authors state that "Tourist behaviors and choices have significant impacts on energy use [and on the environment in general], … [that is why] proper tourist decisions are important to enhance energy efficiency [and environmental responsibility] and realize sustainable tourism development" (p. 295). Derived from the previous approach, it is assumed that the behaviors and choices of tourists must be based on sustainability criteria.

There is a close relationship between sustainable tourist behavior and experiential landscape in PNAs. In this regard, it is considered that "tourism behavior is essential for the protection of the environment and the sustainability of destinations… [therefore] environmentally responsible behavior [is] an important tourism behavior" (Wu et al., 2022, p. 924). In the study How Does Tourist Experience Affect Environmentally Responsible Behavior? (Wu et al., 2022) the authors conclude that "… tourist experiences affect environmentally responsible behavior through attachment to place. The connection with nature is the key factor of environmentally responsible behavior" (p. 924). For this reason, it is necessary to consider the contribution of the experiential landscape to tourist behavior according to the different dimensions of the tourist experience.

Thwaites and Simkins (2007), in their book *Experiential Landscape: An Approach to People, Place and Space*, define the experiential landscape as the holistic way in which relationships are established between people and open spaces, thus being a relationship between the outdoor spaces of daily life and the variety of human experiences that are generated. Under this approach, it is possible to recognize the experiential potential and character of outdoor settings, identify how they change, and create new settings (dynamics and evolution). In this way, applying the experiential approach to the relationships of the subjects with the landscape object of interaction allows for assessing the experiential characteristics during the interpretation and understanding of the visited sites.

A topic of great interest is undoubtedly the spatial relationship of visitors with the landscapes that are the object of observation and interaction, and the process of landscape construction in PNAs. According to Chhetri and Chhetri (2022) "the process of perceiving and experiencing landscapes is widely understood in the extant literature; however, the mappability of perceived landscapes is neither thoroughly investigated nor theorized from a multidisciplinary perspective" (p. 85). According to these authors, perception and cognition are two aspects of great importance in the construction of the landscape since they favor the interpretation of observers and regulating their behaviors.

The process of perceiving and experiencing landscapes in PNAs on the part of the visitors needs to be oriented from a methodological perspective, through which the cognition of the attraction capacity and the affective-motivational relationships with the landscape object of visualization is facilitated. In this sense, it is considered that (Cepollaro & Zanon, 2022)

> landscape education must play a key role … [in which] awareness-raising, cultural initiatives, and training processes require approaches and methods centered on experience, thus shifting from teaching to learning, from a passive to an active role of participants … [in such a way that they] stimulate interest in the landscape and [help] to develop participatory, cooperative and responsible attitudes. (p. 244)

Tourist Behavior in the Post-pandemic Context

The new travel culture in the post-pandemic stage (Çakmak et al., 2023) should promote responsible travel (Kaefer, 2022) and therefore encourage responsible behavior of the new tourist (Eichelberger et al., 2021). This implies greater respect for the environment, making rational use of natural resources and preventing pollution; promoting the sustainable use of raw materials and natural resources; improving our environmental performance; promoting the involvement of all visitors in environmental protection matters; carrying out environmental protection programs; communicating and reporting on environmental management efforts; and preventing the purchase and sale of illegally harvested wood and timber products, among others (Song et al., 2019; Zahoor et al., 2020; Koval et al., 2021). In the new post-pandemic era, new habits are reported in the behavior of visitors of PNAs. According to a report by Hosteltur (2021), there has been a significant concern for the health and security of the destinations to visit, changes in behavior in the way of making reservations, an increase in optimism and the motivation to resume travel to discover natural spaces since there has been a greater diversification in the reasons for the visits, while the increase in environmental awareness and the sustainability of destinations is of great interest. This could impact the reconversion of traditional tourist spaces and the change from irresponsible tourist behavior toward responsible tourist behavior committed to the communities and the destinations visited.

Methodological Framework

The study method was based on qualitative, multidisciplinary, and transversal research principles, using mixed information sources. Empirical methods such as observation, document analysis, diagnostic tests, and work in groups, as well as theoretical methods such as analytical-synthetic, historical-logical, geospatial, and conceptual modeling, were used. The methodological procedure consisted of data collection, analysis and interpretation, and elaboration of conclusions. The study of the behavior of visitors and their relationships with the experiential landscape was based on a field visit to the CMNP with undergraduate tourism students from the Autonomous University of *Ciudad Juárez* (n = 16). The participating students were studying the subjects of Tourism and the Environment and Tourism Planning. The conceptual variables addressed correspond to tourist behavior, the responsible tourist, experiential landscapes, national parks, low environmental tourism, and sustainability. The activity was carried out through a field day in which the students explored the main areas of public use of the park, selected the ecotourist trails to be followed, located the visual observation basins, and took photos of the landscapes with which they felt cognitively and affectively related.

For the development of the practical activity, three questions were elaborated to assess the perception of the tourist potential for recreational use of the undergraduate tourism students who participated in the field day. These were (1) Tell a story based on your own life experience and offer the arguments that generated the emotional bond with the landscape of your choice; (2) Assess the capacity for attraction and motivational-affective relationship with the photographed landscapes according to the valuation sheet (Table 7.1); and (3) How would you rate your landscape experience using the following categories: very satisfied; satisfied; neutral; unsatisfied; very unsatisfied. The valuation sheet for interpreting and understanding the visited sites included the following indicators and perception scales:

- Attraction capacity: functional, ecological, aesthetic, symbolic-representative, cognitive-interpretative, authenticity, accessibility, infrastructure, and interpretive services values. The evaluation of each indicator was carried out at three levels: high (5), medium (3), and low (1).
- Motivational-affective relationship: spectacular (5), very nice (4), nice (3), moderately pleasant (2), unpleasant (1).

Findings

Characterization and Diagnosis of the Current Scenario

The CMNP is located in the northern Mexican state of *Chihuahua*, approximately 48 kilometers north of the city of the same name and state capital. That is why there is a close relationship between sustainable tourist behavior and experiential landscape. The park's creation date was September 1, 1939, by decree published in the Official Gazette of the Federation, covering an area of more than four thousand hectares (SEMARNAT, 2016). Its administration is based on a management program that "has the purpose of constituting the governing instrument of

Table 7.1 Assessment of the attraction capacity and motivational-affective relationship of the photographed landscapes

Experiential landscapes	Attraction capacity	Motivational-affective relationship
	Functional value (5) Ecological value (5) Aesthetic value (5) Symbolic value (5) Cognitive value (5) Authenticity value (5) Accessibility value (4) Infrastructure and services (4)	Spectacular (5) 100% Very nice --- Nice --- Moderatelypleasant--- Unpleasant ---
	Functional value (5) Ecological value (5) Aesthetic value (5) Symbolic value (5) Cognitive value (5) Authenticity value (5) Accessibility value (4) Infrastructure and services (4)	Spectacular (5) 100% Very nice --- Nice --- Moderatelypleasant--- Unpleasant ---
	Functional value (5) Ecological value (5) Aesthetic value (5) Symbolic value (5) Cognitive value (5) Authenticity value (5) Accessibility value (4) Infrastructure and services (4)	Spectacular (5) 100% Very nice --- Nice --- Moderatelypleasant--- Unpleasant ---
	Functional value (5) Ecological value (5) Aesthetic value (5) Symbolic value (5) Cognitive value (5) Authenticity value (5) Accessibility value (4) Infrastructure and services (3)	Spectacular (5) 100% Very nice --- Nice --- Moderatelypleasant--- Unpleasant ---
	Functional value (5) Ecological value (5) Aesthetic value (5) Symbolic value (5) Cognitive value (5) Authenticity value (5) Accessibility value (4) Infrastructure and services (3)	Spectacular (5) 100% Very nice --- Nice --- Moderatelypleasant--- Unpleasant ---

Own elaboration

planning and regulation that establishes the activities, actions, and basic guidelines for the management and administration of the CMNP" (Official Journal of the Federation [DOF], 2016). The designation of *Cumbres de Majalca* as a national park "has as general objective the conservation of wild flora and fauna… due to the environmental services it generates for the region" (CONANP, 2016, p. 13). Derived from the general objective, it is established to conserve wildlife and forest vegetation to avoid erosion and favor the recharge of aquifers; protect the upper parts of the basins and hydrographic sub-basins; and maintain the natural scenic beauty of the landscape components that satisfy the pleasure, distraction, entertainment, leisure, recreation, and diversion of the site, mainly due to the high value of the rock and forest formations that develop in this geographical space (CONANP, 2016).

The geological-geomorphological basement is made up of a relief represented by fractured mountain chains of volcanic origin and modeled by exogenetic dynamic factors, which are separated by intramontane valleys in the form of canyons and large rocky outcrops with a stratified appearance, as well as great variety that gives it spectacular geodiversity. This mountain relief forms the Sierra de Majalca, belonging to the Sierra Madre Occidental of Mexico; it reaches a maximum height of approximately 2800 meters above sea level. The hydro-climatic conditions are of the semi-desert and semi-cold type with altitudinal changes that reach the semi-humid vertical floor. The annual average temperature is close to 14°C, the warmest month is June, and the coldest month is January, which indicates a yearly thermal amplitude of approximately 8°C. The most abundant rainfall occurs in July and August, and the least in April, registering an annual average of close to 178 millimeters and the frequent occurrence of snowfall. From the hydrographic point of view, the park is located in the *Conchos* River basin and comprises three micro-basins corresponding to the *Huerachi* River, *Majalca* Canyon, and *La Fortuna*.

The biogenic characteristics comprise pine-oak forests, oak-pine forests, oak forests, induced forests, xeric scrublands, and natural grasslands, notable for their high biodiversity. The fauna is mainly made up of mammals, reptiles, and birds. Among plants, the most frequent observations correspond to Echinocereus polyacanthus; the Green-flowered Pincushion Cactus (Mammillaria barbata); Coryphantha compacta; Echeveria mucronata;

Dahlia sherffii; Pointleaf Manzanita (Arctostaphylos pungens); Prairie Pricklypear (Opuntia macrorhiza); Mexican Pinyon (Pinus cembroides); Alligator Juniper (Juniperus deppeana); Apache Pine (Pinus engelmannii); and Chihuahuan Pine (Pinus leiophylla) (Naturalista, 2023).

The most frequent sightings of animals correspond to Yarrow's Spiny Lizard (Sceloporus jarrovii); Mexican Jay (Aphelocoma wollweberi); Acorn Woodpecker (Melanerpes formicivorus); American Robin (Turdus migratorius); Rock Squirrel (Otospermophilus variegatus); Northern Flicker (Colaptes auratus); Painted Redstart (Myioborus pictus); Mexican Rose Tarantula (Aphonopelma pallidum); Bridled Titmouse (Baeolophus wollweberi); White-breasted Nuthatch (Sitta carolinensis); White-tailed Deer (Odocoileus virginianus); Zone-tailed Hawk (Buteo albonotatus); and Steller's Jay (Cyanocitta stelleri) (Naturalista, 2023). The most widespread economic activities are tourist-recreational, agricultural, and livestock. Tourism and recreation are particularly interested in carrying out multiple activities such as hiking and free walks, observation of flora and fauna, observation of the geo-forms of the relief, horseback riding, quad biking, rappelling, camping, and sports events. These activities are more intense on weekends, Easter holidays, and July and August holidays. Although there is no official data, it is estimated that approximately 20,000 people enter the CMNP per year, without counting the local inhabitants, who, according to the INEGI 2020 population census, include 157 people in different localities, although to this amount are added around 500 people of floating population (M.C. Recorder, personal communication, 9 March 2023). The infrastructure and furniture, although not enough, include guardhouses; panoramic viewpoints; signaling and signage; health services; some tables, benches, and grills; camping areas; and roads for vehicular traffic.

Following the CMNP Management Program, the following are allowed: camping activities in the Mirador-Camping Areas Public Use subzone; scientific collection; construction of facilities for tourism support services; environmental education; campfires in designated places; filming and photography; scientific research and monitoring; maintenance of built infrastructure; traffic of vehicles on established roads; low environmental impact tourism activities; and sale of food and crafts. Low environmental impact tourism activities are also allowed in the *Mil Castillos-Penas Azules* preservation subzone; and in the subzone of human settlements (CONANP, 2016). The main negative environmental impact factors on the park's environmental systems that represent weaknesses for its strategic positioning are the risks of natural and induced forest fires; prolonged frosts; the presence of pests and forest diseases; uncontrolled grazing; use of all-terrain vehicles; the opening of rough roads that destroy the vegetation, cause loss of soil, and generate dust pollution and noise that is harmful to health; as well as overtourism associated with vacation periods. These massive flows of visitors increase the number of arrivals, occupation of homes located within the park, create a high volume of uncontrolled vehicular traffic, lead to the extraction of firewood, generation of solid waste, degradation of rock formations, and damage to the built infrastructure.

The main strengths of environmental resources and services for the strategic positioning of the CMNP are the great tourist-recreational attractiveness for the inhabitants of the city of *Chihuahua* and nearby regions due to the high biodiversity and geodiversity that give it a high scenic landscape value, the comprehensive natural vegetation cover associated with the vocation of use of the territory; the variety of environmental services that it offers such as rainwater harvesting, which favors the formation of rivers and streams, and the infiltration that it provides to the aquifers that promote agriculture, livestock, as well as the supply to the City of *Chihuahua* and other settlements, at the same time it serves as protection and a source of food for wildlife and allows the capture of carbon dioxide (CO_2) that contributes to reducing global warming. The park also contains a great historical-cultural value in which an archaeological site with manifestations of rock art stands out and is the scene of the culture of the *Conchos* Indians who inhabited northern Mexico. Also to be noted is the existence of medicinal plants used in local self-consumption.

Perception of the Potential of Recreational Tourism Use

The participatory assessment of the tourist-recreational attractions of the CMPN was based on the relationship between affect and cognition generated in the form of stimuli by the landscape components on the behavior of visitors with an experiential approach. To foster the visitor-landscape links, the participants told a story based on their own experience and the arguments that generated the emotional bond with the landscape of their choice, noting the high incidence of scenic panoramic views, open visual fields, and color contrasts. The most used technique for the narration was the personification of the components of the natural landscape and the use of humor. Through this activity, it was possible to recognize the experiential potential and the character of the outdoor settings of the landscapes according to their attractiveness and affective-motivational relationships (Table 7.1). The experiential approach applied to the subjects' relationships with the landscape made it possible to select the five most attractive experiential landscapes in the visiting area. The influence of the stimuli generated by the landscape's components on the visitors' behavior reflects the higher levels of attraction toward the relief and the vegetation as the most expressive perceptible components of the place. The preferred landscapes are of high functional value for recreation and tourism; high representative ecological, aesthetic, and symbolic value. They are characterized by high cognitive value and high natural authenticity. The visitor`s experience must be enriched in terms of accessibility and available infrastructure and services. In general, the assessment of the affective-motivational relationship with the landscape is valued in the category of "spectacular," which indicates a solid and memorable experience.

Finally, the qualification of the experience with the landscape (experiential landscape) was evaluated by 100% of the participants and received the category of very satisfied. During the fieldwork, the students showed environmentally responsible tourist behavior, evidenced in the garbage waste collection in different parts of the park, which other visitors had left (Fig. 7.1). This fact demonstrated an adequate level of landscape environmental education and participation in a tourism activity with a low environmental impact that promotes responsible tourism and visitors that respect the environment and sustainability.

Fig. 7.1 Garbage collection during fieldwork in *Cumbres de Majalca* National Park

Challenges in a Post-pandemic Context and Projection of the Desired Scenario

At an international level, before the COVID-19 pandemic, there were more than 8 billion visits/per year to protected natural areas, of which approximately 80% were concentrated in Europe and North America (Balmford et al., 2015). In the post-pandemic stage, the responsible recovery of nature-based tourism should consider the problem of mass tourism in PNAs due to their fragility and ecological values, paying attention to overcrowding, conflicts and derogatory behavior of tourists, parking and traffic problems, social distancing, health, and hygiene maintenance (McGinlay et al., 2020). This premise of recovery will require the strengthening of long-term resilience in terms of nature-based tourism in PNAs, which will demand the capacity of stakeholders to design innovative experiences for visitors, develop alternative income streams for local communities, address seasonality in a holistic, inclusive, equitable, and adaptable way (Spenceley et al., 2021b); as well as aligning the conservation and human development agendas (UICN-WCPA, 2020), and readapting carrying capacity and environmental planning of spaces for public use, avoiding intensive and expansive tourism models. The tourism allowed in the CMNP, due to its management category and based on the applicable legislation, is low environmental impact tourism.

In response to the question of how a tourism model with low environmental impact can be promoted in the NPCM, it is considered that the desired model of tourism with low post-pandemic environmental effects for this park should reflect a reconceptualization of environmental reductionism in the face of a holistic and integrating approach to the environmental system, in which the requirements of sustainable tourism and preventive management of impacts are implemented, thus reflecting the relationship mechanisms between nature and society in different spatiotemporal scenarios. At the same time, overtourism must be avoided, which is manifested through excessive tourism at certain times of the year, which generates overuse of resources, disturbance of wildlife, extreme concentration of visitors, intensification in the occupation of attractive sites, and congestion of vehicles. All of this means that these behaviors negatively impact visitors' experiences.

The sustainable management of the low-impact tourism development model must be based on the Strategic Framework for Sustainable Tourism in PNA with a perspective of 2030. In this way, a protocol will be adopted for the integration, classification, assessment, and prioritization of information from the CMNP, which allows the identification of the necessary components for tourism use and the minimum conditions for it to be successful, generating a baseline of strategic information that facilitates the decision-making process. At the same time, manuals of Best Environmental Practices in Tourism should be established to promote an experience or intervention that guarantees responsible behavior of visitors, with positive results to prevent, correct, or improve the environmental aspects of integrated tourism development and that contributes to reducing negative environmental impacts. This result will be beneficial since it will contribute to protecting and conserving the environment and its natural resources; foster good relations within the local community and with surrounding communities; promote competitiveness in enterprises due to the improvement in quality and responsibility with the environment and local communities; and raise awareness among consumers, staff, and suppliers about the importance of having a vision of sustainability (SERNATUR, 2011).

Another critical challenge of the new post-pandemic tourism model will be the education of the tourist and their performance as responsible travelers when visiting the CMNP. This strategy will require significant changes in visitor behavior and will positively influence the formation of an enriching experience. That training process must be based on the Global Code of Ethics for Tourism (UNWTO, 2020), which considers self-education, mutual tolerance, and learning of the legitimate differences between peoples and cultures and their diversity as an irreplaceable factor. The education of governments, companies, and communities that can seek it will also be necessary, although the visitor can significantly support this objective in different ways.

The main mechanisms for the prevention and control of impacts should be based on the Regulations for the Management of Visitors supported by studies such as Limit of Acceptable Change and Tourist Carrying Capacity, and on the development of tourism initiatives based on Best Practices, for the benefit of the gateway communities and in the areas of influence of the CMNP. At the same time, Green Marketing strategies should be introduced as part of park management aimed at reducing raw materials use, minimizing the waste generated; promoting responsible consumption as opposed to conventional consumerism; and the commitment to environmental causes as belonging to each visitor in a participatory and collaborative manner, expressed through the positive demonstration effect of one over the other.

The strategic requirements for the consolidation of a tourism development model with low environmental impact and the management of change toward sustainability in natural-based tourism-recreational businesses should promote a sustainable strategic transformation oriented toward family recreation and recreation in such a way that the challenges represented by the weaknesses can be overcome and the strengths identified for the CMNP would be strengthened. These strategies will be appropriate to promote strategic sustainability, sustainable leadership, and responsible environmental behavior of visitors to *Cumbres de Majalca*. For this, the primary strategy to be implemented should consider the following recommendations:

- Implementation of a participatory environmental planning framework that allows the carrying out of a comprehensive strategic planning process
- Strengthening of the conservation subprograms regarding the protection of the environmental system; integrated management of the environmental system; environmental education and interpretation; responsible behavior of visitors, and the strengthening of the culture of participation and integration of all stakeholders for the sustainability of the CMNP
- Control and management of visitor flows and diversification of visitor sites to avoid high-concentration nuclei associated with the main attractions
- Efficient management of solid or liquid waste, avoiding the dumping of materials, substances, or products, such as insecticides, fungicides, and pesticides, which can contaminate soils and bodies of water
- Reconditioning the infrastructure and services in all the polygons where tourist-recreational activities are carried out
- Regulation of vehicular traffic and control of motorized recreation in natural environments, which degrade the environmental system
- Strengthening of environmental monitoring and control mechanisms
- Application for certifications and recognitions that accredit responsible environmental performance

This proposal will contribute to the updating, operationalization, and instrumentation of the Park Management Program. In it, the axes of environmental education and interpretation should be prioritized as the basis for forming the culture of the environmental traveler in national parks, considering a new scenario of tourism recovery in a post-pandemic context. For this, a minimum system of key indicators of sustainable tourism development is proposed to be considered, which includes: protection of resources and attractions; pressure and intensity of use; water and energy consumption; consumption of local and national productions; waste management; social impact; contribution of tourism to the local economy; visitor satisfaction; satisfaction of the local population; and destination safety.

Therefore, for sustainable use in the projected scenario of the CMNP, the development of the Sustainable Tourism Protocol is required; the detailed study of the potential for tourist use and the elaboration of inventory sheets with the evaluation of attractions; determination of the carrying capacity and morphological/functional zoning of public use areas; design of interpretive trails with differentiated themes and suitable for different audiences; determination of scenic-landscape visual basins and their operationalization; and the proposal of a robust program of interpretation and environmental education based on a previous diagnosis of non-formal educational needs.

Conclusion

As a part of the tourist experience, the students satisfactorily undertook the process of group planning and participation in a field day at CMNP, revealing the affect and cognition toward the object of interpretation, as well as the forces that interacted with the environmental stimuli caused by the components of the experiential landscapes visited. In this way, the statement of Wu et al. (2022) was verified when stating that the connection with nature is a key factor of environmentally responsible behavior. It was possible to achieve behavior that made it possible to understand the where, when, and why of the visitors and how the activities carried out influenced their environmentally responsible behavior, perceiving at the same time the complexities of the behavior related to the geospatial dimensions, the time that the activities and consumption of experiences related to the landscapes they observed and interacted with. The study of behavior for sustainable development made it possible to understand that experiential landscapes contributed to landscape environmental education and interpretation and promoted environmentally responsible behavior in the CMNP, which favored the consolidation of sustainable tourism based on the principles of low environmental impact. At the same time, it contributed to the formation of tourists who are increasingly responsible and committed to an environmental cause. It was possible to verify how the theoretical pillars of tourism proposed by the UNWTO (2013) influenced the tourist behaviors, satisfaction, and tourist experiences of the students, who adequately carried out the perceptual assessment of the visited context and its components, especially about politics and governance; capacity development; poverty and social inclusion; and sustainability of the natural and social environment.

This proposal will allow the operationalization and instrumentation of the CMNP Management Program, which will contribute to meeting the new challenges in a post-pandemic context. Developing more scientific research with a post-disciplinary approach to generate new knowledge that serves as a basis for decision-making and managing visitor flows with a minimum impact approach in the public use areas will be advisable. The new post-pandemic projections should be based on the practical operationalization of the concept of low environmental impact tourism and the formation of a culture of experiential landscape interpretation, based on which it will be necessary to plan and schedule activities that promote authentic landscape experiences that strengthen behaviors which are more compatible with the functions of the park. It will be convenient in the future to promote the mapping of the perceived landscapes of the CMPN from a multidisciplinary perspective, integrating the processes of feeling, perceiving and knowing the set of environmental components that are observed, and in the same way, giving particular importance to the construction of mental maps and drawings of the components of the landscape through the establishment of cognitive and affective relationships. As a line of future research, management and social mechanisms for greater resilience and systemic thinking for conservation and responsible use of environmental resources and services in CMNP should be promoted, strengthening the role of tour operators in the experiences of visitors and the pro-environmental responsible behavior of all stakeholders.

The key findings and implications on tourism behavior in the new normal denote the lack of knowledge among visitors, both from a cognitive and affective-motivational point of view, which causes negative attitudes and irresponsible behaviors that cause environmental deterioration and the unsustainability of the tourist use of the territory. The practical implications for practitioners and policymakers consist in the fact that they have scientific information in favor of decision-making processes, which can positively impact the design of strategies that cover behavioral problems that affect good sustainable practices in tourism. The future research directions suggested by the chapter are based on the continuity of this line of research with the participation of all stakeholders, the extension to all groups of visitors to the PNA, and the dissemination of the results for their practical implementation in the medium and long term.

References

Balmford, A., Green, J., Anderson, M., Beresford, J., Huang, C., Naidoo, R., Walpole, M., & Manica, A. (2015). Walk on the wild side: Estimating the global magnitude of visits to protected areas. *PLoS Biology, 13*(2), 1–6. https://www.researchgate.net/publication/272842456_Walk_on_the_Wild_Side_Estimating_the_Global_Magnitude_of_Visits_to_Protected_Areas

Becken, S., & Job, H. (2014). Protected areas in an era of global–local change. *Journal of Sustainable Tourism, 22*(4), 507–527. https://www.tandfonline.com/doi/abs/10.1080/09669582.2013.877913

Çakmak, E., Isaac, R., & Butler, R. (Eds.). (2023). *Changing practices of tourism stakeholders in COVID-19 affected destinations.* Channel View Publications. https://www.channelviewpublications.com/page/detail/?K=9781845418748

Cepollaro, G., & Zanon, B. (2022). The landscape as a learning space. The experiential approach of a 'landscape school' in Trentino, Italy. *Landscape Research, 47*(2), 244–255. https://doi.org/10.1080/01426397.2021.1942442

Chhetri, P., & Chhetri, A. (2022). Theoretical perspectives on landscape perception. In R. S. Singh, B. Dahiya, A. K. Singh, & P. C. Poudel (Eds.), *Practising cultural geographies. Advances in 21st century human settlements.* Springer. https://doi.org/10.1007/978-981-16-6415-1_4

CONANP. (2013). *Programa de Manejo ANPFF Médanos de Samalayuca.* Editado por SEMARNAT. https://www.imip.org.mx/imip/files/planes/PROGRAMA_MANEJO_MEDANOS_DE_SAMALAYUCA.pdf

CONANP. (2016). *Programa de Manejo Parque Nacional Cumbres de Majalca.* Editado por SEMARNAT. https://iefectividad.conanp.gob.mx/i-efectividad/NSMOc/PN%20Cumbres%20de%20Majalca/Contexto%20y%20Planeacion/Uso%20de%20tierra%20y%20aguas%20adyacentes/PM%20Cumbres%20de%20Majalca%202016.pdf

DOF—Diario Oficial de la Federación. (2016, August 8). *Acuerdo por el que se da a conocer el Resumen del Programa de Manejo del Área Natural Protegida con el carácter de Parque Nacional Cumbres de Majalca.* Retrieved May 7, 2023, from https://dof.gob.mx/nota_detalle.php?codigo=5447074&fecha=08/08/2016&print=true%201/61

Eichelberger, S., Heigl, M., Peters, M., & Pikkemaat, B. (2021). Exploring the role of tourists: Responsible behavior triggered by the COVID-19 pandemic. *Sustainability, 13*(11), 1–14. https://doi.org/10.3390/su13115774

Elliot, S. (2014). Behavior, tourist. In J. Jafari & H. Xiao (Eds.), *Encyclopedia of tourism*. Springer. https://doi.org/10.1007/978-3-319-01669-6_17-1

European Commission. (n.d.). *Sustainable development goals*. Retrieved May 6, 2023, from https://commission.europa.eu/strategy-and-policy/international-strategies/sustainable-development-goals/eu-and-united-nations-common-goals-sustainable-future_en

Federal Agency for Nature Conservation. (2020, January 22). *Soft tourism*. Retrieved May 7, 2023, from http://web01.bfn.cu.ennit.de/activities/tourism-and-sports/tourism/ecotourism/soft-tourism/?L=1

Hosteltur. (2021). Sorpresas en el comportamiento del turista postpandemia. Retrieved May 7, 2023, from https://www.hosteltur.com/144781_sorpresas-en-el-comportamiento-del-turista-pospandemia.html

Kaefer, F. (2022). Jamie sweeting on planeterra and the future of responsible travel. In: *Sustainability leadership in tourism. Future of business and finance*. Springer. https://doi.org/10.1007/978-3-031-05314-6_53

Koval, V., Mikhno, I., Udovychenko, I., Gordiichuk, Y., & Kalina, I. (2021). Sustainable natural resource management to ensure strategic environmental development. *TEM Journal, 10*(3), 1022–1030. https://www.ceeol.com/search/article-detail?id=977526

McGinlay, J., Gkoumas, V., Holtvoeth, J., Fuertes, R. F., Bazhenova, E. Y., Benzoni, A., Botsch, K., Martel, C. C., Sánchez, C. C., Cervera, I., Chaminade, G., Doerstel, J., García, C. J., Jones, A., Lammertz, M., Lotman, K., Odar, M., Pastor, T., Ritchie, C. A., Santi, S., Smolej, M., Rico, F. S., Waterman, H., Zwijacz-Kozica, T., Kontoleon, A., Dimitrakopoulos, P. G., & Jones, N. (2020). The impact of COVID-19 on the management of European protected areas and policy implications. *Forests*, (11), 1214. https://www.semanticscholar.org/paper/The-Impact-of-COVID-19-on-the-Management-of-Areas-McGinlay-Gkoumas/1a371ff47bc70ff0400ce66037f48ecb9e3e849c

Naturalista. (2023). *Parque Nacional Cumbres de Majalca—CONANP*. https://www.naturalista.mx/projects/parque-nacional-cumbres-de-majalca-last-accessed-2023/05/07conanp?tab=species

SEMARNAT. (2016, January 9). *Parque Nacional Cumbres de Majalca*. Retrieved May 7, 2023, from https://www.gob.mx/semarnat/articulos/parque-nacional-cumbres-de-majalca

SEMARNAT. (2017, December 17). *Parques Nacionales de México*. Retrieved May 7, 2023, from https://www.gob.mx/semarnat/articulos/parques-nacionales-de-mexico

SERNATUR. (2011). *Manual de Buenas Prácticas, Chile por un Turismo Sustentable. Subsecretaría de Turismo*. https://chilesustentable.sernatur.cl/wp-content/uploads/2013/08/ManualGenericoBaja2.compressed.pdf

Sharmin, F., Sultan, M. T., Badulescu, A., Bac, D. P., & Li, B. (2020). Millennial tourists' environmentally sustainable behavior towards a natural protected area: An integrative framework. *Sustainability, 12*(20), 8545. https://www.mdpi.com/2071-1050/12/20/8545

Song, M., Fisher, R., & Kwoh, Y. (2019). Technological challenges of green innovation and sustainable resource management with large scale data. *Technological Forecasting and Social Change, 144*, 361–368. https://www.sciencedirect.com/science/article/abs/pii/S0040162518311478?via%3Dihub

Spenceley, A., McCool, S., Newsome, D., Báez, A., Barborak, J., Blye, C., Bricker, K., Cahyadi, H., Corrigan, K., Halpenny, E., Hvenegaard, G., King, D., Leung, Y., Mandić, A., Naidoo, R., Rüede, D., Sano, J., Sarhan, M., Santamaria, V., Sousa, T., & Zschiegner, A. (2021a). Tourism in protected and conserved areas amid the Covid-19 pandemic. *Parks*, (27) (Special Issue). https://parksjournal.com/wp-content/uploads/2021/03/Spenceley_et_al10.2305-IUCN.CH_.2021.PARKS-27-SIAS.en_.pdf

Spenceley, A., McCool, S., Newsome, D., Báez, A., Barborak, J. R., Blye, C. J., et al. (2021b). Tourism in protected and conserved areas amid the COVID-19 pandemic. *Parks, 27*, 103–118. https://parksjournal.com/wp-content/uploads/2021/03/Spenceley_et_al10.2305-IUCN.CH_.2021.PARKS-27-SIAS.en_.pdf

The Editors of Encyclopaedia Britannica. (2023, January 23). *National Park*. Retrieved May 7, 2023, from https://www.britannica.com/science/national-park

Thwaites, K., & Simkins, I. (2007). *Experiential landscape. An approach to people, place and space*. Routledge. https://www.routledge.com/Experiential-Landscape-An-Approach-to-People-Place-and-Space/Thwaites-Simkins/p/book/9780415340007#

UICN-WCPA. (2020, December 16). *Sustainable recovery of tourism in protected areas from the Covid-19 pandemic. Transforming vision into action—Oped.*

https://www.unwto.org/covid-19-oneplanet-responsible-recovery-initiatives/iucn-tapas-last accessed-2023/05/07group

UNEP—UNWTO. (2005). *Making tourism more sustainable—A guide for policy makers.* Retrieved May 7, 2023, from https://www.unep.org/resources/report/making-tourism-more-sustainable-guide-policy-makers

UNWTO. (2013). *EU guidebook on sustainable tourism for development.* Retrieved May 7, 2023, from https://www.unwto.org/EU-guidebook-on-sustainable-tourism-for-development#:~:text=Expressed%20simply%2C%20sustainable%20tourism%20can,environment%2C%20and%20host%20communities%E2%80%9D

UNWTO. (2020). *Código Ético Mundial para el Turismo.* Retrieved May 2, 2023, from https://www.unwto.org/es/codigo-etico-mundial-para-el-turismo

UNWTO. (n.d.). *Objetivos de desarrollo sostenible.* Retrieved May 2, 2023, from https://www.unwto.org/es/tourismo-por-los-ods

Wu, D., Li, K., Ma, J., Wang, E., & Zhu, Y. (2022). How does tourist experience affect environmentally responsible behavior? *Sustainability, 14*, 924. https://doi.org/10.3390/su14020924

Zahoor, A., Muhammad, M., Muhammad, N., & Malik, K. (2020). Moving towards a sustainable environment: The dynamic linkage between natural resources, human capital, urbanization, economic growth, and ecological footprint in China. *Resources Policy, 67*, 101677. https://doi.org/10.1016/j.resourpol.2020.101677

Zhang, L., & Zhang, J. (2020). Chapter 13—A systematic review on tourism energy consumption, sustainable tourism, and destination development: A behavioral perspective. In J. Zhang (Ed.), *Transport and energy research* (pp. 295–313). Elsevier. https://doi.org/10.1016/B978-0-12-815965-1.00013-2. https://www.sciencedirect.com/science/article/pii/B9780128159651000132

8

Transcending Beyond Sustainable Tourism Through a 'Well-Being' Perspective: The 'BEST' Sustainability Framework for the New Normal

Zeynep Gulen Hashmi and Sayyeda Zonah

Introduction

The emergence of the COVID-19 pandemic in 2019 has been one of the defining periods of this century, especially for the tourism industry. It has led the tourism industry to reset and rethink a more sustainable future. Unlike any other sector, tourism industry has been hit the hardest (Chaudhary, 2020; Tsai, 2021). The pandemic, coupled with the already prevailing sustainability challenges, has exacerbated all domains of tourism and travel, having harmed the physical, mental, social and economic well-being of the people around the world (Fletcher et al., 2021; Gowreesunkar et al., 2022b). Border closures and lockdowns have led to

Z. G. Hashmi (✉)
National University of Sciences & Technology, Islamabad, Pakistan
e-mail: gulen.hashmi@nbs.nust.edu.pk

S. Zonah
Yasar University, Izmir, Turkey

S. W. Maingi et al. (eds.), *Tourist Behaviour and the New Normal, Volume II*,
https://doi.org/10.1007/978-3-031-45866-8_8

a sharp fall in tourism demand, having affected the well-being of local communities that were heavily dependent on tourism for their survival in particular (Gowreesunkar et al., 2021; Scheyvens & van der Watt, 2021). On the positive side, COVID-19 served as an eye-opener for tourism businesses who saw various business opportunities in the pandemic such as integration of innovative digital technologies in their operations as well as more creativity and sustainability (Sheresheva et al., 2021). Artificial intelligence, internet tours, virtual tours and augmented reality have started to become widespread contributing to more sustainable destination strategies and a more sophisticated tourist experience.

On the demand side, COVID-19 provided tourists and travellers with an opportunity to increase their awareness of sustainability issues, particularly with regard to greater standards of hygiene and sanitation, and heightened awareness of safeguarding biodiversity (Gowreesunkar et al., 2022a). The pandemic led tourists and travellers to adopt new forms of tourism such as nature-based tourism activities, staycations, workcations and six-feet tourism, to name a few (Goel et al., 2021). This change in tourist behaviour was further catalyzed by tourism suppliers who offered information and incentives to the travellers to demonstrate responsible behaviour (Eichelberger et al., 2021). Some popular examples of these sustainability initiatives are stopping open food buffets to limit food waste, encouraging guests to limit their water consumption and minimizing the amount of linen wash in hotels. Thus, the pandemic served as a turning point, enabling tourists to turn away from unsustainable practices and mass tourism behaviour and move towards sustainable tourism (Ioannides et al., 2021).

According to UNWTO, sustainable tourism is defined as '*tourism that takes full account of its current and future economic, social and environmental impacts, addressing the needs of visitors, the industry, the environment and host communities*' (WTO Report, 2002, p. 7). The purpose of sustainable tourism is to make a balance between maintaining cultural integrity, promoting economic benefits, protecting the environment and establishing social justice so as to meet the needs of the host population while emphasizing inter- and intra-generational equity and feasibility in time (Sarfraz et al., 2022). From taxing tourist entries into cities like Venice to shutting down beaches in southeast Asian countries, these sustainable tourism initiatives range from mitigation of tourism's negative impacts to the complete neutralization of these impacts.

Yet, a new concept that goes beyond the concept of sustainable tourism has also emerged. This new form of tourism behaviour is called regenerative tourism and is rooted in adding a positive impact to environment and society at large. From the viewpoint of tourists and travellers, regenerative tourism brings about transformational experiences to tourists whereby they serve a true purpose to the local communities, which in turn makes their holidays become meaningful. This positive feeling and inspiration from their travel experience as well as the assurance that the local cultural heritage and traditions are conserved from one generation to the next, contribute to both personal and societal well-being.

Thus, regenerative tourism is holistic and has a 'well-being' approach. A 'well-being' perspective brings about an understanding of the various systems, interacting components and different stakeholders of the local tourism, as their happiness is the is the ultimate purpose of being sustainable in the tourism industry. It also encourages diversity in the local economic systems to minimize host communities' extreme dependency on tourism. These local communities are included in decision-making to bring value to the communities as well as responsibility towards the environment and its biodiversity. As Hashmi and Muff (2014) posit, it is this interconnectedness between the human systems, resources and the aim of societal well-being that requires holistic thinking.

The tourism sustainability framework, the ***Beyond Sustainable Tourism*** *(BEST)*, aims to contribute to the literature on sustainable tourism, drawing attention to the evolving phases of sustainable tourism and beyond based on new forms of tourism that have emerged from changing tourist behaviour. Tourism businesses and destinations need to manage tourism with an economic concern while simultaneously building competitive advantage through well-being concerns such as sustainable cities and communities (SDG11), and responsible production and consumption (SDG12). Moreover, tourism destinations need to contribute to economic growth through volume while also contributing to quality tourism management. Although a number of sustainability frameworks and models have been developed for the tourism industry, none of these conceptualized works has looked into the evolving phases of sustainable tourism in light of both growth of tourism and quality of tourism. In this respect, the *BEST* serves as an analytical tool to understand and challenge current and emerging tourism behaviour towards sustainability.

More importantly, the *BEST* framework, with its focus on quality of tourism through a 'well-being' concern, can be described as a robust guide to direct future tourism sustainability efforts to benefit society. Tourism leaders can use this sustainability framework to assess evolving tourism behaviour that goes beyond sustainability.

Sustainable Tourism Framework with a 'Well-Being' Touch: The *BEST*

The '***Be****yond* ***S****ustainable* ***T****ourism*' (BEST) sustainability framework, with its 'well-being' lens nested in a holistic worldview, aims to be a practical guide for sustainable destinations that aspire to transcend beyond sustainability by contributing a positive impact to environment and local communities. As an important driver of job creation, tourism serves the needs of travellers and tourists who are in contact with local communities and tourism suppliers. Tourists travel 365 days a year and 24 hours a day, which implies round-the-clock tourism demand in tourist destinations. Since tourism products are intangible and cannot be test-driven in advance, tourists and travellers are extremely sensitive with their choice of travel and accommodation. Moreover, most of the tourism products relate to tourists' experience, which is of a perishable nature. This necessitates driving volume and growth of tourism by tourism businesses and destinations, which might conflict with sustainable tourism behaviour. Finally, the industry faces social challenges such as excessively long working hours, a high turnover rate, a lack of social protection, low wages and gender-based discrimination (ILO, 2022).

Coupled with seasonality, part-time or temporary work, shift and night work as well as vast availability of subcontracting and outsourcing rates, which are common in the industry, there is an urgency to address these well-being challenges to enhance the sustainability of the tourism industry. Highlighting the importance of well-being would facilitate contribution to both economic development and quality of life of society. According to SITF, (2022), well-being entails the following major aspects: overall satisfaction with life (quality of life), education, social justice and health (SITF, 2022). In the context of health, COVID-19 protocols and guidelines helped instil sensitivity and responsibility towards hygiene and sanitation in tourists and travellers. In the context of education, various countries are now focusing on educational tourism as a development strategy in the new normal, with more and more travellers opting for edutourism in the form of educational trips, study tours and in-depth cultural experiences such as homestays. Regarding social justice, in the tourism and hospitality industry, COVID-19 impacted jobs and hours of work—particularly for women (Renaud et al., 2020). Reskilling, upskilling and equality in tourism and hospitality are thus increasingly important post-COVID. This is particularly pertinent for returning workers, including the female workforce. Finally, feeling of happiness and sense of pleasure are at the heart of tourism behaviour. Sustainable behaviour, for instance, is an experience for tourists in which they can pour emotions (Liu et al., 2022).

The BEST framework is inspired by Hashmi and Muff's (2015) S-WELL grid, which highlights the evolution of different types of hotel sustainability with regard to value creation and type of concern for driving sustainability initiatives. The BEST framework presents sustainable tourism as a multi-dimensional construct with two axes, key dimensions of which are: well-being concerns versus economic concerns on one axis; and volume of tourism versus quality of tourism on the other axis. The horizontal axis looks at tourism development in terms of volume and growth of tourism while also considering its contribution to quality of tourism. The vertical axis looks at concerns of evolving tourism behaviour in terms of economic concerns while simultaneously adding positive value through well-being concerns that originate from sustainability challenges. With regard to the nature of the tourism industry and the new forms of tourism that are currently prevalent, the four quadrants of the BEST sustainability framework are *Overtourism*, *Green Tourism*, *Responsible Tourism* and *Regenerative Tourism*. These quadrants depict different and evolving phases of sustainable tourism, which relate to both

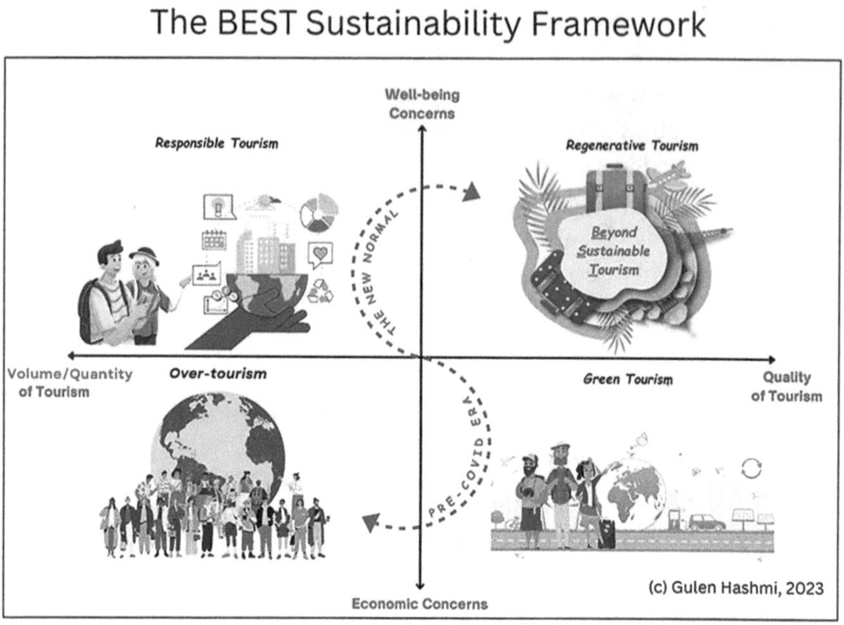

Fig. 8.1 The BEST (**B**eyond **S**ustainable **T**ourism) Sustainability Framework. (Source: Authors' own elaboration)

demand side (tourists and travellers) and supply side (tourism managers and destination planners) of changing tourism behaviour (see Fig. 8.1).

While the quadrant '*Overtourism*' implies the growing numbers of international tourists, crowding and congestion with focus on tourism growth, the quadrant '*Green Tourism*' characterizes tourism behaviour that goes beyond growth of tourism and engages in voluntary green choices and practices. These green initiatives add to enhancing the quality of tourism as there is increased sensitivity regarding the negative impact of tourism on environment and communities. Quadrant '*Responsible Tourism*' is where well-being concerns such as health and hygiene come into play with the ongoing conventional focus on increasing volume and growth of tourism. This phase involves new forms of tourism behaviour such as domestic travel, staycations, workcations and nature-based activities, which focus on mitigating negative impacts on the environment and communities. Finally, quadrant '*Regenerative*

Tourism' is where there is a shift in mindset to make a positive impact on society and environment through a rich and meaningful exchange among various tourism stakeholders, which ultimately add on to the quality of tourism. The pre-COVID period was already characterized by '*Overtourism*' and '*Green Tourism*'. However, with the onset of the pandemic, tourism behaviour started evolving towards '*Responsible Tourism*' and ultimately '*Regenerative Tourism*'.

The following sections elaborate on these four phases of sustainable tourism represented by the BEST sustainability framework. The quadrants challenge destination planners and tourism managers to reflect upon what type of tourism behaviour underlies with regard to type of tourism development and concerns for sustainable tourism behaviour.

The First BEST Quadrant: Overtourism

This is the initial phase of sustainable tourism, which has a mere focus on growth of tourism rooted in economic concerns. As a result of this growth orientation towards tourism, prior to the outbreak of the COVID-19 pandemic, the global tourism sector had witnessed almost constant growth and diversification, with domestic tourism having remained the key driver of the global tourism industry, and international tourism having expanded beyond long-term growth forecasts (UNWTO, 2022). This has led to an overcrowded influx of tourists causing 'overtourism' in high density in many popular destinations. *Overtourism* refers to a scenario when certain residents of a given area believe that tourism has negatively impacted their quality of life, which has resulted in resistance or protests against tourism, visitors, decision-makers or economic interests (UNWTO, 2018). It is a neologism that shows the overcrowding of tourists on a holiday destination. This type of mass tourism causes pollution, the devastation of nature and distress of local populations.

While many tourists want to have an immersive experience and 'live like a local', it impacts the residents by making irreversible changes to their lifestyles, accessibility to facilities and general well-being (Milano et al., 2018). In Iceland, for instance, the government has taken steps to modernize Iceland's parking lots, bridges, and roadways, and increased the number of restrooms and trash cans in rural areas in response to disruptive tourist behaviour (Goodwin, 2017). Similarly, having lost two-thirds of its residents due to overtourism over the past 70 years, Venice presently has a residential population of 57,000, down from 174,000 in 1951. The city is visited by twenty million people a year and devastated also by cruise ships causing pollution.

Famous films that make tourist destinations famous, the ease with which one can reach any corner of the world, and the increase in low-cost air flights and cruises that bring large quantities of tourists, are only some of the causes leading to overtourism. Moreover, demand is rising in some established tourist sites as a result of new tourism phenomena that calls for cheap transportation and short-term rental platforms (Alonso-Almeida et al., 2019). Finally, the impact of the internet and other social media platforms has served to draw attention to few places with significant disruptions such as Palma de Mallorca, Paris, Dubrovnik, Kyoto, Berlin, Bali and Reykjavik (Milano et al., 2018). For instance, the Quarry Bay neighbourhood in Hong Kong has turned into a popular tourist destination, and even banners that prohibit taking pictures or upsetting residents have had no impact. Another example is the influx of Chinese tourists in Barcelona due to top-flight tour operators and increased social media image of the city (Alonso-Almeida et al., 2019).

The above-mentioned negative consequences of overtourism such as destruction of natural ecosystems, increase in amount of waste and escape and malaise of residents have necessitated concerted efforts in the direction of sustainable development and led to the emergence of green tourism behaviour (Korstanje & George, 2020). Thus, the following section focuses on '*Green Tourism*' as the next phase in sustainable tourism, and elaborates on green initiatives that heavily shaped the progression of sustainable tourism behaviour prior to the pandemic.

The Second BEST Quadrant: Green Tourism

Green tourism has long been at the forefront of UN Sustainable Development Goals agenda (UNEP, 2011). The increase in environmental awareness campaigns, in particular, stated the need for a low-carbon era, which led to the consideration of green tourism as the future direction for sustainable tourism development (Sarkar & George, 2018). In this regard, green destinations align with SDG 11 of 'Sustainable Cities and Communities', a good example of which is Bali, a famous tourist destination in Indonesia, which adapted to the green economy framework. Bali revisits general tourism policies to gain economic return through sustainable practices, and reduces negative environmental and social impacts (Law et al., 2016). In Malaysia, for instance, green tourism practices revolve around encouraging the use of green products, introducing waste management techniques such as recycling and providing training programmes on environmental management (The Star, 2010). The Green Tourism Business Scheme in the UK, on the other hand, assures tourists that certain environmental criteria such as energy efficiency, waste minimization and recycling, use of local produce, and support of public transport have been met by the hotels and lodging establishments they choose. Finally, in the context of SDG 13 'Climate Action', the tourism industry also advocates green decision-making in business operations as well as pro-environmental resource management.

Green tourism is concerned with ecological footprint which is further involved with product life cycle as well as practices such as using plastic, lighting, cooling, heating, laundry, shampoo bottles and paper (Pan et al., 2018). Environmental practices such as eco-cuisine, the installation of hinge-activated lighting and the replacement of paper check-in with electronic methods have emerged (Borisenko, 2018). On the side of tourism businesses, the three main incentives for businesses to go green are competitiveness, legitimation, cost and risk reduction as well as good corporate image (Rahman et al., 2012). This, in turn, has implications on tourist behaviour because consumers (tourists) tend to prefer a green hotel over a conventional hotel if they are incentivized and rewarded for using it (Fukey & Issac, 2014). These rewards might range from discounts to reward points for patronizing green hotels.

Green tourists are defined as tourists who behave in an environmentally friendly manner in various tourism contexts when on holiday (Dolnicar & Matus, 2008). They avoid products that have negative impacts related to health of the environment, unnecessary waste, use of materials from endangered species as well as use of large amounts of resources. Green attitudes are increasingly evolving from environmental concerns towards pro-environmental tourism behaviour (Hou & Wu, 2021). Although this change of direction from growth of tourism (overtourism) towards quality of tourism (green tourism) can be considered as a progressive leap in the evolution of sustainable tourism in the pre-COVID-19 period, it is largely rooted in economic concerns where discounts and incentives for tourists, cost reduction and profit maximization for tourism businesses, and competitiveness and marketing for destinations are at play.

The Third BEST Quadrant: Responsible Tourism

The concept of Triple Bottom Line (TBL) is the overarching framework upon which most responsible tourism initiatives are built (Aguiñaga et al., 2018). TBL is the underlying principle for balancing economic, social, and environmental initiatives, and has been widely implemented in India, southeast Asia and Africa under responsible tourism. Health and hygiene measures and certifications such as the *Responsible Travel Code* and the *Responsible Travel Hub* in California constitute other examples of responsible tourism (WTTC Report, 2020). Goodwin (2014) describes responsible tourism as '*making better places for people to live in and better places for people to visit*'. Morrison (2022), on the other hand, highlights it as accepting responsibility for one's actions as a traveller or as a service provider. This is further supported by UN's SDG11 and SDG12, which highlight the importance of delivering responsible tourism at the individual level through ethical consumerism (Burrai et al., 2019). Also, according to the Cape Town Declaration (2002), responsible tourism requires tourists to generate and support economic benefits for local communities, be culturally sensitive especially when entering certain religious grounds, have enjoyable experiences with local communities and protect the cultural heritage to maintain the world's diversity. An example of this is tourists who opt for 'slow' tourism, whereby these responsible tourists stay at a destination for several days and support local businesses and eateries rather than just visit a destination during peak season (Eichelberger et al., 2021).

Indeed, the post-pandemic 'new normal' led to leaning more towards taking responsibility for one's own actions and creating a mindful-driven world for future generations (Gowreesunkar et al., 2021; Stankov et al., 2020). The advent of new consumer segments in tourism that are already drawn to products and services with mindfulness-related themes is evidence for the current growing number of compassionate and responsible travellers. Untact tourism, coined from the longer phrase 'undoing contact' in 2020 relates to one of these mindfulness-related themes as it describes tourist experience or tourism activity designed to facilitate social isolation in outdoor spaces from botanic gardens to golden sandy beaches. Although the term referred to social distancing during the pandemic, it refers to use of non-digitalised untact activities such as outdoor walks or remote nature camping in the new normal. South Korea implemented 'untact' tourism to reinforce eco-tourism in their local communities and promote health and well-being of public (Bae & Chang, 2021). Staycation is another mindfulness-related theme in the new normal. It describes a new tourism segment whereby people take holidays at their own home or in the vicinity rather than visit another place. Staycations ensure quality time and cater to well-being of individuals, enabling them to re-energize at their own pace (Angela, 2023).

As can be seen, responsible tourism is concerned with economic viability, social justice as well as the physical environment, which differentiate it from green tourism that involves mainstream green issues and green concerns. Although responsible tourism behaviour emerges from well-being concerns, and has the three-dimensional concerns of profit, people and planet, it focuses on minimizing negative impacts, and restoring and stimulating sustainable growth in volume. Thus, it is limited to growth of tourism rather than quality of tourism. Considering the pressing need to enhance the quality of tourism for sustainable development, there is a dire need to amplify the positive impact of tourism globally and locally. One way of doing this is to contribute to sustainable societies themselves and this requires taking care of well-being concerns and qualitative tourism development, which are embedded in the concept of 'Regenerative Tourism'. Thus, the final quadrant in the BEST framework elaborates on 'Regenerative Tourism', which is the most advanced phase of sustainable tourism.

The Fourth BEST Quadrant: Regenerative Tourism

Emerging from the need to move from sustainable growth in volume to a more qualitative tourism development with well-being concerns, regenerative tourism is a concept that builds upon the idea of 'building back better'. It aims at adding a positive impact to environment and local communities in a holiday destination. Regenerative tourism is described as sustainable regeneration because it represents an ultra responsible method of travelling and exploring new locations, with the aim of leaving the tourism destination in a better state than before (Ateljevic, 2020; Cheer et al., 2021). It is a concept that goes beyond sustainable tourism as it not only focuses on minimization of adverse impacts of tourism but also maximization of its positive impacts (Bellato et al., 2022). Tourism professionals working with farmers together to restore a desert, or conserving a rainforest with the added value of an ecolodge are some good examples of regenerative tourism.

Regenerative tourism can take various forms. Following the UN's SDG11 and SDG13, the city of Amsterdam introduced 'Doughnut Economics' in April 2020. As a result, the Netherlands Board of Tourism & Conventions (NBTC) unveiled a plan to steer tourists away from Amsterdam by concentrating its marketing and advertising efforts on locations other than the city (CBI, 2022). Similarly, rural tourism has emerged as another mindful practice for regenerative tourism. Traena, a rural town in Norway, has integrated innovative set of facilities including a hotel, museum, cabins for rent, offices and a cultural centre. *Traena 365* will provide services to both locals and tourists, helping to create a thriving local community all year long, promoting local culture and offering visitors unforgettable experiences (Mwesiumo et al., 2022).

Striving together and contributing to social projects is another mindful characteristic of regenerative tourism. One of the most untamed and biodiverse places in the world is Costa Rica's Osa Peninsula, and the Lapa Ros Lodge is situated right in the centre of it all. Many excursions are available for guests to see their restorative work of this lodge which includes 'Twigs, Pigs, and Trash' and 'Local Medicine' trips, which teach about traditional medicinal plants (CBI, 2022). Similarly, in light of Saudi Arabia's 2030 vision, which aims at quality of tourism, the Red Sea and Amaala Projects are ambitious sustainability projects to foster and encourage regenerative growth by integrating training of locals for the future of hospitality and tourism (Dartford, 2021).

There are also some examples of regenerative tourism which are more specific to the hospitality industry. Six Senses Resorts' 'Sustainability and Earth Lab' educates guests about sustainable management of a hotel from construction to operations, which raises their guests' awareness and creates a positive impact. Similarly, the Four Seasons Resort Maui offers its guests a guest-volunteering experience to teach about Maui's past, which includes a session of shoreline clean-up work in collaboration with the Pacific Whale Foundation and the Lahaina Restoration Foundation. By actively engaging with the hotel, guests not only get rewarded for participating but also find a sense of purpose in leaving Maui as a better place than before.

As can be seen, regenerative tourism is not only about bettering local economies, preserving biodiversity and local cultures but also about offering authentic and life-changing experiences to guests who go home with memories, friends and meaningful projects for the environment and society at large. In this regard, regenerative tourism, might be considered as 'beyond sustainable tourism' as well as the next step in sustainable tourism.

Conclusion and Implications

The 'New Normal' of the COVID-19 pandemic shows that the COVID-19 health crisis has not only altered tourist behaviour patterns, but also the sustainable development pattern of the global tourism industry (Gowreesunkar et al., 2021, 2022b). Understanding tourist behaviour and travel habits in terms of changes in tourists' motivations, travel mode and trip frequencies is crucial for healthy development of the tourism industry. Thus, this book chapter contributes to shedding light on how changes in tourism behaviour in the pre-COVID-19 period and the 'new normal', have shaped sustainable tourism development. The chapter showcases an evolving sustainable tourism landscape through the help of the new BEST sustainability framework, which highlights implications of the changing tourism behaviour with a focus on well-being.

Overtourism and green tourism have been the focus of sustainable tourism development until the outbreak of the COVID-19 pandemic, yet there is more to sustainable tourism development than stopping food buffets to limit food waste, encouraging hotel guests to minimize their water consumption and reducing the washing of bed linens. In the post-pandemic 'new normal' we are currently in, tourism behaviour, in terms of both demand and supply side, has been evolving more towards responsible tourism behaviour and regenerative tourism (UNWTO, 2022). Yet, perceptions regarding the meaning of sustainable tourism still differ among tourism professionals, tourists, academia and destination planners. The BEST sustainability framework, in this regard, aims to provide academia, industry practitioners and policy-makers with a multi-dimensional conceptualization to help distinguish among different phases of sustainable tourism originating from changing tourism behaviour involving stakeholders of the tourism industry.

With tourism being the engine of economic development, destinations need to align their sustainability efforts and strategies to well-being concerns, and solve sustainability challenges by going beyond responsible tourism and pro-environmental behaviour that merely aim at minimizing the negative impacts of their actions on the environment and society at large. Although neutralizing tourism's negative impacts on the planet is part of the solution, it is not sufficient to add a positive impact to environment and local communities such as quality of life of residents and indigenous groups in a tourism destination. Going beyond sustainable tourism requires a shift in mindset about what sustainable tourism really means in this new normal. The BEST sustainability framework highlights this emerging shift in mindset, and serves to guide the sustainable tourism domain through a well-being perspective, which takes into account the quality of tourism in addition to its growth. This theoretical contribution is the first attempt to link well-being and happiness to sustainable tourism to bridge the gap between purpose of tourism development and society.

The BEST sustainability framework is a proposed theoretical model, and would require empirical research to confirm extent of its prevalence and applicability in the tourism industry. It is also likely to be replicated in other industries which also have large negative impacts on society such as the mining or oil industry. The framework offers important insights, especially to destination planners and tourism managers who are interested in developing competitive destinations through addressing well-being concerns. Last but not least, the BEST sustainability framework can be considered as a significant contribution to sustainable tourism research, which is gaining more importance in line with the needs of the younger generation and growing environmental and social awareness of our society.

References

Aguiñaga, E., Henriques, I., Scheel, C., & Scheel, A. (2018). Building resilience: A self-sustainable community approach to the triple bottom line. *Journal of Cleaner Production, 173*, 186–196.

Alonso-Almeida, M.-d.-M., Borrajo-Millán, F., & Yi, L. (2019). Are social media data pushing overtourism? The case of Barcelona and Chinese tourists. *Sustainability, 11*(12), 3356. https://doi.org/10.3390/su11123356

Angela. (2023, January 4). *The benefits of staycation trips will surprise you!* global gadding. Global Gadding. Retrieved March 31, 2023, from https://www.globalgadding.com/benefits-of-staycation/

Ateljevic, I. (2020). Transforming the (tourism) world for good and (re)generating the potential 'new normal'. *Tourism Geographies, 22*(3), 467–475. https://doi.org/10.1080/14616688.2020.1759134

Bae, S. Y., & Chang, P. J. (2021). The effect of coronavirus disease-19 (COVID-19) risk perception on behavioural intention towards 'untact' tourism in South Korea during the first wave of the pandemic (March 2020). *Current Issues in Tourism, 24*(7), 1017–1035.

Bellato, L., Frantzeskaki, N., & Nygaard, C. A. (2022). Regenerative tourism: A conceptual framework leveraging theory and practice. *Tourism Geographies, 25*(1), 1–21.

Borisenko, S. (2018). *Tourists' willingness to pay for Green Hotel practices*. Doctoral dissertation, Instituto Politecnico de Leiria, Portugal.

Burrai, E., Buda, D. M., & Stanford, D. (2019). Rethinking the ideology of responsible tourism. *Journal of Sustainable Tourism, 27*(7), 992–1007.

Cape Town Declaration. (2002). Cape Town Conference on responsible tourism in destinations. Retrieved from http://www.capetown.gov.za/en/tourism/Documents/Responsible%20Tourism/Toruism_RT_2002_Cape_Town_Declaration.pdf

CBI. (2022, May 2). *Regenerative tourism: Moving beyond sustainable and responsible tourism*. Retrieved from https://www.cbi.eu/market-information/tourism/regenerative-tourism

Chaudhary, A. (2020). Impact and survival strategy for hospitality industry after COVID-19. *International Journal of Innovative Science and Research Technology, 5*, 489–492.

Cheer, J. M., Ting, H., & Leong, C. M. (2021). Responsible tourism: A new era of responsibility? *Journal of Responsible Tourism Management, 1*(1), 1–17.

Dartford, K. (2021, November 16). Why Saudi Arabia is an inspiring destination for regenerative tourism. *Euronews*. Retrieved March 2023, from https://www.euronews.com/travel/2021/11/16/saudi-arabia-is-this-the-worlds-most-inspiring-regenerative-tourism-initiative

Dolnicar, S., & Matus, K. (2008). Are green tourists a managerially useful target segment? *Journal of Hospitality & Leisure Marketing, 17*(3–4), 314–334.

Eichelberger, S., Heigl, M., Peters, M., & Pikkemaat, B. (2021). Exploring the role of tourists: Responsible behavior triggered by the COVID-19 pandemic. *Sustainability, 13*(11), 5774.

Fletcher, R., Blanco-Romero, A., Blázquez-Salom, M., Cañada, E., Murray Mas, I., & Sekulova, F. (2021). Pathways to post-capitalist tourism. *Tourism Geographies, 27*, 1–22.

Fukey, L., & Issac, S. (2014). Connect among green sustainability and hotel industry: A prospective simulation study. *International Journal of Social, Behavioural, Educational, Economic, Business and Industrial Engineering, 8*(1), 296–312.

Goel, P., Garg, A., Walia, N., Kaur, R., Jain, M., & Singh, S. (2021). Contagious diseases and tourism: A systematic review based on bibliometric and content analysis methods. *Quality & Quantity*, 1–26. https://doi.org/10.1007/s11135-021-01270-z

Goodwin, H. (2014). Responsible tourism and the green economy. In *Green growth and travelism* (pp. 157–168). Routledge.

Goodwin, H. (2017). The challenge of overtourism. *Responsible Tourism Partnership, 4*, 1–19.

Gowreesunkar, V., Maingi, S., Roy, H., & Micera, R. (2021). *Tourism destination management in a post pandemic context: Global issues and destination management solutions* (1st ed.). Emerald Publishing Limited.

Gowreesunkar, V. G., Maingi, S. W., Roy, H., & Micera, R. (2022a). Rebuilding tourism post pandemic–Policy recommendations from global case studies. *Emerald Open Research, 4*(2), 2.

Gowreesunkar, V. G., Maingi, S. W., & Ming'ate, F. L. M. (Eds.). (2022b). *Management of tourism ecosystem services in a post pandemic context: Global perspectives*. Taylor & Francis.

Hashmi, Z. G., & Muff, K. (2014). Rethinking corporate sustainability within the 'well-being' context: The 'COGWHEEL' sustainability framework. *Building Sustainable Legacies, 2*, 67–96.

Hou, H., & Wu, H. (2021). Tourists' perceptions of green building design and their intention of staying in green hotel. *Tourism and Hospitality Research, 21*(1), 115–128.

ILO. (2022). *World employment and social outlook—Trends 2022*. International Labour Organization.

Ioannides, D., Gyimóthy, S., & James, L. (2021). From liminal labor to decent work: A human-centered perspective on sustainable tourism employment. *Sustainability, 13*(2), 851.

Korstanje, M. E., & George, B. P. (2020). Education as a strategy to tackle over tourism for overtourism and inclusive sustainability in the twenty-first century. *Overtourism: Causes, Implications and Solutions*, 341–359. https://doi.org/10.1007/978-3-030-42458-9_18

Law, A., De Lacy, T., Lipman, G., & Jiang, M. (2016). Transitioning to a green economy: The case of tourism in Bali, Indonesia. *Journal of Cleaner Production, 111*, 295–305.

Liu, X., Zeng, Y., He, J., & Li, Z. (2022). Value cocreation research in tourism and hospitality: A comparative bibliometric analysis. *International Journal of Contemporary Hospitality Management, 34*(2), 663–686.

Milano, C., Cheer, J. M., & Novelli, M. (2018). Overtourism: A growing global problem. *The Conversation, 18*(1), 1–5.

Morrison, A. M. (2022). *Tourism marketing in the age of the consumer*. Routledge.

Mwesiumo, D., Halfdanarson, J., & Shlopak, M. (2022). Navigating the early stages of a large sustainability-oriented rural tourism development project: Lessons from Træna. *Norway. Tourism Management, 89*, 104456.

Pan, S. Y., Gao, M., Kim, H., Shah, K. J., Pei, S. L., & Chiang, P. C. (2018). Advances and challenges in sustainable tourism toward a green economy. *Science of the Total Environment, 635*, 452–469.

Rahman, D. H. A. A., Dayang-Affizzah, A. M., & Edman, S. (2012). Tourism and hotels in Sarawak: Economic performance. *Procedia-Social and Behavioral Sciences, 65*, 1020–1026.

Renaud, L. et al. (2020). Covid-19 impacted jobs and hours of work—particularly for women.

Sarfraz, M., Raza, M., Khalid, R., Ivascu, L., Albasher, G., & Ozturk, I. (2022). Coronavirus disease 2019 safety measures for sustainable tourism: the mediating effect of tourist trust. *Frontiers in Psychology, 13*. https://doi.org/10.3389/fpsyg.2022.784773

Sarkar, S. K., & George, B. (2018). Social media technologies in the tourism industry: An analysis with special reference to their role in sustainable tourism development. *International Journal of Tourism Sciences, 18*(4), 269–278.

Scheyvens, R., & van der Watt, H. (2021). Tourism, empowerment and sustainable development: A new framework for analysis. *Sustainability, 13*(22), 12606.

Sheresheva, M., Efremova, M., Valitova, L., Polukhina, A., & Laptev, G. (2021). Russian tourism enterprises' marketing innovations to meet the COVID-19 challenges. *Sustainability, 13*(7), 3756.

Stankov, U., Filimonau, V., & Vujičić, M. D. (2020). A mindful shift: An opportunity for mindfulness-driven tourism in a post-pandemic world. *Tourism Geographies, 22*(3), 703–712.

The Social Impact Taskforce. (2022). *We are grateful to our funders for their financial …—Impact taskforce*. Retrieved April 2023, from https://www.impact-taskforce.com/media/gq5j445w/time-to-deliver-final.pdf

The Star. (2010). 10 hotels in Malaysia now hold Green Award status. http://thestar.com.my/news/story.asp?file=/2010/1/26/nation/5545100&sec=nation

Tsai, M. C. (2021). Developing a sustainability strategy for Taiwan's tourism industry after the COVID-19 pandemic. *PLoS One, 16*(3), e0248319.

UNEP. (2011). Climate change and POPS: predicting the impacts. Report of the UNEP/AMAP Expert Group Secretariat of the Stockholm Convention, Geneva, Switzerland.

UNWTO. (2018). *Overtourism? Understanding and managing urban tourism growth beyond perceptions*. UNWTO.

UNWTO. (2022). International tourism and COVID-19. World Tourism Organization. Retrieved March 30, 2023 from https://www.unwto.org/international-tourism-and-covid-19.

World Tourism Organization. (2002). World Tourism in 2002: Better than expected. Retrieved March 28, 2022 from http://www.world-tourism.org/newsroom/Releases/archives.htm.

World Travel & Tourism Council. (2020). *To recovery & beyond: The future of travel & tourism in the wake of COVID-19*. World Travel & Tourism Council.

9

Ecotourism Practices, Perspectives, and Consumer Preferences, Attitudes, and Expectations: Post-COVID-19 Review

Deborah Kangai, Eliyas Ebrahim Aman, and Árpád Ferenc Papp-Váry

Introduction

Tourism has been an important economic activity in most countries around the world, accounting for 10% of the global Gross Domestic Product (GDP) and 25% of new jobs before the COVID-19 pandemic (WTTC, 2022). It is often touted as a major economic activity that has the greatest contribution to the revenue and quality of life of the indigenous community in a destination. It stimulates economic growth by generating income, employment, investment, and exports and generates

D. Kangai (✉)
Alexandre Lamfalussy Faculty of Economics, István Széchenyi Economics and Management Doctoral School, University of Sopron, Sopron, Hungary

Department of Environmental Studies and Resources Development, Chuka University, Chuka, Kenya
e-mail: deborah.kangai@phd.uni-sopron.hu; deborah.kangai@chuka.ac.ke

S. W. Maingi et al. (eds.), *Tourist Behaviour and the New Normal, Volume II*,
https://doi.org/10.1007/978-3-031-45866-8_9

other valuable benefits, such as the preservation of cultural heritage, improved infrastructure, and local community facilities (UNWTO, 2010; WTTC, 2022). This indicates that the industry has high significance.

Although tourism has numerous benefits for many countries, regions, and communities, including economic ones, the rapid growth of traditional mass tourism has hurt the environment and local cultures in many places (Neto, 2003). These impacts include, but are not limited to, natural resource depletion and environmental degradation in many popular tourism regions. Tourism has the most negative effects on natural resources, ecosystems, and the cultures of local communities (Cobbinah et al., 2017; Gowreesunkar et al., 2022). Sustainable tourism and travel practices such as ecotourism and responsible travel should therefore be developed to reduce the impacts of traditional tourism.

The rise of the COVID-19 pandemic led to a detrimental impact on global industries, with no exception to tourism and ecotourism destinations (Gowreesunkar et al., 2022). It led to a decline in the performance of the international tourism industry in global destinations (Korstanje et al., 2022). According to UNWTO (2021), the COVID-19 pandemic had a negative impact on the tourism industry and caused approximately 74% decline in international visitor numbers in 2020. This is because the pandemic contributed to travel restrictions, border closures, the closure of many tourism facilities, and a decrease in consumer confidence in travel (Korstanje et al., 2022; UNWTO, 2021; WTTC, 2021). About

E. E. Aman
Alexandre Lamfalussy Faculty of Economics, István Széchenyi Economics and Management Doctoral School, University of Sopron, Sopron, Hungary
e-mail: eliyas.ebrahim.aman@phd.uni-sopron.hu

Á. F. Papp-Váry
Faculty of International Management and Business, Budapest Business University, Budapest, Hungary

Alexandre Lamfalussy Faculty of Economics, Lamfalussy Research Center, University of Sopron, Sopron, Hungary
e-mail: papp-vary.arpad@uni-bge.hu; papp-vary.arpad@uni-sopron.hu

96% of worldwide destinations introduced travel restrictions in response to the pandemic, which consequently resulted in a significant decline in tourism activities, with an estimated 70% decline in international tourist arrivals in 2020 (UNWTO, 2020). This resulted in a 5.5% decline in the contribution of the tourism and travel industry to the global GDP and approximately 62 million people (WTTC, 2020).

In the aftermath of the COVID-19 pandemic, the tourism and travel industry remains a key player in global economic recovery (WTTC, 2020) and the achievement of Sustainable Development Goals (SDGs). The multidisciplinary nature of tourism perfectly links it to all the SDGs. It is therefore important to rethink the global practice of tourism in the new normal. In order to meet economic, social, and environmental needs through socio-cultural and ecological integrity, crucial ecological processes, protection of biodiversity and cultural diversity, along with other benefits (Gowreesunkar et al., 2022; Streimikiene et al., 2020), destinations must continue to prioritize the adoption of sustainable tourism and travel practices.

Ecotourism provides alternative and more sustainable livelihood sources to local communities, promotes conservation of natural and cultural resources, and provides ecological experiences to travelers. Ecotourism also provides opportunities for the local community and travelers to participate in the conservation of the environment and biodiversity of the area, and biodiversity in return provides economic incentives to the local community (CREST, 2020; Dorofeeva et al., 2020; Rafa et al., 2021; Ren et al., 2021). The achievement of the aims of ecotourism will however, be more dependent on the motivations of the travelers to the destinations, especially in the post-COVID-19 era. Further, ecotourism must consider the social, economic, and environmental implications of a destination.

The purpose of this chapter is therefore to explore ecotourism, responsible tourism, and environmental tourism while focusing on ecotourism practice, perspectives, and consumer preferences, attitudes, and expectations. The results also provide the readership with grounded theoretical and conceptual foundations on emerging environmental trends and how they influence tourists' behavior toward sustainable ecotourism and responsible tourism development. This would aid the destination managers in providing and devising tactics to improve current and future ecotourism practices.

Literature Review

Despite the linkage between conservation and sustainable management in tourist destinations (Tampakis et al., 2019), there is a need for sustainable tourism policies, coordination of relevant sectoral policies, innovation in tourism products, sustainable mobility, and tools for measuring impacts (UNWTO, 2021). The industry's pursuit of responsible and sustainable tourist practices has paved the path for sustainable tourism development (Aquino et al., 2018), which would support sustainable development.

Brundtland (1987) defines sustainable development as development that meets the needs of the present without compromising the ability of future generations to meet their own needs. Similarly, the concept of sustainable development is employed in sustainable tourism to refer to tourism that enables present-day tourists to enjoy their needs without compromising the ability of future generations of tourists to meet their needs (*Sustainable Tourism for Development Guidebook - Enhancing Capacities for Sustainable Tourism for Development in Developing Countries*, 2013). The core principles of sustainable tourism development are grounded in the coordination of economic, social, and environmental aspects (Dias et al., 2014; OECD, 2001; Shaker & Sîrodoev, 2016). These include providing a high-quality experience for visitors, maintaining the quality of the environment, and improving the quality of life of the host community, on which both the host community and the visitors depend (Ray, 2015). With respect to this, the tourism developers should put emphasis on developing and promoting sustainable tourism practices and consumption.

The increasing demand for sustainable tourism practices led to the emergence of ecotourism in the 1980s. Ecotourism is currently one of the fastest-growing tourism market segments worldwide (Nematpour & Faraji, 2019). It is considered to have a growth rate three times faster than the entire tourism industry (Carvache-Franco et al., 2019). The authors also note that the number of travelers who prefer ecotourism increases by 20% annually, generating about 10 to 20% of the overall global tourism market, thus making it the fastest-growing sector of the tourist industry. Ecotourism is also considered one of the five leading strategic areas for future development in the tourism industry (Dorofeeva et al., 2020). Ecotourism is particularly vital for consistent and sound tourist behavior and the balanced utilization and management of natural resources (Rafa et al., 2021; Ren et al., 2021) and cultural resources. This growth of ecotourism presents excellent opportunities for destinations with natural and cultural resources to tap into them for responsible tourism development.

The COVID-19 pandemic did not only adversely affect the global tourism industry but also ecotourism practice and consumption. Many activities in ecotourism destinations were reduced, while some attraction sites were completely closed. This was catalyzed by the numerous travel restrictions and the concern travelers had for their health and safety. The decrease in the number of visitors adversely impacted the local community's economic welfare due to limited sources of livelihood (*Sustainable Report*, 2021). For example, Galapagos, a popular ecotourism destination, experienced a 97% decline in visitor numbers in April 2020, which negatively affected the community's source of livelihood (*Galapagos Annual Report*, 2020). Similarly, low revenues lowered the amount of money allocated for conservation purposes in Tourism Promotion Zones (TPZ).

Currently, many destinations are recovering from the economic shock of the COVID-19 pandemic (Gowreesunkar et al., 2022). This implies that destinations should focus on increasing the number of visitors and their spending at those destinations. This, along with other factors held constant, would put pressure on the existing infrastructure and other facilities in the destinations (Gössling et al., 2018; Scott & Gössling, 2022). A sustainable approach to recovering from the impact of COVID-19 should therefore be the main focus of destination managers. The proponents of sustainability have sought various methods to promote a shift to sustainable production and consumption for about three decades since the 1992 United Nations Conference on Environment and Development in Rio de Janeiro (Cohen, 2019; Foden et al., 2019). In the

1990s, there was great emphasis on the promotion of cleaner and more efficient processes for manufacturing consumer goods and their intermediary inputs (Hertwich, 2005).

In the early 2000s, "greener" forms of household provisioning were emphasized by educating consumers, designing eco-labels on product packages, and encouraging buyers to make responsible choices (Akenji et al., 2016). Currently, the aim is to achieve absolute reductions in unsustainable consumption levels (Cohen, 2019; Foden et al., 2019) through systemic change of the social and institutional arrangements that propagate modern consumerist habits. This makes ecotourism important since it accounts for the social, economic, and environmental wellbeing of destinations (Kiper, 2013).

Methods

The study started with the identification of the data sources. These included Scopus and Web of Science-indexed journals, Google Scholar, published books, and reports related to sustainable tourism and travel. The reports used were sourced from the United Nations World Tourism Organization, Center for Responsible Travel, Booking.com, World Travel and Tourism Council, Galapagos Conservation Trust, and United Nations News websites. After identifying the data sources, key words in the titles and abstracts were considered to select the articles related to ecotourism, responsible tourism, environmental travelers, environmental trends, and tourists' behavior in the new normal and related terms in many forms. The study focused on studies and reports published in English. The study utilized a total of 43 journal articles, 15 reports, and seven books. After the data sources were selected, the researchers were guided by the following questions to achieve the objectives of the chapter:

- What are the ecotourism practices and perspectives in the new normal?
- What are the trends in ecotourism and responsible tourism in the new normal?
- What are the ecotourists preferences, attitudes, and expectations in the new normal?

Discussion

Ecotourism Practices and Perspectives in the New Normal

Ecotourism is a type of sustainable tourism that focuses on preserving and protecting natural environments while giving visitors the chance to learn and have fun (CREST, 2020). Ecotourism practice and demand have been growing over the years in many global destinations (Das & Chatterjee, 2015), thereby presenting great opportunities for destinations to achieve Sustainable Development Goals. Nature-based and mountain tourism have been identified as the major pillars for sustainable development in rural and mountainous areas, presenting new economic opportunities for environmental protection (Carvache-Franco et al., 2019; Tampakis et al., 2019) through ecotourism.

Sustainable Development Goals (SDGs) can be achieved in many destinations with the help of ecotourism. Destinations could also focus on low-density and eco-sustainable tourism services, which could help lower health risks. This is a great chance for ecotourism, which is a high-value, fast-growing industry with a lot of demand from people with high incomes. Destinations that practice ecotourism include Costa Rica (*From the Field: Costa Rica Points the Way to a Sustainable World*, 2021), the Galapagos Islands, Ecuador (Burbano et al., 2022), Papua (Bohensky et al., 2011), Bhutan, Norway (Gössling et al., 2021), and Kenya, among others. The Costa Rican government, for example, supports ecotourism to improve quality and social impact (Goretti et al., 2021).

In the last few decades, ecotourism and responsible tourism have become more popular as practices and choices. People who travel are now more aware of how tourism affects the environment and want to travel in a more responsible way (Carvache-Franco et al., 2019). Ecotourism is based on the idea that tourism should have fewer negative effects on the environment, economy, and native communities (Fennell, 2017). This means that ecotourism operators should carefully manage their activities to ensure that they do not cause detrimental impacts to the natural environment, economy, or local communities. For example, by limiting the number of people who visit certain areas, limiting access to fragile ecosystems and habitats, and using efficient, low-impact transportation methods like bicycles or electric vehicles (Fennell & Cooper, 2020).

Ecotourism operators are focusing on the health and safety of all stakeholders, including sanitation protocols, protective equipment, and limiting visitors to ensure safety (CREST, 2020). Similarly, outdoor and nature-based activities such as hiking, birdwatching, and wildlife safaris that allow visitors to experience nature while maintaining safe distances are becoming more popular (Fennell, 2017) and support local conservation efforts and involve the local community in the planning and management of ecotourism. The increased interest in eco-friendly facilities such as ecolodges and camping sites has also led to the development of more sustainable accommodations in destinations. Ecotourism has always focused on sustainability, but in the new normal, there is great emphasis on responsible tourism practices such as reducing and recycling waste, supporting local conservation efforts, and conserving water and energy, among others (Aziz et al., 2020), as well as meeting the changing needs of travelers.

Trends in Ecotourism and Responsible Tourism in the New Normal

Responsible tourism refers to a form of tourism that aims to minimize the negative impacts of tourism on the environment, local communities, and cultural heritage while providing positive and authentic travel experiences to visitors as well as maximizing the positive impacts of tourism on these same areas (CREST, 2020). Ecotourism and responsible tourism are closely related since they are grounded on similar principles. There have been several trends in responsible tourism and ecotourism in recent decades.

The trends include the adoption of sustainable tourism practices such as the use of renewable energy sources, reduction of waste and pollution, and conservation of natural resources (Gössling et al., 2021), the promotion of community-based tourism (Gössling et al., 2018) through the provision of accommodation and food, the showcasing of cultural and natural resources (Tampakis et al., 2019), and the use of technology to enhance the tourism experience (Lytras et al., 2010; Navío-Marco et al., 2018). Technology can be used to provide tourists with access to information, facilitate communication, and improve the efficiency of tourism services (Neuhofer et al., 2015; Koh et al., 2023; Xiang et al., 2017), such as making bookings and reservations, revenue management, ticketing, property management, and guest and employee security, among others.

There has been an increasing growth in staycations, especially after COVID-19 (Muritala et al., 2022). Staycations, which refer to vacations spent at home or within one's local area, have become popular due to a number of factors, such as economic uncertainty (APA, 2019), environmental concerns, and the desire for meaningful and authentic travel experiences. Staycations encourage local community and cultural engagement (Doğan & Jelinčić, 2023). These could include visits to local museums, cultural activities, parks, and natural places. Staycations reduce carbon emissions and environmental impact (CREST, 2020), and travelers can save money on local travel and still take a respite (APA, 2019). Staycations can make it hard to disengage from work and other duties and urge people to fall back into usual routines, which can restrict the sense of adventure and exploration that many people associate with travel (Wong et al., 2021). Despite these drawbacks, staycations are an amazing way to relax without overstretching one's budget or harming the environment (APA, 2019).

Another emerging trend includes tourism and climate change, coastal and marine ecotourism, the blue economy, and creative tourism (Castanho et al., 2023). Tourism and global warming are interconnected. Tourism significantly contributes to climate change, and climate change has serious implications for the tourism industry (Rowe, 2019). Greenhouse gas emissions from the tourism industry contribute to global warming and climate change (Gössling & Peeters, 2015). Transportation, lodging, and other tourism-related businesses currently account for about 8% of world greenhouse gas emissions (Lenzen et al., 2018; Scott et al., 2023), and future tourism is expected to increase these emissions (Gössling et al., 2023) due to the expected growth post-COVID-19.

On the other hand, climate change affects the tourism industry in various ways, such as changes in weather (Scott et al., 2019), a rise in sea levels (Becken & Loehr, 2022; Turner et al., 2022), which could affect beach and marine tourism activities, and increased frequency and severity of extreme weather events that could lead to damage to tourism attractions and resources and changes in touristic activities. The world currently requires 30 years to decarbonize the economy for it to stay within the safe boundaries of global warming (Scott et al., 2023). This presents a great challenge to the tourism industry (Gössling & Higham, 2021; Gössling & Schweiggart, 2022), which is still recovering from the shock of COVID-19. Destinations should therefore focus on sustainable tourism practices such as waste reduction and recycling, the use of renewable energy sources, the adoption of transport means that emit low carbon emissions into the atmosphere, and the education of stakeholders (Gössling & Schweiggart, 2022) in the tourism industry on issues and adaptation tactics to climate change.

Consumer Preferences, Attitudes, and Expectations in the New Normal

Responsible tourism requires a change in the mindset of consumers from a focus on consumption and entertainment to one on sustainability and responsible behavior (Gössling et al., 2018). Recent studies indicate that the pandemic has greatly improved tourists' travel behaviors and destination choices, making them more ecologically mindful and changing their inclinations toward inland destinations close to natural surroundings (Kupi & Szemerédi, 2021). Despite the revitalization of tourism facing uncertainties, the changing trends in tourists' behaviors offer great opportunities for the industry to reinvent its practices in favor of sustainable development (Kupi & Szemerédi, 2021). The COVID-19 pandemic has led to a new normal in the global tourism industry, and ecotourism is no exception.

As ecotourists grow more conscious of how their travel and activities affect the ecology, economics, and socio-cultural systems of their local communities, their views and preferences shift (Carvache-Franco et al., 2019). Ecotourists are currently seeking safety and health measures, which also influence the supply side of the industry. There has been more emphasis on social distancing, virtual tours, enhanced cleaning protocols, and contactless check-ins (UNWTO, 2023; CREST, 2020). More so, ecotourists are seeking accommodation facilities that are constructed using sustainable, locally made materials, use green energy sources such as solar, wind, and tidal energy, and have minimal impact on the environment (Carvache-Franco et al., 2019).

Many ecotourists seek out real experiences that let them connect with nature and the people who live there. They like to do nature-based activities like go hiking, watch birds, and look at wildlife, as well as culture-related activities like visiting local markets or taking part in traditional ceremonies and festivals (Anabela et al., 2020; Carvache-Franco et al., 2019). Ecotourists may also prefer to stay in eco-friendly accommodation facilities such as ecolodges or campsites that use renewable energy sources, recycle waste, and minimize their environmental footprint. The COVID-19 pandemic has had a significant impact on changes in consumer preferences for ecotourism. They are more concerned about their health and safety when traveling and therefore seek out ecotourism suppliers that prioritize these factors. According to Rariel (2022) 53% of international travelers now put health and safety before other considerations when choosing a travel destination,

Further, travelers are seeking out sustainable and responsible destinations (Rariel, 2022). The Global Sustainable Tourism Council (GSTC) found in a survey conducted in 2020 that 82% of passengers thought eco-friendly travel was important. Tourists are familiarizing themselves with how their travel decisions influence the environment and local populations. Tourism should support the conservation of natural resources and the local communities' social and economic welfare (Cobbinah et al., 2017). Therefore, consumers' attitudes are also changing toward sustainability. They are seeking destinations that not only support conservation but also those that provide opportunities for them to support the local communities through their travel choices (Anabela et al., 2020). According to a survey conducted by the GSTC in 2020, 72% of travelers believe that tourism should support the conservation of natural and cultural heritage. Rariel (2022) conducted a similar study and found that 67% of respondents believed that it was important to support the local community when traveling.

Consumers expect that destinations should prioritize health and safety measures to ensure safe and amusing experiences (Anabela et al., 2020). According to a survey conducted by the GSTC in 2020, 84% of travelers expect that tourism suppliers will implement health and safety measures in response to a pandemic. They also expect destinations to prioritize sustainability and responsible travel options. Rariel (2022) conducted a study and found that 70% of global travelers believe that travel companies should offer more sustainable travel options.

Conclusion

Ecotourism and responsible tourism have been growing over the years in many global destinations. The growth has been characterized by changing practices, perspectives, and consumer preferences, attitudes, and expectations in the aftermath of the COVID-19 pandemic. The travelers are more aware of the impacts of tourism on the social, cultural, economic, and environmental soundness of destinations. The current practices include the use of low-impact transportation, such as cycling; limiting access to fragile ecosystems; limiting visitor numbers; and focusing on the health and safety of the stakeholders while observing relevant sanitation protocols. The activities include a wide range of outdoor activities such as hiking, nature walking, birdwatching, and photography, among others. There is increased interest in eco-friendly facilities and a focus on responsible tourism practices while meeting the needs of tourists.

The trends in ecotourism and responsible travel include the adoption of sustainable tourism practices such as the use of renewable energy sources, the reduction of waste and pollution, the conservation of natural resources, the promotion of community-based tourism, and the use of technology to enhance the tourism experience. There has been an increasing growth in staycations, increased debate on tourism and climate change, coastal and marine ecotourism, the blue economy, and creative tourism.

The finding from the literature also indicates that the pandemic has greatly improved tourists' travel behaviors and destination choices, making them more socio-culturally, economically, and ecologically mindful while changing their inclinations toward inland destinations close to natural surroundings. Many ecotourists seek out real experiences that let them connect with nature and the local people and prefer to stay in eco-friendly accommodation facilities. They prefer destinations that prioritize health and safety over other factors when choosing a travel destination. Further, travelers are seeking out sustainable and responsible destinations. The consumer's attitudes are also changing toward sustainability, and the expectations are that destinations will prioritize health and safety measures to ensure safe and amusing experiences and prioritize sustainability and responsible travel options.

The study also notes that destinations are recovering from the shock of the pandemic. In the wake of attracting more visitors to attraction sites, the authors recommend that destination managers take appropriate measures to ensure the sustainability of ecotourism and responsible tourism in light of the new trends in consumer behavior.

References

Akenji, L., Bengtsson, M., Bleischwitz, R., Tukker, A., & Schandl, H. (2016). Ossified materialism: Introduction to the special volume on absolute reductions in materials throughput and emissions. *Journal of Cleaner Production, 132*, 1–12. https://doi.org/10.1016/j.jclepro.2016.03.071

American Psychological Association. (2019). Stress in America: Stress and current events. Stress in America TM survey. https://www.apa.org/news/press/releases/stress/2019/stress-america-2019.pdf

Anabela, M. S., Carmen, M., Karel, H., & Alessandro, R. (2020). Behavioral changes in tourism in times of Covid-19: Employment scenarios and policy options. *Research Papers in Economics*. https://doi.org/10.2760/00411

Aquino, R. P., Lueck, M., & Schänzel, H. (2018). A conceptual framework of tourism social entrepreneurship for sustainable community development. *Journal of Hospitality and Tourism Management, 37*, 23–32. https://doi.org/10.1016/j.jhtm.2018.09.001

Aziz, N., Mihardjo, L. W., Sharif, A., & Jermsittiparsert, K. (2020). The role of tourism and renewable energy in testing the environmental Kuznets curve in the BRICS countries: Fresh evidence from methods of moments quantile regression. *Environmental Science and Pollution Research, 27*(31), 39427–39441. https://doi.org/10.1007/s11356-020-10011-y

Becken, S., & Loehr, J. (2022). Tourism governance and enabling drivers for intensifying climate action. *Journal of Sustainable Tourism.* https://doi.org/10.1080/09669582.2022.2032099

Bohensky, E., Butler, J. N., & Mitchell, D. G. (2011). Scenarios for knowledge integration: Exploring ecotourism futures in Milne Bay, Papua New Guinea. *Journal of Marine Biology, 2011*, 1–11. https://doi.org/10.1155/2011/504651

Brundtland, G. H. (1987). Our common future-call for action. *Environmental Conservation, 14* (4), 291–294. Cambridge University Press. https://doi.org/10.1017/s0376892900016805

Burbano, D. V., Valdivieso, J. A., Izurieta, J. C., Meredith, T. C., & Ferri, D. (2022). "Rethink and reset" tourism in the Galapagos Islands: Stakeholders' views on the sustainability of tourism development. *Annals of Tourism Research Empirical Insights, 3*(2), 100057. https://doi.org/10.1016/j.annale.2022.100057

Carvache-Franco, M., Peiro-Signes, A., & Carrascosa-López, C. (2019). Segmentation and motivations in eco-tourism: The case of a coastal national park. *Ocean & Coastal Management, 178*, 104812. https://doi.org/10.1016/j.ocecoaman.2019.05.014

Castanho, R. A., Santos, C., & Couto, G. (2023). Creative tourism in islands and regional sustainable development: What can we learn from the pilot projects implemented in the Azores territory? *Land, 12*(2), 498. https://doi.org/10.3390/land12020498

Center for Responsible Travel. (2020). *The case for responsible travel: Trends & statistics 2020.* https://www.responsibletravel.org/wp-content/uploads/sites/213/2021/03/trends-and-statistics-2020.pdf

Cobbinah, P. B., Black, R., & Thwaites, R. (2017). *Ecotourism implementation in the Kakum Conservation Area, Ghana: Administrative framework and local community experiences* (pp. 129–148). Routledge EBooks. https://doi.org/10.4324/9781315205205-9

Cohen, M. J. (2019). Introduction to the special section: Innovative perspectives on systems of sustainable consumption and production. *Sustainability: Science, Practice and Policy, 15*(1), 104–110. https://doi.org/10.1080/15487733.2019.1703331

Das, M., & Chatterjee, B. (2015). Ecotourism: A panacea or a predicament? *Tourism Management Perspectives, 14*, 3–16. https://doi.org/10.1016/j.tmp.2015.01.002

Dias, N., Curwell, S., & Bichard, E. (2014). The current approach of urban design, its implications for sustainable urban development. *Procedia. Economics and Finance, 18*, 497–504. https://doi.org/10.1016/s2212-5671(14)00968-x

Doğan, E., & Jelinčić, D. A. (2023). Changing patterns of mobility and accessibility to culture and leisure: Paradox of inequalities. *Cities*. https://doi.org/10.1016/j.cities.2022.104093

Dorofeeva, A., Shamayeva, N. P., & Nyurenberger, L. (2020). Ecotourism as a factor of the development of off-season tourism in southern Russia. *IOP Conference Series, 421*(6), 062016. https://doi.org/10.1088/1755-1315/421/6/062016

Fennell, D. A. (2017). *Tourism ethics*. Channel View Publications.

Fennell, D. A., & Cooper, C. (2020). *Sustainable tourism: Principles, contexts and practices*. Channel View Publications.

Foden, M., Browne, A., Evans, D. M., Sharp, L., & Watson, M. (2019). The water–energy–food nexus at home: New opportunities for policy interventions in household sustainability. *The Geographical Journal, 185*(4), 406–418. https://doi.org/10.1111/geoj.12257

From the Field: Costa Rica points the way to a sustainable world. (2021, November 19). *UN News*. https://news.un.org/en/story/2021/10/1103812

Galapagos Annual Report. (2020). Retrieved March 30, 2023, from https://reports.glpg.com/annual-report-2020/en/.

Goretti, M., Leigh, L. Y., Babii, A., Cevik, S., Kaendera, S., Muir, D., Nadeem, S., & Salinas, G. (2021). Tourism in the post-pandemic world. Departmental Paper, 2021(002), 1. https://doi.org/10.5089/9781513561905.087

Gössling, S., Balas, M., Mayer, M., & Sun, Y. (2023). A review of tourism and climate change mitigation: The scales, scopes, stakeholders and strategies of carbon management. *Tourism Management, 95*, 104681. https://doi.org/10.1016/j.tourman.2022.104681

Gössling, S., & Higham, J. P. (2021). The low-carbon imperative: Destination management under urgent climate change. *Journal of Travel Research, 60*(6), 1167–1179. https://doi.org/10.1177/0047287520933679

Gössling, S., & Peeters, P. (2015). Assessing tourism's global environmental impact 1900–2050. *Journal of Sustainable Tourism, 23*(5), 639–659. https://doi.org/10.1080/09669582.2015.1008500

Gössling, S., & Schweiggart, N. (2022). Two years of COVID-19 and tourism: What we learned, and what we should have learned. *Journal of Sustainable Tourism, 30*(4), 915–931. https://doi.org/10.1080/09669582.2022.2029872

Gössling, S., Scott, D., & Hall, C. M. (2018). Global trends in length of stay: implications for destination management and climate change. *Journal of Sustainable Tourism, 26*(12), 2087–2101. https://doi.org/10.1080/09669582.2018.1529771

Gössling, S., Scott, D., & Hall, C. M. (2021). Pandemics, tourism and global change: A rapid assessment of COVID-19. *Journal of Sustainable Tourism, 29*(1), 1–20. https://doi.org/10.1080/09669582.2020.1758708

Gowreesunkar, V., Maingi, S., & Mingate, F. (2022, September 30). *Status of the tourism ecosystem services: Marking more realities. management of tourism ecosystem services in a post pandemic context: Global perspectives.* Routledge & CRC Press. http://www.routledge.com/9781032248080/

Hertwich, E. G. (2005). Life cycle approaches to sustainable consumption: A critical review. *Environmental Science & Technology, 39*(13), 4673–4684. https://doi.org/10.1021/es0497375

Kiper, T. (2013). *Role of ecotourism in sustainable development.* InTech EBooks. https://doi.org/10.5772/55749

Koh, L. Y., Wu, M., Wang, X., & Yuen, K. F. (2023). Willingness to participate in virtual reality technologies: Public adoption and policy perspectives for marine conservation. *Journal of Environmental Management, 334*, 117480. https://doi.org/10.1016/j.jenvman.2023.117480

Korstanje, M. E., Seraphin, H., & Maingi, S. W. (2022). *Tourism through troubled times: Challenges and opportunities of the tourism industry in 21st century.* Emerald Publishing Limited.

Kupi, M., & Szemerédi, E. (2021). Impact of the COVID-19 on the destination choices of Hungarian tourists: A comparative analysis. *Sustainability, 13*(24), 13785. https://doi.org/10.3390/su132413785

Lenzen, M., Sun, Y., Faturay, F., Ting, Y., Geschke, A., & Malik, A. (2018). The carbon footprint of global tourism. *Nature Climate Change, 8*(6), 522–528. https://doi.org/10.1038/s41558-018-0141-x

Lytras, M., De Pablos Patricia, O., Damiani, E., & Diaz, L. (2010). *Digital culture and E-Tourism: Technologies, applications and management approaches: Technologies, applications and management approaches.* IGI Global.

Muritala, B. A., Hernández-Lara, A., & Sánchez-Rebull, M. (2022). COVID-19 staycations and the implications for leisure travel. *Heliyon, 8*(10), e10867. https://doi.org/10.1016/j.heliyon.2022.e10867

Navío-Marco, J., Ruiz-Gómez, L. M., & Sevilla-Sevilla, C. (2018). Progress in information technology and tourism management: 30 years on and 20 years after the internet—revisiting Buhalis & Law's landmark study about eTourism. *Tourism Management, 69*, 460–470. https://doi.org/10.1016/j.tourman.2018.06.002

Nematpour, M., & Faraji, A. (2019). Structural analysis of the tourism impacts in the form of future study in developing countries (case study: Iran). *Journal of Tourism Futures, 5*(3), 259–282. https://doi.org/10.1108/jtf-05-2018-0028

Neto, F. R. (2003). A new approach to sustainable tourism development: Moving beyond environmental protection. *Natural Resources Forum, 27*(3), 212–222. https://doi.org/10.1111/1477-8947.00056

Neuhofer, B., Buhalis, D., & Ladkin, A. (2015). *Technology as a catalyst of change: Enablers and barriers of the tourist experience and their consequences* (pp. 789–802). Springer EBooks. https://doi.org/10.1007/978-3-319-14343-9_57

OECD, D. (2001). *The DAC guidelines: Strategies for sustainable development: Guidance for development cooperation*. Development Cooperation Committee, Organisation for Economic Co-operation and Development, Paris. Retrieved March 16, 2023, from www.OECD.org

Rafa, N., Nuzhat, S., Uddin, S. M. N., Gupta, M. K., & Rakshit, R. (2021). Ecotourism as a forest conservation tool: An NDVI analysis of the Sitakunda Botanical Garden and Ecopark in Chattogram. *Bangladesh. Sustainability, 13*(21), 12190. https://doi.org/10.3390/su132112190

Rariel. (2022). Booking.com sustainable travel report 2020. *GSTC*. Retrieved March 30, 2023, from https://www.gstcouncil.org/booking-com-sustainable-travel-report-2020/

Ray, N. (2015). *Emerging innovative marketing strategies in the tourism industry*. IGI Global.

Ren, L., Li, J., Li, C., & Dang, P. (2021). Can ecotourism contribute to ecosystem? Evidence from local residents' ecological behaviors. *Science of the Total Environment, 757*, 143814. https://doi.org/10.1016/j.scitotenv.2020.143814

Rowe, A. (2019). Sustainability-ready evaluation: A call to action. *New Directions for Evaluation, 2019*(162), 29–48. https://doi.org/10.1002/ev.20365

Scott, D., & Gössling, S. (2022). A review of research into tourism and climate change—Launching the annals of tourism research curated collection on tourism and climate change. *Annals of Tourism Research, 95*, 103409. https://doi.org/10.1016/j.annals.2022.103409

Scott, D., Hall, C. M., & Gössling, S. (2019). Global tourism vulnerability to climate change. *Annals of Tourism Research, 77*, 49–61. https://doi.org/10.1016/j.annals.2019.05.007

Scott, D., Hall, C. M., Rushton, B., & Gössling, S. (2023). A review of the IPCC Sixth Assessment and implications for tourism development and sectoral climate action. *Journal of Sustainable Tourism*, 1–18. https://doi.org/10.1080/09669582.2023.2195597

Shaker, R. R., & Sîrodoev, I. (2016). Assessing sustainable development across Moldova using household and property composition indicators. *Habitat International, 55*, 192–204. https://doi.org/10.1016/j.habitatint.2016.03.005

Štreimikienė, D., Švagždienė, B., Jasinskas, E., & Simanavičius, A. (2020). Sustainable tourism development and competitiveness: The systematic literature review. *Sustainable Development, 29*(1), 259–271. https://doi.org/10.1002/sd.2133

Sustainability Report. (2021). Retrieved March 30, 2023, from https://www.themomentum.com/reports/booking-holdings-2021-sustainability-report.

Sustainable Tourism for Development Guidebook - Enhancing capacities for Sustainable Tourism for development in developing countries. (2013). https://doi.org/10.18111/9789284415496

Tampakis, S., Andrea, V., Karanikola, P., & Pailas, I. (2019). The growth of mountain tourism in a traditional forest area of Greece. *Forests, 10*(11), 1022. https://doi.org/10.3390/f10111022

Turner, R. A., Forster, J., Fitzsimmons, C., & Mahon, R. (2022). Expanding narratives of governance constraints to improve coral reef conservation. *Conservation Biology, 36*(5). https://doi.org/10.1111/cobi.13933

United Nations World Tourism Organisation. (2020). Tourism and COVID-19: UNWTO guidelines to restart tourism. Retrieved March 26, 2023, from https://webunwto.s3.eu-west-1.amazonaws.com/s3fs-public/2020-05/UNWTO-Global-Guidelines-to-Restart-Tourism.pdf

United Nations World Tourism Organization. (2010). World tourism barometer and statistical annex. Retrieved March 26, 2023, from https://www.e--unwto.org/doi/abs/10.18111/wtobarometereng.2010.08.1.

United Nations World Tourism Organization (2021). 2020: Worst Year in Tourism History with 1 Billion Fewer International Arrivals, 28th January News Release, UNWTO, Madrid. Available online: https://webunwto.s3.eu-west-1.amazonaws.com/s3fs-public/2021-01/210128-barometer-en.pdf?VersionId=GaI1QTYG.Ky9LDZ2tlDKc.iRZkinJeuH (Assessed 26th November, 2023).

Wong, I. A., Lin, Z., & Kou, I. E. (2021). Restoring hope and optimism through staycation programs: An application of psychological capital theory. *Journal of Sustainable Tourism, 31*(1), 91–110. https://doi.org/10.1080/09669582.2021.1970172

World Travel & Tourism Council. (2020). *Travel & tourism: Global economic impact & trends 2020*, London, UK. Retrieved May 21, 2023, from https://wttc.org/Portals/0/Documents/Reports/2020/Global%20Economic%20Impact%20Trends%202020.pdf?ver=2021-02-25-183118-360

World Travel and Tourism Council. (2021). Retrieved May 21, 2023, from https://wttc.org/Research/Economic-Impact

World Travel and Tourism Council. (2022). Economic impact reports. Retrieved March 26, 2023, from https://wttc.org/Research/Economic-Impact

World Tourism Organization. (2023). The End of COVID-19-related Travel Restrictions – Summary of findings from the COVID-19-related Travel Restriction reports, UNWTO, Madrid, https://doi.org/10.18111/9789284424320

Xiang, Z., Du, Q., Ma, Y., & Fan, W. (2017). A comparative analysis of major online review platforms: Implications for social media analytics in hospitality and tourism. *Tourism Management, 58*, 51–65. https://doi.org/10.1016/j.tourman.2016.10.001

10

Post-COVID-19 Era: Possibilities for Responsible Tourism

Darius Liutikas, Edis Kriaučiūnas, Viktorija Baranauskienė, and Gintarė Pociūtė-Sereikienė

Introduction

There is no doubt that the COVID-19 pandemic has affected all economic sectors, and tourism has become one of the most affected areas. Researchers in the field of tourism immediately began to examine not only the impact of the COVID-19 pandemic (e.g., Hall et al., 2020; Higgins-Desbiolles, 2020; Škare et al., 2021; Seyfi & Hall, 2020; Gössling et al., 2021), the measures applied to reduce the spread of virus (e.g., Capano et al., 2020; Cheng et al., 2020; Moon, 2020; Hale et al., 2022; Pociūtė-Sereikienė et al., 2022), but also the scenarios and models of tourism development in the post-COVID-19 period (e.g., Brouder, 2020; Haywood, 2020; Zhang et al., 2021; Lew et al., 2022; Sharma et al., 2021b; Hussain & Fusté-Forné, 2021; El-Said & Aziz, 2022).

D. Liutikas (✉) • E. Kriaučiūnas • V. Baranauskienė • G. Pociūtė-Sereikienė
Department of Regional and Urban Studies, Institute of Sociology, Lithuanian Centre for Social Sciences, Vilnius, Lithuania

S. W. Maingi et al. (eds.), *Tourist Behaviour and the New Normal, Volume II*,
https://doi.org/10.1007/978-3-031-45866-8_10

A number of studies in tourism have investigated crisis management and tourism resilience after the COVID-19 pandemic (e.g., Higgins-Desbiolles, 2020; Sharma et al., 2021b; Reddy et al., 2020; Rastegar et al., 2021b). Great attention was also directed towards behavioural changes of the travellers (e.g., Pappas, 2021; Dogramadjieva & Terziyska, 2022; Kırlar-Can & Ertaş, 2022; Shin et al., 2022). Such studies discuss tourism trends and possible transformation of the tourism sector and proposed renewal scenarios after the end of the COVID-19 pandemic. Some researchers (e.g., Benjamin et al., 2020; Brouder, 2020; Niewiadomski, 2020; Romagosa, 2020; O'Connor & Assaker, 2021; Abbas et al., 2021; Gössling & Schweiggart, 2022) claim that the COVID-19 pandemic was an opportunity to change former trends making tourism more sustainable, responsible, and climate friendly. However, the COVID-19 pandemic makes it possible to look more broadly and to link such different conceptions as resilience, digital innovations, sustainability, accessibility, and new business management ideas into unified tourism future development scenarios.

The key elements of the post-COVID-19 tourism discussion are related to the development of local (domestic) tourism, application of technological innovations (e.g., virtual reality and augmented reality), sustainability and responsibility in travelling. All these aspects are related to the change of tourist behaviour. Even risk perception, as noticed by Gu et al. (2022), should be integrated into analysis of tourist behaviour intention.

Su (2022) noticed that the COVID-19 pandemic is stimulating the market of Information and Communication Technology. Technology Acceptance Model (TAM) has come under new scrutiny (El-Said & Aziz, 2022; Liutikas, 2023). Different studies suggest that possible tourists actively used virtual reality as a travel substitute during and after the COVID-19 pandemic. Scholars (Sharma et al., 2021a; Rastegar et al., 2021a; El-Said & Aziz, 2022; Sarkady et al., 2021) agree that the implementation of innovations poses challenges, but it was positively accepted as the best alternative, especially for the lockdown period. Corbisiero and Monaco (2021) after analysis of the changes in Italians' travel behaviour suggests that virtual travel and augmented reality are not a valid alternative to real travel; however, they agree that the COVID-19 pandemic accelerated the use of new technologies (Corbisiero & Monaco, 2021).

Many authors consider the importance of local communities for the restart of tourism (Higgins-Desbiolles et al., 2021; Corbisiero & Monaco, 2021; Koščak & O'Rourke, 2021). Korstanje and George (2022) discuss the positive impact of COVID-19 on tourism as an opportunity for innovation, sustainability, and performance improvement. They raise such questions as the impact of robots in tourism and hospitality service, prospects of space tourism, the role of information and communication technologies, the future of authenticity conception, and the tourist consciousness at the time of uncertainness (Korstanje & George, 2022). Mensah and Boakye (2021) analyse the case of Ghana and propose that the post-COVID-19 tourism industry could be sustained by the use of digitalization, social media, and a revived domestic tourism industry (Mensah & Boakye, 2021).

The main goal of this chapter is to discuss the changes of tourism behaviour and new possibilities of responsible tourism. Had the COVID-19 pandemic influenced the use of new tourist behavioural models? The possible changes of post-COVID-19 tourist behaviour is analysed using social, environmental, and technological factors. This chapter emphasizes the conceptual transformations of the travelling experience. The chapter tackles such issues as the tourist motivation to behave differently and the speed of adaptation to the innovative changes that appeared in the tourism and leisure management.

Data Collection and Methods

This chapter is based on the assessment of primary research data that was collected during the in-depth questionnaires (quantitative analysis method) and the focus group discussion (qualitative analysis method) of tourism experts located in different Lithuanian cities and regions. The research was carried out during the period February–May 2022.

The questionnaire survey involved state tourism institutions (Tourist Information Centres in regions—TICs) and different types of tourism business enterprises (tour operators, specialized guides, walking/bike-riding/ boating tours organizers, etc.), working in the field of inbound and local tourism. The questionnaire involved 7 closed-ended, 12 semi-closed, and 19 open-ended questions. In general, the questionnaire consisted of 38 questions. In total the 95 (sample size) questionnaires were sent to eligible tourism business enterprises and 57 (sample size) questionnaires were sent to state tourism institutions, located in different Lithuanian cities and regions. As a final sample, 50 TICs and 52 tourism business enterprises' completed questionnaires were received, compiled in SPSS spreadsheets, and analysed.

In February 2022 two focus groups discussions were organized. The main criterion behind the selection of respondents for the focus groups was expertise (work experience, active involvement in tourism business, leading position). The first focus group involved ten tourism business experts including representatives of the Ministry of Economy and Innovation, Lithuanian tourism association, Lithuanian guides union, the head of the tour operator enterprise, guides, and other experts. The second focus group discussion was held online and involved 30 experts representing TICs from all around Lithuania. In our case, the online discussion was the most suitable way to gather together as many representatives as possible even from the most distant regions of Lithuania saving their time and travel expenses. The data collected from focus groups was qualitative; it was transcribed and analysed, and some examples of answers were presented in the article in order to supplement the quantitative survey (questionnaire).

The questioners and the focus group discussion were based on the same question groups in order to collect the widest possible range of views and opinions from different tourism experts on the similar issues. The questions were structured into five main groups: general evaluation of the situation, state (governmental) aid, innovations, local tourism and regions, sustainability, and climate change. In this chapter, the greatest attention was paid to the questions that discussed innovations, sustainability, climate, and tourist behaviour change.

Rethinking Tourist Behaviour After COVID-19: Time for Responsibility? Literature Review

The Cape Town Declaration on Responsible Tourism stated the main attributes of responsible tourism (International Conference on Responsible Tourism in Destinations, 2002). This kind of tourism:

- minimises negative economic, environmental and social impacts;
- generates greater economic benefits for local people and enhances the well-being of host communities, improves working conditions and access to the industry;
- involves local people in decisions that affect their lives and life changes;
- makes positive contributions to the conservation of natural and cultural heritage, to the maintenance of the world's diversity;
- provides more enjoyable experiences for tourists through more meaningful connections with local people, and a greater understanding of local cultural, social and environmental issues;
- provides access for people with disabilities and the disadvantaged;
- is culturally sensitive, engenders respect between tourists and hosts, and builds local pride and confidence.

Same common responsibilities as environmental protection, respect of local cultures, benefit of local communities, conservation of natural resources, and minimization of pollution were also identified in Goodwin and Francis' (2003) often-cited article on ethical and responsible tourism. Koščak and O'Rourke (2021) described responsible tourism as "the type of tourism which sustains and develops local communities in a balanced way for the benefit of future generations" (Koščak & O'Rourke, 2021, p. xii). Goodwin (2016) analysed such aspects of responsible tourism as social responsibility in business, responsible destination marketing, and responsibility in environmental, social, and economic areas.

Such notion and agenda of responsible tourism can play a new role in the post-COVID-19 tourism era. The question is whether the COVID-19 pandemic will encourage a more responsible view at the world and people's influence on it. Koščak and O'Rourke (2021) noticed that locally based sustainable tourism may have a brighter prospect over the medium to longer term (Koščak & O'Rourke, 2021, p. xii). Sharma and Arora (2022) identified such post-COVID-19 dimensions as local tourism and domestic travel and virtual events as a critical success.

Seyitoğlu and Costa (2022) suggested six main categories (financial strategies, travellers' expectations and confidence, coordination and collaboration, employment, (post-)pandemic tourism marketing, and sustainable (post-)pandemic tourism) to include into a scenario planning framework for the post-pandemic tourism in European destinations. Gu et al. (2022) include social, economic, and environmental conditions in the System Dynamics analysis of tourism future. Liutikas (2023) included social, environmental, and technological factors in the analysis of the post-COVID-19 tourism.

United Nations World Tourism Organization (UNWTO) One Planet Sustainable Tourism Programme recommended six lines of action to guide a future of responsible tourism (One Planet…, 2020) in the post-COVID-19 era. These lines are related to the principles of responsible tourism in the areas of public health, social inclusion, biodiversity conservation, climate action, circular economy and governance, and finance. The COVID-19 crisis is also an opportunity to stimulate sustainable consumption and production. The crisis showed that tourism has limits and must be managed. This management has to integrate local supply chains, circularity, resource efficiency, and new sustainable business models (One Planet…, 2020).

Assaf et al. (2022) highlighted a research agenda for post-COVID-19 tourism. Post-COVID-19 research pillars in tourism include consumer behaviour (the impact of COVID-19 on tourist behaviour), demand and performance modelling (at the destination level analysis of spatial effects on various geographical scales, such as country, regions, cities, and rural areas), forecasting, destination and facility management, information technology, quality of life (importance of sustainability and safety), and use of climate-friendly tourism products (Assaf et al., 2022).

Palacios-Florencio et al. (2021) stated that sustainable tourism can become a solution to the crisis caused by COVID-19. Maybe, the most important thing is that a positive tourist's attitude towards the development of sustainable tourism is observed. Sustainable tourism could serve as a driving force of tourism in a post-COVID-19 era whose practices and principles can be applied to all tourist destinations (Palacios-Florencio et al., 2021). Eichelberger et al. (2021) also confirmed that a crisis helped to increase knowledge, which contributes to awareness and determines the changes in a behaviour.

During the COVID-19 pandemic, the tourism sector had become a major beneficiary benefiting from the general economic stimulus and support measures provided by governments (Pociūtė-Sereikienė et al., 2022). The majority of governments of European countries has introduced various business support measures, which were also related to the post-COVID-19 tourism vision. Decision makers in tourism understand the opportunities to rethink the future of tourism and its role in the context of safety, sustainability, and technological development. The developing strategies for the revitalization and transformation of tourism sector discuss ideas on a more responsible and sustainable tourism model (Department for Digital, Culture, Media, & Sport, 2021).

On the other hand, are the currently developed strategies acceptable to the tourists themselves? What kind of new tourism models can ensure a rapid response to unforeseen events in future? Can sustainable and responsible tourism become a key conception of the post-COVID-19 tourism vision? How does the transition to responsibility and sustainability manifest in practice? These questions should be addressed to the representatives of tourism business and tourism experts in Lithuania.

Findings

Organizational Changes of Tourism Business in Lithuania

The COVID-19 pandemic period has forced many tourism service providers to review their products and search for ways to adapt and survive. In most countries, during the COVID-19 pandemic, international tourism was replaced by domestic tourism (Knezevic Cvelbar et al., 2021), which led to the expansion of the domestic tourism market. Tourism organizers who focused on natural tourism within the country became the winners, because they didn't need to reorient their activities. Other representatives of tourism business had to change their products fundamentally or to be satisfied with governmental support and wait for the end of the COVID-19 pandemic and an increase in the number of tourists.

It was noticed that there is a growing need for responsible and sustainable tourism during the COVID-19 pandemic. The representatives of the tourism business distinguished the changing needs of tourists: "There is a greater interest of travellers in the objects of active tourism" (R 65) or "There is a need for small groups of tourists and interest in sustainable tourism" (R 70) and "Travels to nature where there are no restrictions are becoming more popular" (R 22). Respondents also noted that travellers "consume less" (R 60).

In the focus group discussion, it was noted that

> *even before the pandemic, there was a trend of growth of local, natural, eco and rural tourism, and during the pandemic, everything grew, we can say it was the blooming time of homesteads in rural areas and the flowering time of regions.* (Respondent R2, Focus group with actors from tourism business)

Another observation:

> *...if you remember the 2020 shutdown, the only thing that was open was the sightseeing paths... they were more crowded than the big malls.* (Respondent R8, Focus group with actors from tourism business)

Therefore, taking into account the changing needs of travellers, tourism organizers had to change their offered products accordingly. However, respondents from our research noticed that the products of responsible tourism were newly offered by a relatively small number of tourism business representatives. Around one-third of tourism business representatives who took part in the survey were more or less related to the development of responsible tourism products during the COVID-19 pandemic. Half of them offered these products even before the pandemic. Thus, only the rest of the tourism business organizers (14 per cent) stated that they during the COVID-19 pandemic created and offered new products to travellers that can be attributed to responsible tourism. Some answers:

> *We have dozens of amazing routes, some of which we have enriched with the insertion of the elements slow and natural tourism.* (R74, survey of tourism businesses)

> *It was necessary to adapt to the quarantine conditions, for this we created a new service—solo routes. People can hike independently throughout Lithuania. The service was in great demand during the quarantine. We also noticed that during the non-quarantine period, when events were already allowed, there is an audience that does not like mass gatherings, they are happy to choose solo routes.* (R45, survey of tourism businesses)

For some tourism organizers, the COVID-19 pandemic has helped to popularize newly developed products:

> *Excursions in the limestone quarries of Akmenė (Northern part of Lithuania) started to be organized a year ago. We started inviting people before the pandemic. The first years of activity were not intense, later with the increase in popularity it became easier to collect groups of excursionists. The year of the pandemic was one of the most successful, as the majority of Lithuanians travelled around the country and gladly joined our organized trips. We've created some new routes for smaller groups, more focused on less visited areas.* (R43, survey of tourism businesses)

In the focus group discussion, it was noted:

> *…local tourism in Lithuania was in the new-born stage before the pandemic, and now it is already a teenager, the one who jumps out, looks for his own forms, tries to realize how much money it can generate and realize how much influence it can have… Small towns, for example, Anykščiai (North-eastern Lithuania)… the city had 2 catering places, now there are 34 restaurants and other catering places….* (Respondent R8, Focus group with actors from Tourism business)

During the COVID-19 pandemic, the importance of cooperation also became apparent: both between the tourism organizers themselves, between them and local communities, and cooperation with self-governing institutions. From the focus group discussion:

> *I have a very good example. In Klaipėda district (Western Lithuania) there is a route "Fish Road". They made a cluster for the service providers, that is more than 20 service providers. You come and you buy that experience. Business and local government work together and promote the region. And that service package is being expanded…, this is purely community involvement. Let's say one captain can't provide service, another one is sailing, a smaller group of people has arrived, the boat is smaller. That's how that communication happens.* (Respondent R8, Focus group with actors from tourism business)

Therefore, most of the tourism business primarily sought to survive during the COVID-19 pandemic, and the development of responsible tourism products was part of the survival strategy; some successfully applied it, while others chose other ways of survival.

Adaptation to Climate Change and Tourist Behaviour

The topic of climate change has become more relevant in the tourism sector in recent years. The international scientific society discussed the importance of sustainability in tourism, emphasizing the input of tourism service providers and also the tourists themselves (Assaf et al., 2022; Gössling et al., 2021; Gössling & Schweiggart, 2022; Ioannides & Gyimóthy, 2020).

According to the state tourism institutions, the foreign tourists positively accepted the environment-friendly travelling solutions. This was especially essential for cruise ship passengers, who required "clean" transfer transport such as electric buses. Also, some municipalities offered the special routes for tourists traveling by bicycles, which combined different public transport vehicles including trains, buses, and boats. Also, there are companies and municipalities that introduced electric buses, electric water bikes and other ecologic vehicles powered by renewable resources. These innovations appeared not only to mitigate the climate change, but also to attract tourists while offering the alternative clean and "exotic" way to explore the region:

> [...] *it was especially popular in summer—electric water bikes, which were really in high demand among tourists, because they look more exotic, and in general the realization that you are using an ecological tool is really very important.* (Respondent A1, Focus group with actors from TICs)

Even though the survey revealed that there have not been many activities regarding climate change, environment-positive thinking and aspirations, ideas, and suggestions on how to improve tourism activities in a nature-friendly way were mentioned:

> *[...] the aspiration of our district is to continue developing bicycle tourism [...] that visiting tourists come by train, buses, with their own vehicle and continue to ride bicycles.* (Respondent A9, Focus group with actors from TICs)

The discussants of focus group also mentioned that the COVID-19 pandemic was an "expensive" time for the companies; therefore, the implementation of environment-friendly instruments and services was postponed:

> *The introduction of any new product that affects communities costs a lot of money. I think that this environmental impact will grow sooner or later, but the question is how much it will grow. [...] I think that ecology will grow sooner or later, but it costs a lot of money.* (Respondent R4, Focus group with actors from tourism business)

The climate protection trend in the field of tourism is quite clear, but the COVID-19 pandemic slowed down its greater expression. The investments in practices that mitigate the climate change have been postponed, prioritizing the "survival" activities. In general, the results of the research underlined that more environment-friendly activities and tools were introduced in the Western part of Lithuania (sea-region).

Application of New Technologies

The COVID-19 pandemic was the time to step forward and introduce novelties in the tourism sector. The results of our survey show that in order to survive and adapt to the changed environment and new tourist habits, the great majority of tourism business enterprises and state tourism institutions (TICs) introduced technological and non-technological innovations.

The main unifying feature while discussing the new technologies in the tourism sector was the ability to use the services and products avoiding human contact and keeping the distance. Following this statement, the results of our research show that both state institutions and private business enterprises invested in *mobile applications, audio guides, created and developed interactive virtual tours and excursions, digitalized the information and routes* (Fig. 10.1). There was a variety of routes offered for solo or family travellers, usually for outdoor activities (either in nature or in urban environment). Also, the virtual excursions that could be attended from home were created. The newly presented and renewed audio guides became popular tools among independent travellers.

The most often mentioned other technological innovations were associated with *developed electronic and information technology services.* One of the first steps was the transfer of workplaces towards home and adaptation to remote work. Since life has moved to virtual online space, the offered services had to adapt to the changed circumstances. Therefore, the enterprises and TICs were working on expanding the online services such as online stores, travel sales, ticket reservations, and online games (e.g., interactive games introducing city history). In order to avoid human contact, information terminals, interactive boards, QR code scanning machines, self-service terminals allowing the guests independently to check-in at the hotels, energy-saving and more user-friendly remote room environment management systems were installed.

The results of the survey showed that new products were in demand and boosted domestic tourism. Also, our research underlined that transformation of the tourism sector and introduction of technologies was one of the most essential factors allowing to survive the turbulent times. The use of new technologies is a way to improve the competitiveness and to provide new opportunities for tourists. Digital technologies change the way tourists behave and tourism experience in general.

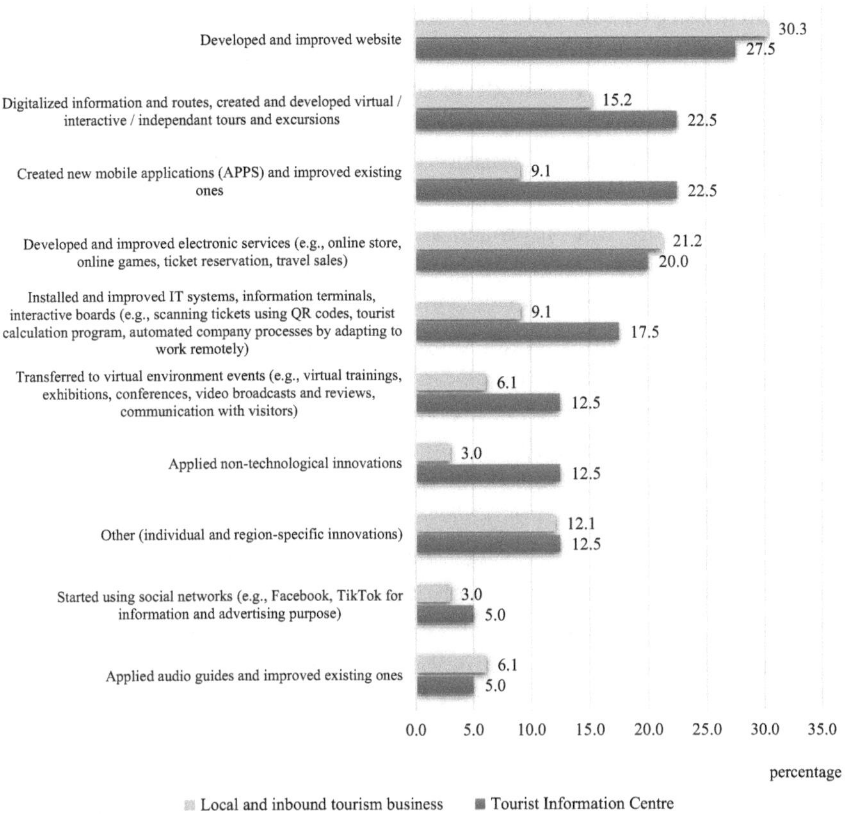

Fig. 10.1 Applied technological and non-technological innovations (since 16 March 2020, the beginning of the COVID-19 pandemic), calculated from positive answers (total 40; TIC 40 and 33 tourism business enterprises). (Source: Authors' own calculations based on a survey of state tourism institutions and local and inbound tourism business)

Discussion: New Normal or Return to Past

The COVID-19 pandemic has changed people's needs, behaviours, business processes, and government policies in all parts of the world. This situation is often called the "new normal" (Carroll & Conboy, 2020). Some researchers argue that some of these changes are permanent rather than temporary (Smolka & Heugens, 2020). Travellers have also realized that consumerism can lead to global crises, so more and more people could choose responsible and environmentally friendly travel.

Domestic tourism as the drive of recovery has been emphasized by UNWTO (UNWTO, 2020). Kriaučiūnas et al. (2023) noticed that sustainable and resilient rural tourism could become an attractive alternative to mass tourism. The innovative business models related to tourism in rural areas include the approaches of sustainability, green and circular economy, and digitalization. In general, recreation and tourism activities in the rural areas become active and significant agents of environmental, economic, and social change; they can revitalize and reorganize local economies to supplement income of farming, craft, and service sectors and to provide opportunities to re-evaluate natural and cultural resources of landscape (Hall et al., 2016).

Liutikas (2023) proposed the concept of re-tourism driven by societal and technological changes and based on improved knowledge and application of innovations. It involves adaptation and mitigation of negative impact, flexibility and dynamic, change and transformation. Later developed *RegTour* model includes the assessment of domestic tourist destinations and the application of innovations and investment in sustainable and environmentally friendly ways (Fig. 10.2).

The main aspect of this model is that it includes societal and technological changes, elements of resilience for the future crisis, flexibility and transformation, and such elements as sustainability, adaptability, education, and innovations. The authors based on research believe that the *RegTour* model is an undeniable reality of "new normal". The recovering tourism will no longer be the same as it was before the COVID-19 pandemic, but at the same time new opportunities for tourism business development are opening up. The main aspects of the *RegTour* model were also confirmed by the respondents who took part in the study:

> First of all, people again want to travel in nature, to consume organic natural food, and to enjoy products of slow tourism. It will be possible to include for inbound tourism some of the best products that were very popular among local tourists in the 2020 and 2021 seasons. (R 74, survey of tourism businesses)

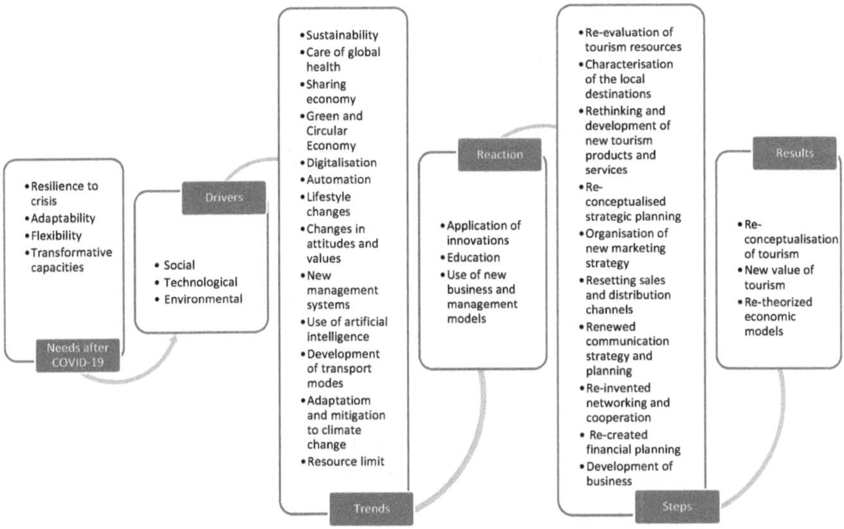

Fig. 10.2 *RegTour* model of tourism development. (Source: Authors' own work based on research)

Although there are those who doubt and believe that the pandemic time will be forgotten and post-COVID-19 time will be returning to the past—everything will be approximately as it was before the COVID-19 pandemic:

> ... *in terms of both that slow tourism and those communities, yes, there is that necessity to include those new products and everything else into those routes, but there's a problem because it's not going to be mass tourism, it's not going to be big tourist flows. Finally, it will be something like 2–5% of all those travels. What do we see today? Everyone went on tours again and... again people will be part of mass tourism, and will forget all sustainable tourism and ecotourism. Mass tourism will continue to develop. Such is life, as they say. Will that slow tourism come, or will that community tourism step in?* (Respondent R4, Focus group with actors from tourism business)

However, during the COVID-19 pandemic, it was noticed that Lithuania in the "new normal" conditions can win precisely because of the increasing requirements for safety and ecological responsibility (Kaip keisis turizmas…, 2020). The tourists were more focused on non-mass tourism in Lithuania. There are no world-famous tourist facilities in the country; the country is sparsely populated with many areas of natural tourism resources. Such a more nature-oriented tendency was already formed before the COVID-19 pandemic and may increase in the future. The participants of our study predict similarly:

> *Lithuania and other Baltic countries can offer what in many cases a post-pandemic traveller would want—new and undiscovered tourist regions, wonderful natural scenery, uncrowded tourist centres, good price-quality ratio.* (R32, survey of tourism businesses)

Such forecasts allow us to look positively at the future of tourism, seeing that the COVID-19 pandemic, in addition to negative effects, also has positive effects and that the "new normal" using principles of *RegTour* model may in some respects be of higher quality than before. How does this model redefine the concept of responsible tourism in the post-pandemic context? There are several important aspects that can further enhance the use of responsible tourism principles and practices: empowerment of community engagement, focus on health and safety, creation of effective destination management and planning strategies, implementation of personal and communal carbon offsetting actions and adoption of digital technologies and innovation.

Conclusions

Various new challenges may threaten the social, economic, or environmental sustainability of tourism businesses, as well as challenge the health or security of the different world regions or the entire world. Like the rest of the world, Lithuanian tourism service providers had to adapt to changing conditions. Our study shows that the trends in tourism development during the COVID-19 pandemic were similar to other countries: international tourism was replaced by local tourism, and travellers became more oriented to safety and ecology. The trend of responsible tourism and the supply of virtual tourism products has been increasing. The results of our research prove the importance of innovations during the COVID-19 pandemic and post-pandemic period. The research results indicated that the tourism business and state tourism institutions offered new products and services (IT products as virtual routes, mobile applications, exhibitions, audio guides, etc.) and facilitated new organizational models related to technological development and the development of the new tourism destinations (Pociūtė-Sereikienė et al., 2022).

Responsible tourism is related to re-shaping our mindset and preferences. The authors provided examples of how future trends in tourism correlate to the lessons learned from the COVID-19 pandemic. One of the insights that can ensure a rapid response to unforeseen events in the future is sustainable, technologically advanced, and community-based tourism development (*RegTour* model). *RegTour* model and the ideas of responsible tourism that it brings go in line with the sustainable development goals (SDG) presented by the United Nations (United Nations, 2023). Responsible tourism is directly connected to the SDGs concerning the climate (SDG 13) and life on land (SDG 15) and in water (SDG 14). Also, we believe that the great attention in future tourism sector will be paid on responsible consumption (SDG 12) and clean energy (SDG 7).

The research findings indicate that tourists were inclined to use new tourism services and products during the COVID-19 pandemic. The new products and services were essential for the tourism and leisure sector to survive, and the majority of them continue to be used in the post-COVID reality and provide opportunities to be more resilient and less dependent on unexpected circumstances in the future. New technological innovations and new business models let new attractiveness criteria for the local tourism destinations develop (Kriaučiūnas et al., 2023). Tourists were more engaged in the local destinations and new local tourism products.

In general, the discussion with tourism experts allows us to assume that tourists and their behaviour are changing. For instance, the hiking tourists prefer off-road routes. Therefore, the state tourism institutions instead of paved roads prepared and adapted the natural forest trails (without changing the natural forest cover) and included into the hiking route maps. Also, the tourists started to respect the nature more. The respondents agreed that tourists became more environment-conscious and leave less litter and even collect those that are left by others.

The increased need for security, the need for nature-oriented tourism, the need for ecological and responsible tourism, the expanded virtual services that complement real trips have formed the "new normal". All participants in the tourism process—travellers, tourism organizers, and state institutions—will have to understand it and adapt to it.

Despite some scepticism about the future of sustainable tourism and the possible return of mass tourism, the authors of this chapter believe that responsible tourism will undoubtedly remain as one of the main aspects of tourism development direction, determining ideas in the "new normal". Responsible tourism was already increasing before the pandemic, and the need for it only intensified during the COVID-19 pandemic. This trend will be determined both by the increasing promotion of these ideas among tourism service providers and by the spread of sustainable lifestyle among travellers.

Future research directions could be related to the analysis of evidence-based data on the use of the principles of responsible tourism and practical examples addressed associated with responsible tourism development. An application of longitudinal research approach is necessary in the studies of tourist behaviour. So, research on responsible tourism needs to be repeated in the near future to explore how the ideas and thoughts of responsible and sustainable approach during the COVID-19 pandemic have transformed patterns of tourist behaviour after the COVID-19 crisis.FundingThis paper presents results of the research project "Changes and prospects in the tourism sector following the COVID-19 pandemic". This work was supported by the European Regional Development Fund (project no 13.1.1-LMT-K-718-05-0009) under grant agreement with the Research Council of Lithuania (LMTLT) funded as the European Union's measure in response to the COVID-19 pandemic.

References

Abbas, J., Mubeen, R., Iorember, P. T., Raza, S., & Mamirkulova, G. (2021). Exploring the impact of COVID-19 on tourism: Transformational potential and implications for a sustainable recovery of the travel and leisure industry. *Current Research in Behavioral Sciences, 2*, e100033. https://doi.org/10.1016/j.crbeha.2021.100033

Assaf, A. G., Kock, F., & Tsionas, M. (2022). Tourism during and after COVID-19: An expert-informed agenda for future research. *Journal of Travel Research, 61*(2), 454–457. https://doi.org/10.1177/00472875211017237

Benjamin, S., Dillette, A., & Alderman, D. H. (2020). "We can't return to normal": Committing to tourism equity in the post-pandemic age. *Tourism Geographies, 22*(3), 476–483. https://doi.org/10.1080/14616688.2020.1759130

Brouder, P. (2020). Reset redux: Possible evolutionary pathways towards the transformation of tourism in a COVID-19 world. *Tourism Geographies, 22*(3), 484–490. https://doi.org/10.1080/14616688.2020.1760928

Capano, G., Howlett, M., Jarvis, D. S. L., Ramesh, M., & Goyal, N. (2020). Mobilizing policy (in)capacity to fight COVID-19: Understanding variations in state responses. *Policy and Society, 39*(3), 285–308. https://doi.org/10.1080/14494035.2020.1787628

Carroll, N., & Conboy, K. (2020). Normalising the "new normal": Changing tech-driven work practices under pandemic time pressure. *International Journal of Information Management, 55*(3), e102186. https://doi.org/10.1016/j.ijinfomgt.2020.102186

Cheng, C., Barceló, J., Hartnett, A., Kubinec, R., & Messerschmidt, L. (2020). COVID-19 government response event dataset (*CoronaNet v1.0*). Retrieved April 5, 2022, from https://www.coronanet-project.org

Corbisiero, F., & Monaco, S. (2021). Post-pandemic tourism resilience: Changes in Italians' travel behavior and the possible responses of tourist cities. *Worldwide Hospitality and Tourism Themes, 13*(3), 401–417. https://www.emerald.com/insight/content/doi/10.1108/WHATT-01-2021-0011/full/html

Department for Digital, Culture, Media & Sport. (2021). *The tourism recovery plan*. Department for Digital, Culture, Media & Sport. https://assets.publishing.service.gov.uk/government/uploads/system/uploads/attachment_data/file/992974/Tourism_Recovery_Plan__Web_Accessible_.pdf

Dogramadjieva, E., & Terziyska, I. (2022). One year later: Shifts and endurances in travel intentions of Bulgarian residents in the time of pandemic. *European Journal of Tourism Research, 32*, 1–30. https://doi.org/10.54055/ejtr.v32i.2704

Eichelberger, S., Heigl, M., Peters, M., & Pikkemaat, B. (2021). Exploring the role of tourists: Responsible behavior triggered by the COVID-19 pandemic. *Sustainability, 13*, 1–14. https://doi.org/10.3390/su13115774

El-Said, O., & Aziz, H. (2022). Virtual tours a means to an end: An analysis of virtual tours' role in tourism recovery post COVID-19. *Journal of Travel Research, 61*(3), 528–548. https://doi.org/10.1177/0047287521997567

Goodwin, H. (2016). *Responsible tourism. Using tourism for sustainable development* (2nd ed.). Goodfellow Publishers. https://doi.org/10.23912/978-1-910158-84-5-3101

Goodwin, H., & Francis, J. (2003). Ethical and responsible tourism: Consumer trends in the UK. *Journal of Vacation Marketing, 9*(3), 271–284. https://doi.org/10.1177/135676670300900306

Gössling, S., & Schweiggart, N. (2022). Two years of COVID-19 and tourism: What we learned, and what we should have learned. *Journal of Sustainable Tourism, 30*(4), 915–931. https://doi.org/10.1080/09669582.2022.2029872

Gössling, S., Scott, D., & Hall, S. M. (2021). Pandemics, tourism and global change: A rapid assessment of COVID-19. *Journal of Sustainable Tourism, 29*(1), 1–20. https://doi.org/10.1080/09669582.2020.1758708

Gu, Y., Onggo, B. S., Kunc, M. H., & Bayer, S. (2022). Small Island Developing States (SIDS) COVID-19 post-pandemic tourism recovery: A system dynamics approach. *Current Issues in Tourism, 25*(9), 1481–1508. https://doi.org/10.1080/13683500.2021.1924636

Hale, T., Anania, J., Andretti de Mello, B., Angrist, N., Barnes, R., Boby, R., Cameron-Blake, E., Cavalieri, A., Di Folco, M., Edwards, B., Ellen, L., Elms, J., Furst, R., Green, K., Kira, B., Luciano, M., Majumdar, S., Nagesh, R., Petheric, A., Phillips, T., Pott, A., Sampaio, J., Tatlow, H., Webster, S., Wood, A., Zha, H., Zhang, Y., & Wade, A. (2022). Variation in government *responses to COVID-19. BSG-WP-2020/032. Version 13.0* [Blavatnik School of Government Working Paper]. Oxford. https://www.bsg.ox.ac.uk/sites/default/files/2022-04/BSG-WP-2020-032-v13.pdf

Hall, D., Roberts, L., ir Mitchell, M. (2016). Tourism and countryside: Dynamic relationships. In D. Hall, L. Roberts, ir M. Mitchell (Eds.), New directions in rural tourism (pp. 3–19). Routledge. https://doi.org/10.4324/9781315248097

Hall, C. M., Scott, D., & Gössling, S. (2020). Pandemics, transformations and tourism: Be careful what you wish for. *Tourism Geographies, 22*(3), 577–598. https://doi.org/10.1080/14616688.2020.1759131

Haywood, K. M. (2020). A post COVID-19 future – Tourism re-imagined and re-enabled. *Tourism Geographies, 22*(3), 599–609. https://doi.org/10.1080/14616688.2020.1762120

Higgins-Desbiolles, F. (2020). Socialising tourism for social and ecological justice after COVID-19. *Tourism Geographies, 22*(3), 610–623. https://doi.org/10.1080/14616688.2020.1757748

Higgins-Desbiolles, F., Bigby, B. C., & Doering, A. (2021). Socialising tourism after COVID-19: Reclaiming tourism as a social force? *Journal of Tourism Futures, 8*(2), 208–219. https://doi.org/10.1108/JTF-03-2021-0058

Hussain, A., & Fusté-Forné, F. (2021). Post-pandemic recovery: A case of domestic tourism in Akaroa (South Island, New Zealand). *World, 2*(1), 127–138. https://doi.org/10.3390/world2010009

International Conference on Responsible Tourism in Destinations. (2002). *The Cape Town Declaration*. Cape Town. https://responsibletourismpartnership.org/cape-town-declaration-on-responsible-tourism/

Ioannides, D., & Gyimóthy, S. (2020). The COVID-19 crisis as an opportunity for escaping the unsustainable global tourism path. *Tourism Geographies, 22*(3), 624–632. https://doi.org/10.1080/14616688.2020.1763445

Kaip keisis turizmas po pandemijos: Lietuvai šviečiasi ne tokia ir bloga ateitis. (2020, September 29). *Delfi*. https://www.delfi.lt/keliones/naujienos/kaip-keisis-turizmas-po-pandemijos-lietuvai-svieciasi-ne-tokia-ir-bloga-ateitis.d?id=85331227

Kırlar-Can, B., & Ertaş, M. (2022). Traveling in the time of risk: The impact of vaccination on Turkish travellers. *European Journal of Tourism Research, 31*, e3112. https://doi.org/10.54055/ejtr.v31i.2469

Knezevic Cvelbar, L., Antonucci, B., Cutrufo, N., Marongiu, L., Rodrigues, M., & Teoh, T. (2021). Research for TRAN committee – Relaunching transport and tourism in the EU after COVID-19 – Tourism sector. *European Parliament, Policy Department for Structural and Cohesion Policies*, Brussels.

Korstanje, M. E., & George, B. (2022). *The nature and future of tourism: A post-COVID-19 context*. CRC Press. https://doi.org/10.1201/9781003277507

Koščak, M., & O'Rourke, T. (2021). *Post-pandemic Sustainable tourism management: The new reality of managing ethical and responsible tourism*. Routledge. https://doi.org/10.4324/9781003153108

Kriaučiūnas, E., Liutikas, D., & Burneika, D. (2023). The impact of Covid-19 on tourism and accommodation sector in Lithuanian rural areas. *Tourism: An International Interdisciplinary Journal* (in print).

Lew, A. A., Cheer, M. J., Brouder, P., & Mostafanezhad, M. (Eds.). (2022). *Global tourism and COVID-19. Implications for theory and practice*. Routledge. https://doi.org/10.4324/9781003223252

Liutikas, D. (2023). Post COVID-19 tourism: Transformations of travelling experience. In K. Dube, N. Godwell, & N. Swart (Eds.), *COVID-19, tourist destinations and prospects for recovery* (pp. 277–301). Springer Nature. https://doi.org/10.1007/978-3-031-22257-3_15

Mensah, E. A., & Boakye, K. A. (2021). Conceptualizing post-COVID 19 tourism recovery: A three-step framework, 2021. *Tourism Planning & Development, 20*(1), 37–61. https://doi.org/10.1080/21568316.2021.1945674

Moon, M. J. (2020). Fighting against COVID-19 with agility, transparency, and participation: Wicked policy problems and new governance challenges. *Public Administration Review, 80*(4), 651–656. https://doi.org/10.1111/puar.13214

Niewiadomski, P. (2020). COVID-19: From temporary de-globalisation to a re-discovery of tourism? *Tourism Geographies, 22*(3), 651–656. https://doi.org/10.1080/14616688.2020.1757749

O'Connor, P., & Assaker, G. (2021). COVID-19's effects on future pro-environmental traveler behavior: An empirical examination using norm activation, economic sacrifices, and risk perception theories. *Journal of Sustainable Tourism, 29*(1), 89–107. https://doi.org/10.1080/09669582.2021.1879821

One Planet Sustainable Tourism Programme. (2020). One Planet vision for a responsible recovery of the tourism sector. https://webunwto.s3.eu-west-1.amazonaws.com/s3fs-public/2020-06/one-planet-vision-responsible-recovery-of-the-tourism-sector.pdf

Palacios-Florencio, B., Santos-Roldán, L., Berbel-Pineda, J. M., & Castillo-Canalejo, A. M. (2021). Sustainable tourism as a driving force of the tourism industry in a post-Covid-19 scenario. *Social Indicators Research, 158*, 991–1011. https://doi.org/10.1007/s11205-021-02735-2

Pappas, N. (2021). COVID19: Holiday intentions during a pandemic. *Tourism Management, 84*, e104287. https://doi.org/10.1016/j.tourman.2021.104287

Pociūtė-Sereikienė, G., Baranauskienė, V., Liutikas, D., Kriaučiūnas, E., & Burneika, D. (2022). Challenges of the tourism sector in Lithuania in the context of COVID-19 pandemic: State aid instruments and the efficiency of the tourism business support. *European Spatial Research and Policy, 29*(2), 235–270. https://doi.org/10.18778/1231-1952.29.1.01

Rastegar, N., Flaherty, J., Liang, L. L., & Choi, H. C. (2021a). The adoption of self-service kiosks in quick-service restaurants. *European Journal of Tourism Research, 27*, e2709. https://doi.org/10.54055/ejtr.v27i.2139

Rastegar, R., Higgins-Desbiolles, F., & Ruhanen, L. (2021b). COVID-19 and a justice framework to guide tourism recovery. *Annals of Tourism Research, 91*, e103161. https://doi.org/10.1016/j.annals.2021.103161

Reddy, M. V., Boyd, S. W., & Nica, M. (2020). Towards a post-conflict tourism recovery framework. *Annals of Tourism Research, 84*, e102940. https://doi.org/10.1016/j.annals.2020.102940

Romagosa, F. (2020). The COVID-19 crisis: Opportunities for sustainable and proximity tourism. *Tourism Geographies, 22*(3), 690–694. https://doi.org/10.1080/14616688.2020.1763447

Sarkady, D., Neuburger, L., & Egger, R. (2021). Virtual reality as a travel substitution tool during COVID-19. In W. Wörndl, C. Koo, & J. L. Stienmetz (Eds.), *Information and communication technologies in tourism 2021* (pp. 452–463). Springer.

Seyfi, S., & Hall, C. M. (2020). COVID-19 pandemic, tourism and degrowth. In C. M. Hall, L. Lundmark, & J. Zhang (Eds.), *Degrowth and tourism: New perspectives on tourism entrepreneurship, destinations and policy* (pp. 287–313). Routledge.

Seyitoğlu, F., & Costa, C. (2022). A scenario planning framework for (post-) pandemic tourism in European destinations. *European Planning Studies, 30*(12), 2554–2574. https://doi.org/10.1080/09654313.2022.2045571

Sharma, A., & Arora, S. (2022). *The effect of COVID-19 pandemic on tourism and hospitality industry – A review*. In A. Sharma, A. Hassan, & P. Mohanty (Eds.), *COVID-19 and the tourism industry: Sustainability, resilience and new directions*. Routledge. https://doi.org/10.4324/9781003207467

Sharma, A., Shin, H., Santa-María, M. J., & Nicolau, J. L. (2021a). Hotels' COVID-19 innovation and performance. *Annals of Tourism Research, 88*(6), 1–21. https://doi.org/10.1016/j.annals.2021.103180

Sharma, D. G., Thomas, A., & Paul, J. (2021b). Reviving tourism industry post-COVID-19: A resilience-based framework. *Tourism Management Perspectives, 37*, e100786. https://doi.org/10.1016/j.tmp.2020.100786

Shin, H., Nicolau, J. L., Kang, J., Sharma, A., & Lee, H. (2022). Travel decision determinants during and after COVID-19: The role of tourist trust, travel constraints, and attitudinal factors. *Tourism Management, 88*, e104428. https://doi.org/10.1016/j.tourman.2021.104428

Škare, M., Soriano, R. D., & Porada-Rochon, M. (2021). Impact of COVID-19 on the travel and tourism industry. *Technological Forecasting and Social Change, 63*, 1–14. https://doi.org/10.1016/j.techfore.2020.120469

Smolka, K. M., & Heugens, P. P. (2020). The emergence of protoinstitutions in the new normal business landscape: Dialectic institutional work and the Dutch drone industry. *Journal of Management Studies, 57*(3), 626–663. https://doi.org/10.1111/joms.12540

Su, C. J. (2022). Post-pandemic studies in tourism and hospitality. *Service Business, 16*, 413–416. https://doi.org/10.1007/s11628-022-00495-8

United Nations. (2023, May 9). *Sustainable development goals*. https://sdgs.un.org/goals

UNWTO. (2020, April 5). *UNWTO highlights potential of domestic tourism to help drive economic recovery in destinations worldwide*. https://www.unwto.

org/news/unwto-highlights-potential-of-domestic-tourism-to-help-drive-economic-recovery-in-destinations-worldwide

Zhang, H., Song, H., Wen, L., & Liu, C. (2021). Forecasting tourism recovery amid COVID-19. *Annals of Tourism Research, 87*, e103149. https://doi.org/10.1016/j.annals.2021.103149

11

Ecotourism Principles, Responsible Travel, and Building a Sustainable Post-pandemic Destination Kenya

Joseph K. Muriithi and Philip Ireri

Introduction

Tourism is a global phenomenon that impacts many people and places. Moreover, the tourism sector involves many actors including tourists, host communities, government officials, industry managers, facility employees, and civil society members. The multiplicity of actors with different interests forms the basis of developing aspects of actors' responsibility to both destinations and the host communities (Bramwell et al., 2008). Therefore, the philosophy behind the responsible tourism concept implies that all actors in tourism should take responsibility for their actions in the destinations they travel to (Eichelberger et al., 2021). This sense of actors' responsibility is the foundation of the concept of responsible tourism and underlines its importance (Mihalic, 2016). Responsible

J. K. Muriithi (✉)
Department of Environmental Studies and Community Development, Kenyatta University, Nairobi, Kenya

P. Ireri
Essibility Africa, Nairobi, Kenya

S. W. Maingi et al. (eds.), *Tourist Behaviour and the New Normal, Volume II*,
https://doi.org/10.1007/978-3-031-45866-8_11

practices among actors are important in ensuring the sustainability of destinations (Mathew & Sreejesh, 2017). Consequently, over the years, there has been a growing awareness of the importance of promoting responsible principles and practices in many travel destinations across the world. Furthermore, there have also been growing debates on the importance of emphasizing responsible values in tourism since the 1970s (Mihalic, 2016).

Responsible tourism is focused on the natural and social environments and contexts around which the travel phenomenon happens (Mihalic, 2016). Generally, it focuses on environmental conservation and promoting community development in the destinations (Fennell & de Grosbois, 2021). The key characteristics of responsible tourism also revolve around the two ideas of the natural and social environment in that responsible tourism seeks to minimize negative socioeconomic and environmental impacts while providing more economic benefits to local communities to improve their well-being. In this regard, responsible tourism is close to other related concepts such as sustainable tourism, responsible travel, green tourism, ethical tourism, and ecotourism that espouse similar ideals and values and which demonstrate behaviour change among actors in the destinations (Camilleri, 2016; Mihalic, 2016).

Despite the wide acceptance of these concepts and the values they espouse, responsible tourism faces sustainability challenges where there is more emphasis on growth that ignores the fairness, ecological limits, human benefits, and sustainable futures of tourism (Higgins-Desbiolles, 2018). Many destinations experience overtourism, and host communities do not get a fair share of benefits from tourism. There is slow adoption of circularity measures in tourism accommodation facilities, and biodiversity in nature-based tourism destinations continues to be lost. There is also the ever-present challenge of climate change impacts which continue to threaten tourism destinations. However, the greatest threats to the development of sustainable tourism in the world today remain behavioural and management factors like overtourism which represent the single greatest threat to sustainable tourism (Mihalic, 2020; Goodwin, 2017).

In Africa, Kenya is a leading destination for nature-based and wildlife tourism. The country has other attractions like beach tourism in the coastal region and cultural tourism products and attractions spread across the country. As a destination, Kenya also faces many tourism-destination-related challenges like biodiversity loss and community benefit issues that threaten the sustainability of her tourism (Muriithi, 2022a). While practical interventions to promote sustainable and responsible tourism practices and principles are in place in Kenya, there has been a prevalence of these challenges in many of the tourism destinations. For example, since the introduction of the certification of tourism facilities to promote sustainable practices among tourism hotel facilities, many unsustainable practices continue to persist in many destinations in Kenya (Muriithi, 2014). This chapter examines responsible tourism concerns in Kenya by focusing on the application of ecotourism principles and practices and their impacts on the post-pandemic context. More specifically, the chapter addresses three key questions. In what ways has ecotourism certification influenced the development of responsible tourism in Kenya? How have Kenyan tourism destinations embraced responsible tourism principles and practices? Lastly, how has the COVID-19 pandemic affected how responsible tourism principles in Kenya's tourism are applied?

Literature Review

Ecotourism, Responsible Travel Principles, and Practices

The term ecotourism was first used in 1983 by Mexican conservationist and architect Hector Ceballos-Lascurain who later defined it as visits to relatively undisturbed natural areas to admire, enjoy, and study their scenery, wildlife, and culture (Ceballos-Lascurain, 2006). The objectives of ecotourism are to reduce tourism's negative impacts on people and the environment; support preservation of natural areas; ensure the sustainable spread of benefits; and empower residents (Thompson et al., 2018). The principles of ecotourism include minimization of the impacts of tourism; beneficial experience for all involved; deriving of financial benefits from environmental conservation; empowerment and financial benefits for residents; expansion of cultural and environmental awareness; and increasing awareness of host's environmental, social, and political status (Bricker, 2017). They also cover the engagement of tourists in destinations' environmental conservation efforts; embracing environmentally friendly (green) tourism practices in waste management and other areas; minimization of the use of non-renewable resources; and empowerment of residents (Ibrahim et al., 2019).

Ecotourism is the answer to the challenges of mass tourism and other forms of travel that are associated with environmental degradation, short-term economic benefits, and unsustainable resource use (Dimitriou, 2017). It yields various benefits including creating employment through green jobs, reducing poverty, growing local economies, and supporting the conservation of a destination's natural and cultural heritage (Bricker, 2017). It however faces various challenges including some businesses engaging in ecotourism "lite" whereby they make large eco-friendly claims after adopting only a few cosmetic practices (Thompson et al., 2018). Even in Kenya, practitioners have been found to use the "eco" prefix for selfish promotional and financial reasons rather than a genuine concern for environmental sustainability (Atieno & Njoroge, 2018). There is therefore a wide gap between ecotourism principles and their practical application that emanates from the challenge of managing visitor impacts in fragile destinations and the multiplicity of stakeholders and interests involved among other factors (Dimitriou, 2017).

Responsible Travel and Role of Actors

Responsible tourism is a concept that became popular in the 1980s and represents travel that addresses the social-economic and environmental challenges of tourism including climate change (Camilleri, 2016). It encompasses the attitudes and actions of tourism actors as far as the environmental and social-economic aspects of the business are concerned (Mondal & Samaddar, 2021). Responsible tourism is not equivalent to sustainable tourism since the former considers the industry's micro aspects while the latter deals with those at the macro level (Ting et al., 2020). Responsible tourism therefore covers how sustainable tourism policies and strategies are translated into behaviour and action by suppliers, consumers, governments, and other tourism stakeholders (Mihalic, 2016). Responsible tourism actors include tourists, tour operators, tourism businesses, local residents, destination marketing organizations (DMOs), governments, and NGOs (Dimitriou, 2017; Mondal & Samaddar, 2021). All tourism stakeholders have a role to play in promoting the industry's positive economic, environmental, social, and cultural outcomes (Mihalic, 2016).

Responsible tourism practices include when tourism enterprises embrace green technologies, water conservation, energy conservation, renewable energy, effective waste management, pollution mitigation, and offsetting of carbon dioxide emissions, all of which can have a positive impact on the facility's bottom line and the society in general (Camilleri, 2016). Governments promote responsible tourism through legislation, regulatory frameworks, and financial incentives like tax breaks, standardization, and guidelines (Camilleri, 2016). Local communities' roles include ensuring equity in the distribution of the benefits of responsible tourism among their members, a role that Kenyan communities have attained limited success in due to several internal and external constraints (Ireri et al., 2020; Ireri et al., 2023). Tour guides in Kenya also play important roles including educating visitors on responsible behaviour during safaris, though they also engage in irresponsible actions including offroad driving and breaching the minimum permitted distances when viewing wild animals (Kabii et al., 2019). Indeed, attaining responsibility in tourism requires customized, concerted, and selfless action by all stakeholders (Ting et al., 2020).

Tourism Destinations in Kenya

There are three approaches by which tourism destinations are defined: the classical approach which looks at them as spatial geographical areas that meet certain criteria; the systems approach which takes into consideration the different systems and actors that interact in complex and adaptive ways; and the currently evolving smart approach that recognizes the role of ICT in providing information and knowledge about tourism products (Jovicic, 2019). We have retained the geographical approach in this chapter as this is the focal point where tourism occurs. Ecotourism Kenya indicates on their website that they have certified 81 tourist accommodation facilities from five destinations: Coast, Laikipia/Samburu, Masai Mara/South Rift, North Rift, and Tsavo/Amboseli (Ecotourism Kenya, n.d.). This categorization, however, clearly leaves out other destinations where ecotourism facilities are found including in and around Nairobi City. Magical Kenya, the country's official tourism promotion website that is managed by Kenya Tourism Board (KTB), identifies the following as Kenya's tourist destinations: Central Highlands, Coastal Kenya, Northern Kenya, Rift Valley, Southern Kenya, and Western Kenya, in addition to Cities and Intimate Encounters (KTB, n.d.). Table 11.1 provides details of these destinations:

Responsible Tourism in Post-pandemic Contexts

The World Health Organization (WHO) declared the COVID-19 disease a global pandemic in March 2020 (WHO, 2020). On one hand, the pandemic had positive impacts on tourism including a reduction in visitor numbers and associated negative impacts on the environment even in popular destinations that may have been hitherto experiencing overtourism (Spenceley, 2021). The economic slowdown that was associated with the pandemic resulted in reduction in atmospheric pollution, with nitrogen dioxide levels reducing by up to 30% in some parts of China (Cassidy, 2020). The growth in augmented reality (AR) and virtual reality (VR) tours promoted responsible behaviour and practices before physical visits to destinations (Ting et al., 2020). On the other hand, it had negative impacts on nature-based tourism including increase in poaching and other wildlife threats as poor people sought alternative means of livelihoods and donor incomes for combating these activities dwindled (Lindsey et al., 2020; Spenceley, 2021). The outbreak's impacts were more severe in Latin America and Africa (Waithaka et al., 2021). In the Kenyan tourism industry, there was massive loss of jobs and drastic decline in revenues including park entry fees and taxes, with businesses losing up to 80% of their incomes (Ondicho, 2021).

Table 11.1 Kenya's tourism destinations

Destination	Attractions
Central Highlands	Mount Kenya, Mount Kenya National Park, Meru National Park, Aberdare National Park, Private and Community ranches including Lewa Wildlife Conservancy, Ngare Ndare Forest, Ol Pejeta Conservancy, and Il Ngwesi, Thompson Falls in Nyahururu,
Coastal Kenya	Diani Beach, Nyali beach, Watamu Beach, Wasini Island, White sandy beaches, Kisite Mpunguti Marine Park, Haller Park, Mida Creek, Watamu Marine Park, Fort Jesus, Lamu, Malindi, luxury hotels and resorts
Northern Kenya	Samburu, Reteti Elephant Sanctuary, Lake Turkana, Chalbi Desert, Mt. Ololokwe and South Horr
Rift Valley	Lake Nakuru National Park, Lake Naivasha, Lake Bogoria hot springs, Lake Baringo, Hell's Gate National Park, Lake Elementaita, Menengai Caldera, Mount Longonot, Kerio Valley,
Southern Kenya	Maasai Mara National Reserve, Amboseli National Park, Chyulu Hills, Tsavo East National Park, Tsavo West National Park, Mzima Springs, Shompole Conservancy, Naboisho Conservancy
Western Kenya	Kakamega Forest Reserve, Kericho tea plantations, Lake Victoria, Kit Mikayi and Thimlich Ohinga sacred stones, Mount Elgon, Jaramogi Oginga Odinga Mausoleum, Rusinga Island,
Other destinations	-Cities including Nairobi, Mombasa and Kisumu -Intimate encounters that cover Forests, Lakes, Orphanages, Highlands and Valleys.

Source: Kenya Tourism Board

Tourists have demonstrated increased responsibility in attitudes and behaviour in the post-pandemic period (Eichelberger et al., 2021). Indeed, a study of 1096 Finnish travellers revealed that responsibility was one of the factors they would consider in the post-pandemic period (Ritalahti & Ali, 2022). The United Nations World Tourism Organization (UNWTO) has also recommended different lines of action to ensure tourism recovery is responsible, and these include rendering targeted support to vulnerable sections of the community, monitoring and reporting on carbon dioxide emissions, and shifting towards circular economy processes including re-purposing, refurbishing, and reusing of products (UNWTO, 2020). This is critical as tourism accounts for 8% of global carbon dioxide emissions and the industry's growth is higher than the available decarbonization technologies (Lenzen et al., 2018).

The pandemic has provided all actors in the sector with an opportunity to reset tourism into a more sustainable path (Ioannides & Gyimóthy, 2020). It has brought into question the numbers-led tourism growth model advocated by agencies like the UNWTO and the businesses that are profiting from these forms of mass tourism (Gössling et al., 2020). The recovery period is characterized by a higher reliance on domestic visitors even for tourism businesses that previously targeted foreign markets, a trend that risks degrading destinations due to the high numbers that are usually associated with this category of visitors (Spenceley, 2021). There is also a risk of local communities and other stakeholders losing confidence in the future of wildlife and nature-based tourism and converting critical wildlife areas into other uses (Lindsey et al., 2020; Spenceley, 2021). This is important for Kenya, a country that has seen a rapid increase in wildlife conservancies in recent years as reported on the website of the Kenya Wildlife Conservancies Association (KWCA, n.d.). Proponents of wildlife conservancies hold that they provide employment and support the preservation of wildlife and natural areas while critics like Ogada (2019) contend that they are a neo-colonial landgrab that threatens local communities' identity, culture, and livelihoods by denying them access to ancestral land, pasture, and water. In the post-pandemic period, even the benefits of responsible tourism have been questioned as it has emerged that it is not possible to determine whether the Kenyan eco-certification programme has led to an increase in the number of visitors for participating enterprises and tourism actors are now calling for a more dynamic and accountable audit process to guard against greenwashing (Diazzi, 2022).

Research Methodology

This chapter uses a qualitative research approach to undertake content analysis of information and key informant information to analyse the steps taken in Kenya to enhance responsible travel. A desk study reviewed website materials and documents which then underwent content analysis while phone interviews with key informants from various stakeholder categories enabled us to understand how the COVID-19 pandemic affected travel in various tourism destinations in Kenya. In this regard, nine key informant interviewees who form a part of the responsible tourism chain in Kenya were interviewed. These consisted of two proprietors of responsible tourist accommodation facilities; two destination-based tour guides; one Destination Management Organization (DMO) representative; one tour operator; one manager of a community-owned tourist enterprise; one county government tourism officer; and one manager of a public forest that enjoys high levels of tourist visitation. These informant interviews yielded important information that allowed us to understand how the key principles of sustainable tourism and responsible travel were affected by the pandemic and how the future of responsible travel in Kenya's various destinations is likely to be. In terms of data collection and analysis, the key informant interviews were recorded using a smart phone and later transcribed for in-depth analysis. Analysis proceeded by taking several rounds of reading the transcripts while noting important emergent messages from different informants. The messages were then collated and categorized according to the similarity of themes or messages according to the objectives of the study. At this point, the messages used were then used as the take-home findings from the interviews and were triangulated with other literature reviews in the course of writing this paper.

Results and Discussion

Certification and Infusion of Responsible Tourism Principles and Practice

Certification and Responsible Tourism

Certification is a way of ensuring sustainability and is also seen as a management tool and recognition award to businesses that comply with a given criterion (Conaghan & Hanrahan, 2010). In Kenya, the better-known certification programme is known as the Eco-rating Certification Scheme which certifies tourism accommodation facilities (Muriithi, 2014). The voluntary scheme is managed by Ecotourism Kenya, a civil society organization that brings together tourism sector players including the private sector, community-based organizations, and individuals to promote responsible tourism practices in the country. Consequently, the goals of the certification scheme align well with the principles and practices of responsible and sustainable tourism. As Esparon et al. (2014) has pointed out, certification schemes are important sustainable tourism management tools and therefore the Eco-rating Certification Scheme was introduced in 2002 to infuse sustainable tourism values and practices in the conduct of tourism business in Kenya.

In terms of its application, certification works by rating or categorizing tourism facilities in terms of how they respond to environmental and socioeconomic concerns in the destinations in which they operate. In this sense, a hotel facility like a lodge or camp subjects itself to assessment and evaluation on a set criterion initially outlined in a questionnaire before an on-site assessment is made. In this sense, the Kenya Eco-rating Certification is performance-based, the opposite of the process-based certification when determining how eco-labels are awarded. According to Chester and Crabtree (2002), performance-based certification assesses facilities on what they have accomplished in practice based on a set of environmental, economic, social, and cultural criteria. On the other hand, process-based schemes award certification-based statements of commitment to certain ecotourism ideals by the applicants.

Implementation of Eco-rating Certification in Kenya

The eco-rating certification scheme is voluntary and certifies facilities spread across the country. Currently, 81 facilities are listed to be certified in different eco-rating categories (Ecotourism Kenya, n.d.). Table 11.2 summarizes the current number of eco-rated facilities in Kenya. This is a fair number of facilities that can serve as a critical mass of tourism actors in promoting the adoption and infusion of responsible tourism practices given that the certification criteria confirm that eco-rated facilities have

Table 11.2 Number of eco-rated facilities in Kenya (April 2023)

No.	Tourism region	Eco-rating category	Sub-total	Total
1	Maasai Mara	Gold	25	40
		Silver	14	
		Bronze	1	
2	Amboseli/Tsavo	Gold	4	9
		Silver	5	
		Bronze	-	
3	Coast Region	Gold	4	12
		Silver	8	
		Bronze	-	
4	Laikipia/Samburu	Gold	7	12
		Silver	4	
		Bronze	1	
5	North Rift	Gold	4	8
		Silver	4	
		Bronze	-	
	Total			81

Source: Ecotourism Kenya (https://ecotourismkenya.org/eco-rated-facilities)

to meet standards that ideally provide them with marketing the environmental and social, economic and cultural practices they implement depending on the level (gold, silver or bronze) of certification. The award of an eco-label allows certified facilities to "green market" their business as responsible businesses by displaying the award logo or eco-label online, in exhibitions, and on vehicles. Spenceley (2019) has pointed out that a key motivation is to seek certification for marketing and promoting facilities. However, according to Honey (2008), green marketing risks enabling the eco-labels to be used in "green washing" a situation where facilities use green labels without being independently assessed as having met the set criteria. Furthermore, as Diazzi (2022) points out, tourism certification should also strive to guard against greenwashing. Perhaps, this is the cause of negative public perceptions of the role of certification as a method of infusing responsible tourism principles and practices as captured by a key informant below:

> *...you might, for instance, find that accommodation facilities are certified ecotourism-wise, but you will be surprised that some of their practices are not adhering to the ecotourism principles...you will go and see them in those attrac-*

tion [web]sites that have been marketed online but in practice, they are not willing to adhere 100% to the practice of sustainable tourism. KI2

Table 11.2 reveals one important point. Half of the eco-rated facilities are from the Maasai Mara tourism circuit. This is revealing given that the Maasai Mara area is one of the most visited tourist destinations which in the past had caused concerns about overtourism (Maingi, 2019). Other studies (Waithaka, 2004) have also decried the unsustainable ecological and human threats facing the Maasai Mara region and equally a potential risk to sustainable tourism development in the country.

Impacts of Certification on Responsible Tourism

An important question often asked is what certification of tourism accommodation facilities has been able to achieve in terms of contributing to the realization of sustainable tourism goals. A number of positive achievements have been noted. To start with, through the eco-rating scheme, it has been possible to raise awareness among the industry players on the need to embrace sustainable practices. Awareness creation is an important factor in entrenching responsible tourism among the tourism actors both locally and abroad. Importantly, the certification has promoted the participation and inclusion of local people in sustainable tourism development. As studies have shown, the inclusion of people in tourism is a key sustainability pillar in the journey of realizing sustainable development (Ritalahti & Ali, 2022; Saufi et al., 2014).

Besides these positive attributes, there are also persisting challenges that present difficulties in assessing certification positively. As previously suggested, many public and community expectations have not been able to be met. Questions have been asked about equity and fairness in the distribution of benefits. In many instances, benefit-sharing arrangements have been deemed to favour private tourism investors, especially where tourism partnership agreements are skewed in favour of the investor. However, the argument only goes as far as sharing monetary returns from tourism without considering the in-kind community support rendered to communities including social services and infrastructure support like the building of classrooms and health centres. This kind of benefit was noted especially during the COVID-19 pandemic period when some tourist facilities were kind enough to provide community support as captured by the sentiments of a key informant:

> *Our destination has been known to promote sustainable tourism for a while. Most of the facilities have taken up sustainable practices. Whether it is how we grow food, whether it is how we source our water, how we construct infrastructure in the facilities…Many of the facilities did not lay off staff during COVID-19. This is highly commendable for them to have taken up sustainable practices. Whether it is how we ing loans to stay afloat….* KI4

Another prevalent challenge in some destinations is increased impacts on the landscape caused by the problem of poor conservation planning. For example, in the Maasai Mara area, the problem of fencing persists by preventing wildlife from accessing zoned conservation areas within the conservancies for their exclusive use. The consequence of the fencing menace acts as a blockage that poses a potential challenge of a decline in wildlife numbers that serve as the key tourist attractions in the conservancies (Veldhuis et al., 2019; Løvschal et al., 2017). Furthermore, other community members graze their cattle in conservation areas, way beyond the allowed areas. This often creates bad relations between the tourism investors and the community members, but there is a mechanism agreed upon in the lease agreements on how to address such issues when they arise.

Furthermore, an examination of the eco-rated facilities reveals the challenges of using a certification programme to promote responsible tourism. To begin with, the Ecotourism Kenya website shows that 43 facilities are "Gold-rated", 35 are "Silver-rated", and only 2 are "Bronze-rated". Many facilities have therefore attained the highest level of certification possible and may have no further incentive to strive to practice responsible tourism. Secondly, the "Gold-rated" level for instance has facilities that range from small, tented conservancy-based camps like the 9-tent Naboisho Camp, to large urban-based brick-and-mortar entities like the 74-room Serena Beach Resort & Spa. In addition to the practical challenge of applying the same criteria to certify such a wide range of facilities, this raises the likelihood of the scheme inadvertently promoting mass tourism simultaneously with promoting ecotourism. Finally, the Ecotourism Kenya website reports that the Eco-rating Scheme was launched in 2002. For the long time that the programme has been in existence, there is a very low number of certified facilities. This suggests that it has had a limited impact on responsible tourism practices in the country's tourism industry as a whole.

Transitioning Nature and Wildlife Tourism Towards Responsible and Sustainable Tourism?

Nature and wildlife are the main tourist attractions in Kenya. Traditionally, tourism destinations revolve around the network of national parks and reserves that dot the country. There is also a portion of beach tourism along the Kenyan coast. Over the years, however, many protected areas have faced many challenges related to human activities in the destinations. For example, Waithaka (2004) identified ecological and anthropogenic threats affecting the Maasai Mara ecosystem. These threats by implication affected the development of sustainable tourism in the Maasai Mara destination. They included blockage and destruction of wildlife migratory corridors, habitat fragmentation, land sub-division, land tenure change, and crop farming. These challenges have become the precursor to the establishment of community wildlife conservancies model. The model looks at rangelands as potential areas for co-existence between people and wildlife with tourism being a sustainable livelihood means for communities and the accommodation of multiple land uses including tourism and pastoralism.

Furthermore, popular destinations such as the Maasai Mara National Reserve were also laden with other challenges like overtourism (Maingi, 2019). Questions have been raised on the best model of creating co-existence between local communities and wildlife mediated through the promotion of sustainable tourism (Western et al., 2015). In recognition of the fact that in Kenya majority of wildlife lives outside officially designated protected areas borders (Western et al., 2009) early ideas about the possibilities of wildlife conservancy-based tourism started taking shape.

Therefore, conservancies have emerged as the transitioning model for the development of more responsible and sustainable tourism where local communities benefit, tourism investors and the ecosystem benefit, and biodiversity eventually is conserved.

Wildlife conservancy-based tourism in Kenya is seen as an ideal form of sustainable tourism because of the perceived observance of environmental, socioeconomic, and cultural criteria on which they are certified as eco-lodges. The conservancy-based tourism model brings landowners and tourism investors together to manage the land collectively. A conservancy is created when a group of landowners with individual land title deeds comes together to collectively manage their small parcels of land as a bigger conservation area managed together by tourism investors through a lease agreement. This means that landowners who have formed a conservation area lease out their conservancy to investors who establish lodges and camps on the conservancy in return for tourist visitor fees and monthly lease fees (Muriithi, 2022b).

Nowadays, conservancy-based tourism has been the driving force in the convergence of conservation and community development in Kenya. Through lease agreements signed between landowners and tourism investors to develop ecotourism in the conservancies, community members receive monthly lease fees, employment opportunities, development, and infrastructure support through such things as hospitals and classrooms, and community capacity building through entrepreneurship training. With the development of the conservancy-based tourism model across the country's rangelands new tourism destinations that focus on responsible tourism are being established. Conservancies create and bring much better benefits to local people, the environment, and tourism investors than under the previous laissez-faire protected area tourism model in the conventional parks and reserves which was under state control and management.

On the transition to more responsible travel, especially in most wildlife tourism destinations, evaluation of benefits from the wildlife conservancy-based tourism model suggests a much better application of responsible tourism principles and practices. A few examples suffice. With the continued establishment of more conservancies, landowners now recognize the value of conserving and protecting wildlife. Due to the regular incomes generated through lease fees, landowners now link the monthly incomes with wildlife protection. Furthermore, land lease income provides more reliable livelihood opportunities than traditional pastoralism it has come to replace. Compared to the previous community conservation dividend-sharing arrangements of park fees from county governments, land lease fees were unfairly distributed with many cases of misappropriation of park fees by officials reported in the past (Thompson et al., 2009). Additionally, since conservancy land also serves as areas for local community residency, they also serve as destinations for cultural tourism. They are located in areas with rich cultural heritage and allow communities to showcase their traditional customs, cultural practices, and traditional knowledge. They, therefore, provide opportunities for demonstrating cultural sensitivity and responsibility by providing cultural tourism products and lifestyles of the local communities. Therefore, tourists gain useful insights into the local people's culture and lifestyles besides watching wildlife majority of which lives in the conservancies. Conservancies therefore help preserve and celebrate local cultures and promote cultural exchange between tourists and communities, therefore, fostering mutual respect and understanding. Respect for local culture is one criterion used in assessing tourism facilities when they seek to be eco-certified. Moreover, conservancies host few visitors at any one time than conventional parks and reserves. There are restrictions on the number of vehicles allowed within the wildlife zones in the conservancies and this helps reduce the environmental impacts in the core conservation areas. There is also widespread adoption of sustainable practices in the lodges like the use of renewable energy like solar for heating purposes, sustainable use and reuse of water, and waste management practices (Muriithi, 2014). All these practices support environmental responsibilities as tourism in the conservancies continue to grow, contrary to what has been the practices in parks and reserves.

Effects of the Pandemic on Implementation of Responsible Tourism Principles

As the literature review suggests, the two overarching principles of ecotourism and responsible travel to destinations are environmental conservation and support for the community's socioeconomic and cultural well-being. Sentiments expressed by key informants from the tourism industry reflect the sustainability uncertainty of Kenya's tourism destinations during times of crisis. The COVID-19 pandemic affected how tourism is organized and managed because of international travel restrictions. At the height of the pandemic, many destinations experienced limited environmental pollution and exploitation of natural resources. This allowed biodiversity regeneration as a result of fewer tourist visits to key destinations. However, the post-pandemic recovery period in Kenya coincided with the transition to junior secondary school under the new Competency-Based Education (CBC) system that emphasizes the teaching of environment-related studies subjects like biology, geography, tourism, and agriculture which has enhanced school tours and excursions to various tourism destinations. This increase in visitations combined with visitations by other members of the public has seen increased pollution levels in many publicly accessible destinations, especially through littering by food wrappings. Positively, in some other destinations, dustbins were provided as a way of managing waste littered by visitors. In this regard, it is important to enhance environmental awareness creation as part of promoting responsible tourism through public education using signage in various destinations as well as through tour guiding. This was well captured by two key informants thus:

> ...*Litter is dumped indiscriminately. The problem is caused by domestic tourists. There is a need for more awareness and warnings informing tourists not to litter everywhere....* KI1

> ...*When a visitor comes, you explain to them the rules of the destination. That they should not throw their litter into the forest or the ocean. Explain to them first. Secondly, educate them on the benefit of the forests and the threats to forests by littering....* KI6

These sentiments underscore the importance of visitor education, which is captured in studies on responsible tourism (Mondal & Samaddar, 2021; MacLeod, 2016) as a way of promoting responsible travel. Furthermore, littering with poor disposal of plastics in destinations officially designated as protected areas is actually against the law. In 2020, at the height of COVID-19, the government banned the littering of plastic materials in all protected areas through regulation (Government of Kenya, 2020). This policy was aimed at promoting responsible travel in Kenya's protected areas which are the main tourism destinations. Another boost to the principle of responsible travel was a government policy decision to secure the jobs of all rangers including community rangers working in parks and conservancies across the country (Muriithi, 2022b). Rangers play an important role in monitoring environmental crimes such as poaching in protected areas and conservancies, at a time of crisis when there is little monitoring and surveillance of the conservation area that attracts tourists. As the literature reviewed has indicated, government interventions by way of policy formulation are important measures (Camilleri, 2016; Mihalic, 2016) as they contribute towards responsible and sustainable tourism through legislation, policies, and regulations.

In terms of the pandemic's effects on the application of social-economic principles of responsible travel, a negative picture also emerges but with positive prospects for local tourism from the views of key informants. This is captured in the words of key informants:

> *...In early 2019, it started to pick up. In 2020, the pandemic struck. In 2021 it was still a downturn. So, I encountered challenges immediately after I opened the business and I have not recovered yet. So, I have suffered, many losses. Even my colleagues, their companies have similarly suffered...we haven't recovered...we hope things will change....* KI5

> *...it is both yes and no on the effects of the pandemic. On the effect on the community, we saw a rise in domestic tourism. And domestic tourism within where communities live. If it is people in Nairobi, there is the uptake of tourism within Nairobi...the pandemic opened a lot of opportunities for local tourism.* KI8

The impacts of the pandemic on the community relate to how principles and practices of responsible tourism touch on the social and economic aspects of the local people. Many local community members who depended directly on aspects of tourism like employment or selling items to international tourists were hard hit with their livelihood options suddenly coming to a stop. Additionally, community development support provided by many tourism investors, in particular in the conservancies came to an abrupt end. Regarding the sustainability of socioeconomic practices of responsible tourism, it is suggested that tourism actors support communities in diversifying livelihoods, therefore, reducing too much dependence on tourism. Recognition of the role of local tourism in supporting tourism is also well appreciated by key informants. The implication is a suggestion for government to develop strategies to promote and strengthen local tourism as a support system for responsible travel.

The outbreak of the COVID-19 pandemic opened new tourism destinations and new forms of tourism. Key informants indicated that proximal natural areas like hills suddenly became attractive destinations during the pandemic and also during the post-pandemic period as people sought relief from the lengthy lockdown. They further reported that the pandemic increased homestays as preferred areas to travel to. All this came with a challenge to responsible tourism as one key informant reported:

> *... Another issue is the entry of a lot of quacks. Quacks in this way. Even as we call it local tourism, if somebody decides to give a service, you give something that is not standard. So, you are competing with people without standards. You know initially, we were controlled by things like classification. It could be that a place is for about three-star, four-star, or five-star. But now the things that we used to have, those you are calling five-star, most of them collapsed. You see. So now those that have remained are worse off.* KI9

The pandemic and post-pandemic period, therefore, resulted in the opening up of new pristine areas for tourism, and an increase in unregulated tourism businesses. Both of these are bound to hurt responsible tourism in Kenya. The rise in unregulated business only means the possibility of increased unsustainable practices. This though is contrary to what the literature had suggested that in the post-pandemic era, actors ought to have developed increased responsibility, positive attitudes, and behaviour (Eichelberger et al., 2021).

This study has identified several implications for practice to ensure that responsible travel principles are observed at the destination level. For example, this study highlights the importance of enhancing community engagement and benefit-sharing in ecotourism practice as a way of promoting responsible travel. This can be achieved through initiatives that ensure full engagement with the communities, building their capacities, and adopting fair revenue-sharing models. In addition, this study underscores the importance of tourism actors to prioritize environmental sustainability and biodiversity conservation measures in tourism destinations. Actors need to implement robust environmental management practices including habitat protection, responsible wildlife viewing practices, waste reduction, sustainable water management, and energy efficiency. Putting into practice these measures contributes to minimizing human impacts on the environment at the destination level. Furthermore, it is also important to address greenwashing practices and promote transparency in the certification processes. This study recommends the establishment of clear guidelines and regulations for certification and eco-labelling that allow rigorous enforcement through keen monitoring and evaluation processes that enhance credibility and transparency in tourism certification.

This study also recommends three areas for further research. To start with, the study suggests the need to focus research on evaluating the effectiveness of the existing ecotourism certification and eco-labelling schemes in Kenya. This should include assessing their impacts on businesses, consumer (tourists) perceptions, and trust in the schemes, as well as identifying opportunities for improvement in terms of transparency, credibility as well as alignment with international best practices. Additionally, the study recommends research be conducted on innovative strategies for responsible travel. This can focus on practices and strategies for promoting responsible travel in different destinations. This can include possibilities of examining the potential of digital platforms and virtual experiences to minimize the environmental impacts of tourism. A final research proposal can focus on exploring the long-term impacts of COVID-19 and other forms of crisis that affect responsible travel in

Kenya especially because Kenya is prone to most of these disruptive occurrences. This research focus can seek to examine sustainable recovery strategies, changing visitors' behaviour and preferences as well as impacts of disruptions on community livelihoods and environmental conservation efforts.

Conclusion

This chapter examines how ecotourism and responsible tourism principles are applied in Kenya with a focus on three areas: changing characterization of destinations, the eco-certification process, and the impact of the pandemic on responsible travel principles and practices. This study provides additional evidence on efforts to implement responsible tourism practices in Kenya's tourism destinations and the challenges encountered.

Regarding certification, there is an ambivalent attitude towards the process by many observers. There is a strong feeling among some actors that certification is done to promote and market individual businesses instead of focusing on the core criteria over which they are certified regarding care for the environment and concern for the local communities. It is however noted that concerns over certification are not a consequence of the pandemic but a direct outcome of greenwashing desires by tourism actors. Hence, eco-certification has been seen as a discreet drawback to the realization of responsible and sustainable tourism in the country. Concerning the destinations and practice of responsible travel, the emergence of wildlife-based conservancies is perceived as providing an ideal model of the practice of responsible tourism. While an alternative model is suggested, the challenge is about what happens to mainstream protected area tourism and the handling of unsustainable practices. The proliferation of hotels with high numbers of accommodation facilities of the nature of brick-and-mortar suggests the persistence of unsustainable practices from the previous era of mass tourism. Furthermore, with a significant proportion of protected areas tourism continuing to be under government management, there are questions related to inefficiencies in the implementation of responsible travel practices. It is also notable that the private sector has been in the lead in promoting responsible tourism. However, questions about private-sector greenwashing practices also cause a dent in the pursuit of responsible tourism. In sum, there remain sustainability gaps that need to be addressed for Kenya to be a responsible and sustainable travel destination.

Finally, the pandemic had contrasting outcomes regarding the overarching responsible travel principles of environmental conservation and community development. On the one hand, it enabled the regeneration of biodiversity in many tourist destinations because of low travel in several destinations. There has also been a relaxation in observing important environmental considerations like visitor numbers and increased littering therefore compromising on the key sustainability principles. On the other hand, in regard to socioeconomic responsibilities, a fall in tourism revenues during the pandemic suppressed local communities' ability to earn their livelihoods with many facilities seeking to recover from financial losses occasioned by the COVID-19 pandemic. As a way forward in both cases, there is a need for enhanced awareness, responsible travel education, and capacity building on the obligations of different actors to adhere to the principles and practices of responsible travel.

References

Atieno, L., & Njoroge, J. M. (2018). The ecotourism metaphor and environmental sustainability in Kenya. *Tourism and Hospitality Research, 18*(1), 49–60.

Bramwell, B., Lane, B., McCabe, S., Mosedale, J., & Scarles, C. (2008). Research perspectives on responsible tourism. *Journal of Sustainable Tourism, 16*(3), 253.

Bricker, K. (2017). The international ecotourism society. *Travel and Tourism Research Association: Advancing Tourism Research Globally, 11*.

Camilleri, M. A. (2016). Responsible tourism that creates shared value among stakeholders. *Tourism Planning & Development, 13*(2), 219–235.

Cassidy, E. (2020, March 26). *How to find and visualize Nitrogen Dioxide satellite data*. Retrieved March 5, 2023, from https://www.earthdata.nasa.gov/learn/articles/find-no2-data

Ceballos-Lascurain, H. (2006). Arg. Hector Ceballos-Lascurain: The architect of ecotourism. *ECOCLUB: International Ecotourism Monthly, 85*, 2–4.

Chester, G., & Crabtree, A. (2002). Australia: The nature and ecotourism accreditation programme. In M. Honey (Ed.), *Ecotourism & certification – Setting standards in practice* (pp. 161–185). Island Press.
Conaghan, A., & Hanrahan, J. (2010). Global conformity of indicators for eco-certification programmes. *Contemporary Issues in Irish and Global Tourism and Hospitality, 93*.
Diazzi, S. (2022). *Ecotourism certification for accommodation facilities: A qualitative evaluation of the Eco-Rating Scheme by Ecotourism Kenya*. Master's Thesis, University of Porto.
Dimitriou, C. K. (2017). From theory to practice of ecotourism: Major obstacles that stand in the way and best practices that lead to success. *European Journal of Tourism, Hospitality and Recreation, 8*(1), 26–37.
Ecotourism Kenya. (n.d.). *Eco rated facilities*. Retrieved March 25, 2023, from https://ecotourismkenya.org/eco-rated-facilities/
Eichelberger, S., Heigl, M., Peters, M., & Pikkemaat, B. (2021). Exploring the role of tourists: Responsible behavior triggered by the COVID-19 pandemic. *Sustainability, 13*(11), 5774.
Esparon, M., Gyuris, E., & Stoeckl, N. (2014). Does ECO certification deliver benefits? An empirical investigation of visitors' perceptions of the importance of ECO certification's attributes and of operators' performance. *Journal of Sustainable Tourism, 22*(1), 148–169.
Fennell, D. A., & de Grosbois, D. (2021). Communicating sustainability and ecotourism principles by ecolodges: A global analysis. *Tourism Recreation Research*, 1–19.
Goodwin, H. (2017). The challenge of overtourism. *Responsible Tourism Partnership, 4*, 1–19.
Gössling, S., Scott, D., & Hall, C. M. (2020). Pandemics, tourism and global change: A rapid assessment of COVID-19. *Journal of Sustainable Tourism, 29*(1), 1–20.
Government of Kenya. (2020). Ministry of environment and forestry-implementation plan for the ban of single use plastics in protected areas. Retrieved March 28, 2023, from https://www.tourism.go.ke/wp-content/uploads/2020/05/Action-Plan-Single-Use-Plastics-Ban-in-Protected-Areas-February2020.pdf
Higgins-Desbiolles, F. (2018). Sustainable tourism: Sustaining tourism or something more? *Tourism Management Perspectives, 25*, 157–160.
Honey, M. (2008). *Ecotourism and sustainable development: Who owns paradise?* Washington, D.C. Island press.

Ibrahim, I., Zukhri, N., & Rendy, R. (2019). From nature tourism to ecotourism: Assessing the ecotourism principles fulfillment of tourism natural areas in Bangka Belitung. *Society, 7*(2), 281–302.

Ioannides, D., & Gyimóthy, S. (2020). The COVID-19 crisis as an opportunity for escaping the unsustainable global tourism path. *Tourism Geographies, 22*(3), 624–632.

Ireri, P., Kung'u, J. B., & Muriithi, J. K. (2020). Distribution of the benefits of Ngare Ndare Forest Trust and Il Ngwesi Group Ranch ecotourism enterprises in Kenya. *African Journal of Hospitality, Tourism and Leisure, 9*(5), 1178–1192.

Ireri, P., Muriithi, J., & Kung'u, J. B. (2023). An analysis of Kenya's policy on local community participation in tourism. *The International Journal of Interdisciplinary Social and Community Studies, 18*(1), 51.

Jovicic, D. Z. (2019). From the traditional understanding of tourism destination to the smart tourism destination. *Current Issues in Tourism, 22*(3), 276–282.

Kabii, F., Wandaka, J., Wamathai, A., & Jilo, N. (2019). The role of tour guides in promotion of sustainable tourism practices in Kenya. *Journal of Tourism and Hospitality Management, 7*(2), 40–47.

Kenya Tourism Board (KTB). (n.d.). *Places to go*. Retrieved March 5, 2023, from https://magicalkenya.com/places-to-go/

KWCA. (n.d.). Conservancies: Transforming communities while safeguarding our iconic wildlife. Retrieved March 5, 2023, from https://kwcakenya.com/conservancies/status-of-wildlife-conservancies-in-kenya/

Lenzen, M., Sun, Y. Y., Faturay, F., Ting, Y. P., Geschke, A., & Malik, A. (2018). The carbon footprint of global tourism. *Nature Climate Change, 8*(6), 522–528.

Lindsey, P., Allan, J., Brehony, P., Dickman, A., Robson, A., Begg, C., et al. (2020). Conserving Africa's wildlife and wildlands through the COVID-19 crisis and beyond. *Nature Ecology & Evolution, 4*(10), 1300–1310.

Løvschal, M., Bøcher, P. K., Pilgaard, J., Amoke, I., Odingo, A., Thuo, A., & Svenning, J. C. (2017). Fencing bodes a rapid collapse of the unique Greater Mara ecosystem. *Scientific Reports, 7*(1), 1–7.

MacLeod, N. (2016). Self-guided trails–A route to more responsible tourism? *Tourism Recreation Research, 41*(2), 134–144.

Maingi, S. W. (2019). Sustainable tourism certification, local governance and management in dealing with overtourism in East Africa. *Worldwide Hospitality and Tourism Themes, 11*(5), 532–551.

Mathew, P. V., & Sreejesh, S. (2017). Impact of responsible tourism on destination sustainability and quality of life of community in tourism destinations. *Journal of Hospitality and Tourism Management, 31*, 83–89.

Mihalic, T. (2016). Sustainable-responsible tourism discourse—Towards 'responsustable' tourism. *Journal of Cleaner Production, 111*, 461–470.

Mihalic, T. (2020). Conceptualising overtourism: A sustainability approach. *Annals of Tourism Research, 84*, 103025.

Mondal, S., & Samaddar, K. (2021). Responsible tourism towards sustainable development: Literature review and research agenda. *Asia Pacific Business Review, 27*(2), 229–266.

Muriithi, J. K. (2014). The role of certification in the enhancement of sustainable *Ecotourism Development in Kenya*. In *Le Kenya Dans la 3e Révolution Tourism: Au au-delà du safari*. L'Harmattan.

Muriithi, J. K. (2022a). Role of conservancy-based tourism in the management of natural ecosystem services in the Maasai Mara, Kenya. In *Management of tourism ecosystem services in a post pandemic context* (pp. 11–29). Routledge.

Muriithi, J. K. (2022b). Sustainability perspectives of the wildlife conservancy–based tourism model in a post-pandemic context in Kenya. In *Tourism through Troubled Times: Challenges and opportunities of the tourism industry in 21st century* (pp. 181–202). Emerald Publishing.

Ogada, M. (2019, June 27). Decolonising conservation: It is about the land, stupid. *The Elephant, 27*.

Ondicho, T. (2021). Impact of COVID-19 on tourism in Kenya: Strategies for recovery. *Kenya Policy Briefs, 2*(1), 45–46.

Ritalahti, J., & Ali, A. (2022). Finnish tourists' responsible travel intentions in times of COVID-19: Implications for travel intermediaries. In *Towards sustainable and resilience tourism futures: Insights from the International Competence Network of Tourism Research and Education (ICNT)*. Springer.

Saufi, A., O'Brien, D., & Wilkins, H. (2014). Inhibitors to host community participation in sustainable tourism development in developing countries. *Journal of Sustainable Tourism, 22*(5), 801–820.

Spenceley, A. (2019). Sustainable tourism certification in the African hotel sector. *Tourism Review, 74*(2), 179–193.

Spenceley, A. (2021). *The future of nature-based tourism: Impacts of COVID-19 and paths to sustainability*. Luc Hoffmann Institute.

Thompson, B. S., Gillen, J., & Friess, D. A. (2018). Challenging the principles of ecotourism: Insights from entrepreneurs on environmental and economic sustainability in Langkawi, Malaysia. *Journal of Sustainable Tourism, 26*(2), 257–276.

Thompson, D. M., Serneels, S., Kaelo, D. O., & Trench, P. C. (2009). Maasai Mara—Land privatization and wildlife decline: Can conservation pay its way? In *Staying Maasai? Livelihoods, conservation and development in East African rangelands* (pp. 77–114). Springer.

Ting, H., Jean, L. X., Meng, L. C., Cheah, J. H., & Cheer, J. M. (2020). Editorial—Responsible tourism: A call to action for turbulent times. *Asian Journal of Business Research, 10*(2).

UNWTO. (2020). *One Planet vision for a responsible recovery of the tourism sector*. Retrieved March 2, 2023, from https://webunwto.s3.eu-west-1.amazonaws.com/s3fs-public/2020-06/one-planet-vision-responsible-recovery-of-the-tourism-sector.pdf

Veldhuis, M. P., Ritchie, M. E., Ogutu, J. O., Morrison, T. A., Beale, C. M., Estes, A. B., Mwakilema, W., Ojwang, G. O., Parr, C. L., Probert, J., & Wargute, P. W. (2019). Cross-boundary human impacts compromise the Serengeti-Mara ecosystem. *Science, 363*(6434), 1424–1428.

Waithaka, J. (2004). Maasai Mara—An ecosystem under siege: An African case study on the societal dimension of rangeland conservation. *African Journal of Range and Forage Science, 21*(2), 79–88.

Waithaka, J., Dudley, N., Álvarez, M., Arguedas Mora, S., Chapman, S., Figgis, P., … & Wong, M. (2021). Impacts of COVID-19 on protected and conserved areas: A global overview and regional perspectives.

Western, D., Russell, S., & Cuthill, I. (2009). The status of wildlife in protected areas compared to non-protected areas of Kenya. *PLoS One, 4*(7), e6140.

Western, D., Waithaka, J., & Kamanga, J. (2015). Finding space for wildlife beyond national parks and reducing conflict through community-based conservation: The Kenya experience. *Parks, 21*(1), 51–62.

WHO. (2020, March 11). *WHO Director-General's opening remarks at the media briefing on COVID-19 – 11 March 2020*. Retrieved March 2, 2023, from https://www.who.int/director-general/speeches/detail/who-director-general-s-opening-remarks-at-the-media-briefing-on-covid-19%2D%2D-11-march-2020

Part III

Economic Sustainability and Governance Trends and Tourist Behavior in the New Normal (SDG 1, 16 and 17)

12

Trends in Post-pandemic Tourism in Developing Countries: The Case of Ghana

Francis Boadu and Silvia Fernandes

Introduction

Digital marketing can be defined as the process of providing access to product or delivering service and post service to customers using technological devices. It also uses electronic media to promote products and services (Yasmin et al., 2015). Digital marketing in Africa has been documented in reference to digital statistics such as internet use and access to social platforms like Facebook, Twitter, LinkedIn, Instagram, Pinterest, etc. Studies and theories around this subject have proven that there is a radical change toward digitalization whereby consumers are searching more and more on the internet. The availability of digital market has widened the scope of potential customers' search and has also granted access to the global market. Digital marketing uses platforms of

F. Boadu
Faculty of Economics, University of Algarve, Faro, Portugal

S. Fernandes (✉)
Faculty of Economics & CinTurs, University of Algarve, Faro, Portugal
e-mail: sfernan@ualg.pt

S. W. Maingi et al. (eds.), *Tourist Behaviour and the New Normal, Volume II*,
https://doi.org/10.1007/978-3-031-45866-8_12

communication, promotion, brand awareness for consumer interaction where a feedback mechanism is fundamental. Businesses can reach more customers, even those beyond the region considered, creating a tool which has equipped them to communicate, sell and promote their products and services. On the other hand, it has developed consumers' access to their preferred items. Content development, customer capital, integrated marketing and communications, user experience, and blockchain technology are considered important areas for digital marketing in Africa. Digital media, online presence and content, and mobile advertising (Ofosu et al., 2013) are developments that can give Africa the chance to promote its tourist attractions. However, there are difficulties specific to the application of digital marketing in its tourism industry (Yasmin et al., 2015).

In the case of Ghana, tourism management is sponsored by the Ministry of Tourism. This body was created in 1993 to promote, develop, and coordinate Ghana's tourism agenda. In 2003, the ministry was renamed Ministry of Tourism and Modernization of the Capital City, being charged to develop Accra (the national capital) into a modern international city. But in 2009, the government reused the original name whose task is the policy making for Ghana's tourism sector. The implementation agency is known as Ghana Tourist Board (GTB), and the body for private tourism organizations and their partners is Ghana Tourism Federation (GHATOF). Two additional important partners are the Ghana Tourist Development Company and the Hotel Catering and Tourism Training Centre (NTMS, 2009–2012). The Ghana Tourism Authority (GTA) established in 2016 a digital platform that had a database of all tourist destinations. This platform acted as a convenient access point for travelers looking for details on destinations, such as location, lodging options, area, general description, and so on.

According to the UNWTO statistics, foreign tourist arrivals increased from 62.7 million in 2017 to 67 million in 2018, indicating that African tourism industry is expanding (UNWTO, 2018). Additionally, reports from the World Travel and Tourism Council (WTTC) and the African Travel and Tourism Association (ATTA) show that $194.2 million were generated by tourism in Africa representing about 8.5% of GDP (ATTA, 2019; WTTC, 2019). The government or the competent authorities can take steps to expand the travel industry. These steps include establishing preferential rates, creating a college for tourism training, and launching marketing campaigns to promote Ghana's attractions. Advertisements and exclusive travel offer can be published on social media. These means can assist clients in gaining access to valuable first-hand knowledge and in expanding their travel options. Travel deals can now reach a wider audience of potential passengers. The travel firms can now use a variety of marketing strategies to get their name out there, connect with a larger audience, share special offers, and easily publish information.

Mobile electronic devices also have contributed for these trends, as they give easy access to the internet. Interacting with customers and other businesses has never been easier. Sharing information globally has become much simpler since the use of social media. Also, many management choices have been impacted by online reviews, and consequent steps to improve goods and services have been considered.

In addition, tracking consumer interactions with products or services, from the initial inquiry to the end of the service, is optimized. The obtained data can be used to examine patterns of consumption, costumer's behavior, and services that provide a more personalized experience. The availability of straightforward bookings and immediate transactions is another useful aspect of digital marketing tools. People no longer need to wait a few days for a hotel or travel company to respond to their requests. Now, consumers can quickly search and select a variety of options that satisfy their needs.

Research Problem

It has been observed that factors influencing the performance of domestic tourism in the country considered include variables like limited awareness, low income, inadequate information, low media usage, marketing quality challenges, and lack of qualified people. In this context, digital marketing can bring a great change due to its role and impact. It is dominating the tourism industry today and, as a result, many organizations developed digital marketing channels to enhance their offers. They are also attentive to mobile digital marketing and content marketing tools.

Africa can be described as having a rich heritage that attracts a lot of tourists to the continent. Although there are challenges to its tourism sector, these positive trends envision a prospective growth if the necessary engagement process is done. Since its independence, Ghana is striving to become a unique and worthwhile tourist destination. As a result, the country must brand itself to the world through travel and trade innovations (Oxford Business Group, 2017). The improved branding will attract more visitors and research work to be held in Ghana. Sites of attraction should be well documented and structured and meet standards of quality. Tourist sites recognized as secure and peaceful are nowadays highly appreciated across the globe. Thus, the purpose of this study is to identify opportunities available in digital marketing for tourism, having the West African country Ghana as a case study. The underlying research design adopted a quantitative approach for data analysis. Questionnaires were used for gathering these data from the public, with the aim of understanding the effects of digitalization from the perspective of tourists and tourism-based businesses. It then presents recommendations that might help companies take advantage of its relevance, popularity, and effectiveness. The following research questions were considered as a guide:

1. Which opportunities are available in digital marketing and tourism?
2. Which dimensions are used to change the operations in tourism?
3. Which strategies have been pursued to operate in this continuous online trend?

Significance of the Study

The digital market has greatly impacted the sphere of selling and marketing goods/services. In Africa, digital statistics in the areas of internet and social media reveal an increasing use. This can have a positive impact on business if it is well considered and adopted (Mkwizu, 2019). Due to the global shift toward the digital sphere, social media marketing campaigns are conducted almost always. If a sector wishes to expand its market share, social media can serve as the front office of its digital business. However, engaging well the audience and motivating it is crucial. Social media can be a way to leverage a brand and advertise it to the world. Today, when people visit a website or a social network, they expect to be guided to learn and interact more with the products/services offered. Digital marketing can be a means of promoting brands around the world, as well as an interesting showcase of travel offers. The use of videos and stories can create a more vivid experience of the featured places or activities. This study, in answering those questions, will serve as a guide for policy makers to identify the gaps that could be solved through the adoption of digital marketing in tourism. Every business, including tourism, has been greatly impacted by the ease of access to the internet and related applications to meet client needs (Gupta, 2019). This study can also serve as a reference for future research in evaluating the opportunities that digital world offers to other sectors.

Methodology

Research Design

This study used a research approach to data collection, analysis, and observations to describe and explain aspects related with the relationship between digital marketing and tourism decision making. This study applied quantitative methods to answer the research questions for the awareness of digital channels' use and how it can affect marketing and bring changes to the tourism operations in Ghana. This is then discussed on how to gather new information and operate in a continuous changing world. The need of consistency in replies influenced the decision to use a survey as the instrument to gather data from the sample population because fixed-response questions help to reduce variability in the results.

Source of Data and Target Population

The responses from the residents of Kumasi, a suburb of the Ashanti region in Ghana, provided the dataset for this investigation. The questionnaire was designed with Google forms and was rolled out to collect responses. Surveys provide a simpler way to code, analyze, and interpret information (Malhotra & Birks, 2007). A sample is a subset of elements in the entire population that provides an almost accurate representation. Bless et al. (2013) however argue that the advantages of studying a whole population outweigh those of studying a sample because of data reliability and validity. Nonetheless, the use of sampling saves time and cost and provides attention to peculiar individual cases. Scott and Morrison (2007) affirmed that sampling refers to the selection of a subset of persons or things from a larger population, with the intention of representing the population. The choice of sample participants depends on the judgment in limiting the selection of subject participants to those who have been exposed to the phenomenon under investigation. Thus, the target population was restricted to residents of Kumasi, who have embarked on tourism in its various forms throughout Ghana. This ensured enough and relevant information within the limited period available for the study. A total sample size of 62 respondents was considered.

Research Instruments

The key aim of this study is to determine the influence of social and digital media in the choice of tourism destinations. In using a questionnaire, a lot of information can be obtained from a desired number of people over a time frame, easy to analyze objectively and not prone to provide biased information as compared to interviews. It is cost-efficient and practical and allows respondents to remain anonymous. For ensuring the research neutrality and validity, the questions asked were closely related to those goals. A total of 11 questions made up the survey: the first and second questions focused on the respondents' demographics, while the third through sixth questions asked about their use of social media, and the final part of the survey asked about opinions on acquisition of data and needs of respondents in terms of safety.

Data Collection

In research the most important process is the collection of data which can be obtained through primary or secondary sources. Data collection enables access into the inner recesses of group life, organizational structure, bureaucratic processes as well as motivations for individual behavior (Yin, 2011). This study adopted the use of primary data, and the survey was administered in September 2022. The research design administered the questionnaires through online platforms for data collection, devoid of time delays. The researchers should make sure not to get involved in influencing the judgements of the respondents. Thus, clear instructions on how to answer the questions were provided in the survey. Respondents' participation was voluntary and done with the appropriate courtesy.

Data Analysis

According to Marshall and Rossman (2012), data analysis is basically the process of bringing order and structure to make meaning of data collected. In analyzing the data, the interpretive paradigm was adopted since the work wants to understand how participants are managing the influence of digitalization in tourism. Interpretivism focuses on identifying the complex nature of social phenomena and building facts around it. Its purpose is to understand and interpret the occurrence of events, experiences, and social structures, as well as the unique values and principles people attach to these phenomena (Collis & Hussey, 2009; Rubin & Babbie, 2009). This method is favorable for this study because the research wanted to know more about how respondents, who represent the tourism industry, understand digital marketing and its impact on tourism activities. The respondents were pre-informed that the results could be given to whom showed interest in them. Respondents were also made aware that they could pull out from the research and assured that their feedback remained confidential.

Results

Examining the questionnaire, other studies, and documents about tourism in Ghana (and throughout Africa), this study aims to have insights into how digitalization and digital marketing are affecting business and marketing plans. This section is divided into the following subsections: data collecting, demographics, various preferences, and summary. The data used here are named as dataset-A, to be compared with another related study whose sample is named dataset-B.

Dataset-A's Overview

A total of 62 questionnaires were received from the residents of Kumasi, Ghana. The google form's link was shared to many individuals to participate in the survey. However, the number of responses was influenced by respondents' knowledge and interest in tourism.

Demographics

In the total number of respondents, 45% are females and 55% are males, with 49% in the age range of 18–25 years, 46% in the range of 26–35 years, and 5% above 35 years.

Sample Preferences

The respondents were asked about their use of digital channels, the willingness to gather information about tourist sites before taking the final decision, and the media through which they acquire/gather information on tourist attractions, as well as their preferred source of data.

Frequency of Digital Channel Use

Out of the responses, 73% use digital channels 'very often', while 24% use them 'often' and 3% of the sample 'rarely' use them.

Means of Getting Information

The respondents were then asked about how they get the information, where 87% refer the option 'over the internet', 8% refer 'at tourist sites', 3% opt for 'in the destination', and 2% refer 'by phone'.

Preferred Source of Data

In the sample, 60 respondents chose 'social media' (97%), 11 respondents opted for 'television' (18%), 3 referred 'newspapers' (5%), and 7 chose billboards (11%) as their preferred source of data on tourist destinations (see Fig. 12.1).

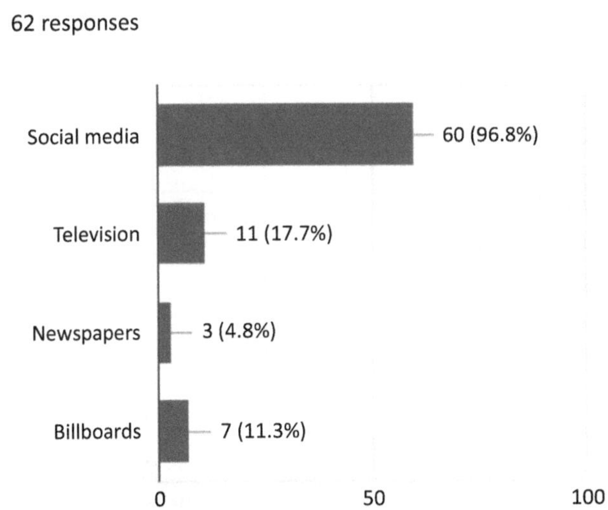

Fig. 12.1 Preferred source of data on destinations

Data Acquisition

In this variable (with a Likert scale), respondents were asked about the difficulty in acquiring data about destinations in Ghana where 2 agreed that it is 'very easy', 11 referred 'easy', 33 respondents chose 'moderate', 8 respondents consider that it is 'difficult', and other 8 agreed it is 'very difficult' to acquire data on tourist destinations in Ghana (see Fig. 12.2).

Satisfactory Information

Then the respondents were asked about the level of satisfaction with the information provided by digital channels on tourist attractions/destinations where 5 of them agree that it is 'very satisfactory', 12 answered 'satisfactory', 27 respondents chose 'moderate', and 16 are dissatisfied, while are very dissatisfied with this aspect in Ghana (see Fig. 12.3).

Digital Marketing Adoption

The respondents were asked about their opinion if tourism firms use digital marketing where 19% referred 'Yes' while 57% stated 'No', and the remaining 24% chose 'Maybe'.

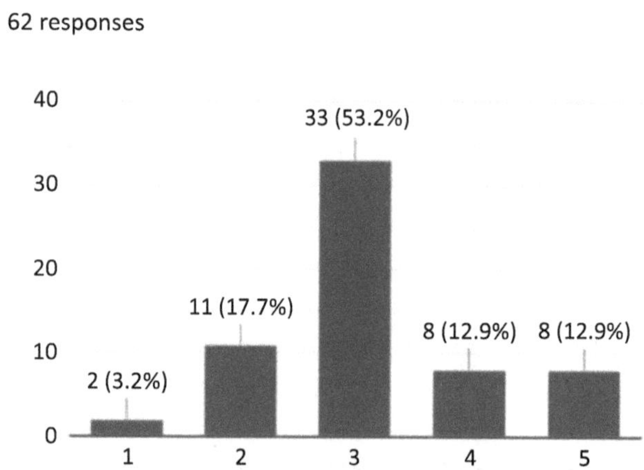

Fig. 12.2 Data acquisition on destinations

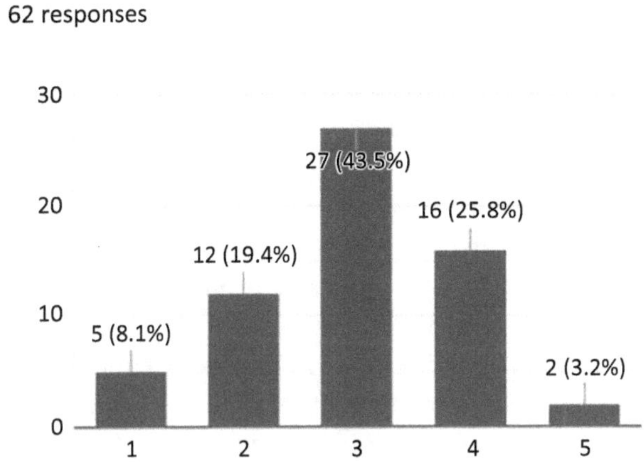

Fig. 12.3 Satisfactory information on tourist destinations

Safety of Tourist Destinations

Out of 56 responses, it was emphatically stated by 84% of the sample that they prioritize safety at the tourist attraction/destination, whereas 11% chose 'Maybe' and 5% referred 'No'.

Reason of Using Digital Channels

With a high percentage (of 93%) the surveyed firms prefer using digital channels to gather information about tourist destinations before travel, while 5% chose 'Maybe' and 2% stated 'No' on this aspect.

Dataset-B's Overview

The dataset-B study (Dzisi & Ofosu, 2014) was also applied to Ghana's firms. It revealed a positive relationship between marketing strategies and the performance of small and medium-sized enterprises (SMEs). Other studies have found a strong correlation between marketing strategies and performance in terms of growth in revenue, improved efficiency, and connection with customers (David et al., 2013; Azeez, 2019). SMEs mostly used traditional forms of marketing—mainly television and radio, newspaper and magazine, banners, and billboards—to attract customers. They also participate in fairs/exhibitions and use word-of-mouth to foster a long-term relationship with their customers. Only a few SMEs deployed technology-based marketing strategies in reaching their customers. Facebook was the social network most used, compared to corporate websites and emails. Thus, dataset-A (from the present study) shows an evolution compared to dataset-B in the side of firms. It reveals an increased use of digital marketing by firms/institutions of tourism and destination management. Its adoption did boost firm sales revenue and performance due to its wider coverage than traditional means. Thus, making digital marketing a part of the educational curriculum in Ghana is a requirement (Ocansey, 2022). And learn research methods and theory on technology acceptance/use and related issues (Dhar & Gorlin, 2013; Venkatesh et al., 2016). Firms and stakeholders should turn it a priority in organizing training procedures and encouraging participation. Technology-based service innovation leads to service value, tourist revisits, and experience sharing (Kotoua & Ilkan, 2017; Preko et al., 2022). Managers and other stakeholders of Ghana's tourism and hospitality sectors should invest in training their staff to deliver excellent experiences. Managers should also examine the latest use of digital platforms to understand the future trends (Aninkorah et al., 2017).

Discussion and Findings

In this study, the aim was understanding participants' perspectives on how social media and digital marketing influence tourist decisions and the operation of tourism-oriented businesses. The results made it clear that the internet is the major and largest source of information about tourism in Ghana. Therefore, social media, as its largest segment, has influence since interested people tend to rely on digital media to acquire information on tourism. The research purpose was achieved through the analysis of a questionnaire, literature review, and a comparison between two datasets/studies. The following objectives/questions were used as a guide to accomplish its aim:

1. Which opportunities are available in digital marketing for tourism?
2. Which dimensions are used to change the operations in tourism?
3. What strategies are pursued to operate in this continuous online trend?

Opportunities and Strategies

One very important opportunity digital marketing can offer to Ghana's tourism sector is by helping to reach a wider range of customers. Primarily, advertisement on tourist destinations was done through traditional media such as newspapers, door-to-door campaigns, banners, paraphernalia, etc. With the reduction of newspaper patronage and limited coverage of door-to-door campaigns, the contact audience is restricted to a few people and a few square miles. Also, this form of advertising is relatively expensive and time consuming. Through the adoption of digital marketing, tourism industry would have their products/services exposed to a larger customer base. Ghana tourism sector serves as an important source of employment and income. As a result, the growth of tourism sector would be a great boost to the country's economy.

Targeted Marketing

The drive of targeted marketing is to identify a group/segment who usually belong to a shared demographic and is highly probable to purchase a certain good or service. Although sometimes challenging to execute, this strategy is highly productive. Through social media polls, data can be gathered and then analyzed to predict what potential customers are interested and their shared characteristics. After this, potential customers can be segmented and then actively advertised, by matching them to tourist products of their preference.

Online Payment Options

Furthermore, by creating and maintaining a centralized website for Ghana's tourism sector, that advertises all its locations and facilities, an option for online payment could be added. This would be particularly useful for tourists who do not live in Ghana because they would be able to plan and make payments before arriving there, in their own currency. Another advantage is that it would be easier for the Tourism Ministry, policy makers, managers of tourist locations, and other key stakeholders to regulate pricing and keep track of the revenue.

Easy Access to Facilities

By creating a regulated website for Ghana's tourism, available accommodation facilities could be advertised. As a digital marketing tool, such a portal would contain not only information and pricing details, but also other accommodation aspects (hotels, guest houses, resorts, etc.). This is important, especially for non-Ghana tourists, to assure a good experience in the destination. Such initiative would help induce a sense of security, relieving tourists from the stress related to personally finding accommodation and target facilities.

Conclusion

This study focuses on the opportunities of digital marketing and tourism in Africa, having Ghana as a case study. It was observed that digital marketing has a greater influence now. From an interesting comparison between two studies in Ghana—the present work with dataset-A and a previous study's dataset-B—it is acknowledged that the tourist destination institutions and tourism firms have not adopted the use of digital marketing in the same pace than individuals. Tourism is a risky activity and so customers need much information to remove all doubts and reduce perceived risks. Social media can then be used to create massive awareness about destinations and facilitate less expensive communication among stakeholders. Additionally, in advertising destinations, social media marketers must make sure that the truth is communicated. Destination management institutions should ensure safety, cleanliness, and security of the sites. Any update should be timely communicated, as this can be a competitive advantage of a destination.

Implications for Practice

This study emphasizes how crucial digital marketing is to boost tourism in Ghana. The results show that the country's tourism industry is still underusing digital marketing. Therefore, to reach a larger audience and increase tourist flow, tourism businesses and policymakers need to invest more in digital media/marketing. Tourism organizations may efficiently engage potential tourists and give them a seamless travel experience by using digital platforms such as social media, websites, and mobile apps. The firms should start refining their revenue model toward online and social media presence, through models like advertising, live events and other forms of digital content. They should leverage their reputation with ethical treatment or service provision standards to enhance their brand. The internet and social networks can be used to create massive awareness about destinations and facilitate less expensive communication among stakeholders to create a viral market for their products and services. Given that tourists prioritize new factors, destination management businesses should be able to provide safety, security, and cleanliness at places and facilities. And any improvement should be accurate and promptly reported to the audience.

Recommendations for Further Research

In Ghana, effective marketing techniques still include door-to-door sales and word-of-mouth, banners, branded merchandise, and radio/television advertisement. But firms must be highlighted about digital marketing plans. Non-traditional marketing tactics are profitable investments because they remove geographical and other barriers to company interactions, enabling a variety of stakeholders to access and share information, resources, and services. Also, this study is limited to the residents of Kumasi, and to the literature review and descriptive statistics as research techniques. A larger sample size, probably comparing tourist locations between a developed and a developing nation, could be useful to provide more insight into the impact of digital marketing. Because it gives relevant signs both in terms of opportunities and challenges of its efficient use. Further research can deepen the factors that affect the choice of destinations and the business models that firms can adopt to cope with this era of digitalization. The focal point of this study is limited to the perspectives of tourist-centered institutions on how digital media/marketing is influencing decision making. A broader scope, considering governance interventions, could also be approached.

References

Aninkorah, B., Bakah, J., Dankwah, S., Falke, A., & Codjia, M. (2017). *The influence of the internet and social media on tourism in Ghana, Project Tourism & Social Media*. University of Ghana Business School, Department of Marketing and Entrepreneurship.

ATTA. (2019). *Analysis of Africa's tourism market for April 2019*. African Travel and Tourism Association.

Azeez, R. (2019). Is tourism a sustainable haven for economic growth in north African countries? An evidence from panel analysis. *Munich Personal RePec Archive, 93449*. https://doi.org/10.2139/ssrn.3323486

Bless, C., Smith, C., & Sithole, S. (2013). *Fundamentals of social research methods: An African perspective* (5th ed.). Juta.

Collis, J., & Hussey, R. (2009). *Business research: A practical guide for undergraduate and postgraduate students* (3rd ed.). Palgrave Macmillan.

David, C., Neeraj, A., Derek, K., John, W., Michael, R., Ruba, B., Julia, C., & Julio, B. (2013). *Lessons on technology and growth from small-business leaders*. The Boston Consulting Group.

Dhar, R., & Gorlin, M. (2013). A dual-system framework to understand preference construction processes in choice. *Journal of Consumer Psychology, 23*(4), 528–542. https://doi.org/10.1016/j.jcps.2013.02.002

Dzisi, S., & Ofosu, D. (2014). Marketing strategies and the performance of SMEs in Ghana. *European Journal of Business and Management, 6*(5), 102–111.

Gupta. (2019). *Digital Marketing* (2nd ed.). McGraw Hill.

Kotoua, S., & Ilkan, M. (2017). Tourism destination marketing and information technology in Ghana. *Journal of Destination Marketing & Management, 6*(2), 127–135. https://doi.org/10.1016/j.jdmm.2017.04.007

Malhotra, N., & Birks, D. (2007). *Marketing research: An applied approach.* Pearson Education.

Marshall and Rossman. (2012). Designing qualitative research. *Organization Management Journal, 8*(3), 193–195. https://doi.org/10.1057/omj.2011.23

Mkwizu, K. (2019). Digital marketing and tourism: Opportunities for Africa. *International Hospitality Review, 34*(1), 5–12. https://doi.org/10.1108/IHR-09-2019-0015

NTMS. (2009–2012). *Non-tariff measures to trade: Economic and policy issues for developing countries.* https://unctad.org/system/files/official-document/ditctab20121_en.pdf

Ocansey, T. (2022). *Digital marketing and the performance of cashew exporting firms in Ghana.* Thesis. University of Ghana.

Ofosu, D., Gyanewa, M., & Boadi, I. (2013). The influence of short message service (SMS) on the purchase behaviour of young consumers in a tertiary institution in Ghana. *Review of Integrative Business & Economics Research, 2*(2), 276–303.

Oxford Business Group. (2017). *The report: Ghana 2017.* https://oxfordbusinessgroup.com/ghana-2017

Preko, A., Amoako, G., Dzogbenuku, R., & Kosiba, J. (2022). Digital tourism experience for tourist site revisit: An empirical view from Ghana. *Journal of Hospitality and Tourism Insights*, 2514–9792. https://doi.org/10.1108/JHTI-10-2021-0294

Rubin, A., & Babbie, E. (2009). *Research methods for social work* (7th ed.). Cengage Learning.

Scott and Morrison. (2007). *Key ideas in educational research.* Continuum.

UNWTO. (2018). *Tourism highlights.* Retrieved July 14, 2019, from https://www.e-unwto.org/doi/pdf/10.18111/9789284419876

Venkatesh, V., Thong, J., & Xu, X. (2016). Unified theory of acceptance and use of technology: A synthesis and the road ahead. *Journal of the Association for Information Systems, 17*(5), 328–376. https://doi.org/10.17705/1jais.00428

WTTC. (2019). *Travel and tourism global economic impact and trends 2019.* World Travel and Tourism Council.

Yasmin, A., Tasneem, S., & Fatem, K. (2015). Effectiveness of digital marketing in the challenging age: An empirical study. *International Journal of Management Science and Business Administration, 1*(5), 69–80. https://doi.org/10.18775/ijmsba.1849-5664-5419.2014.15.1006

Yin, R. (2011). *Qualitative research from start to finish*. Guilford.

13

Protected Areas and Tourism in the New Normal: An Analysis of Temporal and Scalar Tensions of Sustainable Tourism Governance in the Finnish Arctic

Aapo Lundén and Alix Varnajot

Introduction

Crises have long been a factor of disruption in the global tourism industry (Cheer et al., 2021). Tourism is extremely sensitive and susceptible to wars, pandemics, and natural disasters (Sönmez et al., 1999). Despite continuous global growth, several crises in recent decades have significantly impacted tourism (Bonham et al., 2006). The ash cloud from the eruption of the Eyjafjallajökull volcano in Iceland, for example, caused substantial disruptions to European air transportation in the spring of 2010, leading to millions of stranded passengers (Bolić & Sivčev, 2011). Climate change and its consequences are understood as a global crisis, with significant economic and environmental disruptions for the industry (Hall, 2010). Recently, the ongoing war in Ukraine and the

A. Lundén (✉) • A. Varnajot
University of Oulu, Oulu, Finland
e-mail: Aapo.Lunden@oulu.fi; Alix.Varnajot@oulu.fi

S. W. Maingi et al. (eds.), *Tourist Behaviour and the New Normal, Volume II*,
https://doi.org/10.1007/978-3-031-45866-8_13

COVID-19 pandemic have also challenged tourism with international, regional, and local travel restrictions (Brouder, 2020; Clark et al., 2022; Gössling et al., 2020).

Generally, the COVID-19 pandemic forced tourism operators to promptly adapt and look for new markets, increasing the role and share of domestic tourism in some cases. On the one hand, the pandemic created what has been termed an "anthropause" (Searle et al., 2021), referring to an opportunity for destination (eco)systems to recover from intense tourism. However, on the other hand, lockdowns and limits to international travel forced people to "turn back to nature", to spend time in local and national green spaces and to engage in leisure activities in closer destinations, especially in countries with well-established green leisure infrastructures, such as the Nordics. This underlines the role of green spaces, such as protected areas and national parks, in providing a temporary push for domestic tourism markets (Brouder, 2020). Thus, this chapter focuses on the governance and sustainability challenges brought by the temporary growth of tourism in protected areas.

Globally, the Nordics represent a key region for tourists seeking nature-based activities and experiences (Fredman et al., 2021). In parallel, nature-based tourism is one of the fastest growing tourism segments in the world (Wearing et al., 2009), taking place in various types of protected areas. According to the International Union for Conservation of Nature, the standard definition of a protected area is "[an] area of land and/or sea especially dedicated to the protection and maintenance of biological diversity, and of natural and associated cultural resources, and managed through legal or other effective means" (Dudley & Stolton, 2008, p. 9). Protected areas can be further categorised according to their main features, biological values and purpose from strict conservation and wilderness areas (Category I) to areas managed for sustainable resource use (Category V). There is an active debate on the practical meanings and management of protected areas, as well as on how to balance cultural and ecological priorities (see Brechin et al., 2010). However, this chapter focuses on Finnish national parks because they usually and historically reflect the main ideas of the universal spirit and the definition of protected areas. Thus, in principle, nature-based tourism in protected areas represents an integral part of achieving Sustainable Development Goals (SDGs) (Snyman & Bricker, 2019).

This chapter critically examines the impact of these recent disruptions on tourism in protected areas, particularly those generated by the COVID-19 pandemic and climate change. First, we explore the emergence of domestic crowding and overtourism in protected areas (e.g., national parks) induced by these global crises, with a focus on Finland and the Nordics in general. The Nordics represent a relevant case because outdoor culture and access to nature (e.g., *friluftsliv, everyman's right*) are traditional parts of the Nordic lifestyle. Second, we look into these issues further by discussing the long-term effects of both climate change and the COVID-19 pandemic, and how in the Anthropocene, global crises and abstract processes can have direct impacts at regional and local scales (Varnajot & Saarinen, 2021). Thus, our goal is to explore the consequences and governance issues posed by immediate crowding and long-term climate change in protected areas. By doing this, we aim to challenge some prevailing notions of sustainable governance in tourism within these areas, rooted in short-term, targeted management strategies, exemplified by crowding due to COVID-19.

This chapter is divided into three sections. The first section reviews key literature on sustainable tourism and the growth of protected area tourism, where the COVID-19 pandemic has recently played a crucial role in tourist behaviours. Second, we explore the complexity of protected area tourism governance and its temporal, scalar, and consequential nature, regarding how short-term visitor management development may provide broad governance challenges in the long-term. Finally, we discuss the potential inherent challenges of sustainable protected area tourism in the Arctic by evaluating the broader temporal and scalar implications of current dominant protected area tourism governance ideals. We conclude by linking the discussion to the SDGs and some of the challenges they potentially bring to protected area tourism governance in climate-vulnerable destinations.

COVID-19 and the Return to Nature

Access to nature and its commercialisation are closely intertwined with outdoor leisure activities and nature-based tourism, but their origins often stem from socioeconomic and industrial motivations. For example, in Finland, everyman's rights were initially introduced to allow landless people to practise slash-and-burn farming (Lehtonen et al., 2007). However, their interpretation remains vague, posing challenges in defining the boundaries between the public and private use of nature (e.g., tourism and commercial berry picking) (Tuulentie & Rantala, 2013). Moreover, the popularity of second homes in Finland and the Nordics at large can be linked to rapid urbanisation, which has led people to seek a reconnection with nature (Löfgren, 2002). This growing interest in green spaces has been accompanied by the expansion of other industries. For instance, the development of extractive industries in peripheral areas was accompanied by the construction of road networks, electricity grids, and other facilities that could later be utilised for nature-based leisure and tourism activities (Sandell, 2007). The creation of the Abisko National Park in Sweden, for example, followed the establishment of the train connection for ore transportation from Kiruna to Narvik, Norway. In Finland, forestry has also contributed to improved access to the most remote parts of the country through road building. Thus, access to Nordic nature, notably in Finland, arises from a blend of factors like enhanced infrastructure, societal advancements, economic growth, private transport accessibility, and industrialisation, which cultivated a strong affinity for natural environments and the incorporation of everyman's right into the Nordic way of life.

The impact of the pandemic on protected areas' visitor rates varied significantly across countries (McGinlay et al., 2020) depending on various factors, like distance to major cities or local measures implemented to contain the spread of the virus. In general, nature-based tourism provides relatively risk-free options for leisure, including the control of social distancing and transport in countries with high-level public outdoor and protected areas. Therefore, it is no surprise that after strict lockdown measures were lifted, visitor rates increased, and crowding of protected areas was observed in countries such as the USA (Jacobs et al., 2020; Volenec et al., 2021), Sweden (McGinlay et al., 2020), Norway (OECD, 2022), and Finland (Metsähallitus, 2022). Thus, in general, such a surge of visitors made protected areas' governance more complex and required shifting priorities to visitor control in many countries, as health measures were added to the basic tasks of protected areas (McGinlay et al., 2020)

The Nordics was a specific region where outdoor activities became especially popular after the COVID-19 lockdown, leading to a 63–75% increase according to mobile data collected in different types of outdoor areas (Ritchie, 2020). At the same time, in regions like Arctic Finland, with some of the country's most iconic and popular national parks, foreign tourism decreased, especially in ski resorts located in the vicinity of national parks (e.g., Levi and Ruka ski resorts located nearby Pallas-Ylläs and Oulanka national parks, respectively). Figure 13.1 shows the role of domestic tourism, national parks, and broader tourism impacts in Finland, as well as how the cessation of international tourism and mobility influenced visitation in Finnish national parks. Thus, the domestic turn to national parks provided compensation for local tourism operators in ski resorts.

Nevertheless, this surge in the number of visitors to protected areas has created new governance challenges. According to McGinlay et al.'s analysis (2020), four key challenges were especially evident in the aftermath of the COVID-19 pandemic in protected areas: overcrowding ("escape to nature"), changes in visitor profiles (first-time visitors), unexpected and unfavourable visitor behaviour (e.g., littering, wear-and-tear) and rising conflicts with different visitor groups (e.g., day visitors and hikers). In the next section we explore such "novel" governance challenges and their potential short- and long-term implications from a scalar and temporal perspective. In doing so, we question whether symbiotic governance ideals require re-evaluation from a more nuanced and broader environmental perspective. We also examine the need for a similar re-evaluation of local-level governance solutions.

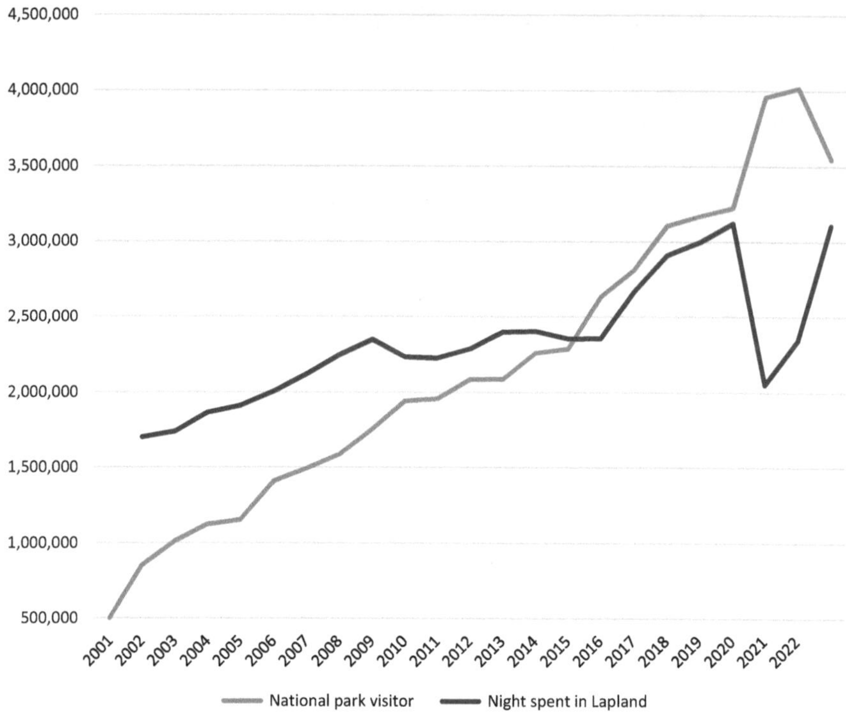

Fig. 13.1 Visitor trends in Finnish national parks and nights spent in the Lapland region 2001–2022. (Source: Metsähallitus, 2023; Statistics Finland, 2023)

From Loving to Exploitation: Hitting the Limits of Tourism-Conservation Symbiosis in the New Normal?

Protected areas contribute significantly to global and regional economies and provide economic justification for tourism in comparison to other types of land use (Hall, 2006). Prior to the COVID-19 pandemic, protected areas received approximately eight billion visits globally, generating roughly USD 600 billion in direct in-country expenditure and USD 250 billion in consumer surplus (Balmford et al., 2015). Considering the most recent tourism figures, international tourism has rebounded and even suppressed previous visitor rates in some Nordic destinations (Fig. 13.1).

The dominant paradigm of sustainable protected area tourism can be related to the symbiotic relationship between tourism and conservation, referring to the idea that well-managed tourism can support the conservation of natural and cultural resources (Budowski, 1976). Thus, the symbiotic ideal provides an economic justification, compensation, and rationality for the "non-use" of protected areas, while natural areas and iconic landscapes benefit from their exclusion from "normal" use of nature (Puhakka & Saarinen, 2013). For example, in the case of wildlife tourism in many countries, tourism revenue is directly channelled for conservation purposes and management (Newsome, 2020). Here, tourism is a conservation lifeline for protected areas under such governance regimes. However, in the Nordic context, since protected areas are free to access, their funding comes from the government budget (Øian et al., 2018). Thus, economic rationality, in the sense of "nature paying its way" through tourism revenues, is more indirect. In the last 20 years, various research efforts have focused on the Nordic context, with studies examining the impact of protected area tourism on the regional economy (see Saarinen, 2003, 2007; Lundmark et al., 2010), and more recently, on the well-being and public health benefits they provide (Puhakka et al., 2017). These studies have shown how benefits can be measured beyond the direct relationship between protected areas and tourism revenue. Additionally, already due to pre-pandemic overtourism, discussions on soft exclusion and tolls have also been introduced in the Nordic context (Øian et al., 2018; Kaltenborn et al., 2001). Finally, the symbiotic trajectory of tourism and conservation paints a scenario where improved management can mitigate crowding in protected areas. This suggests that the solution to challenges posed by the growth of protected area tourism lies in enhanced and more efficient management.

Protected areas have become iconic destinations that connect and concentrate on other types of tourism activities. In Arctic Finland and Canada, ski resorts have often been established in fells and mountains in the close vicinity of the most popular national parks. In a sense, such areas create a second version of the symbiotic ideal, or rather, a more synthetic version of it, by blending mutual benefits through infrastructure and economic development on a new scale. The appeal of protected areas in these contexts often supports further development of tourism infrastructure to accommodate more visitors, which again requires more management and resources for maintenance, zoning, and controlling visitors. In the same spirit, and tellingly, visitor management, zoning and guiding has become an issue of "crowd" controlling in many countries, requiring managerial interventions based on visitor characteristics, pricing, reservation systems (e.g., lotteries) and digital tools (e.g., route planning based on crowding) (Milman, 2019).

The tourism-conservation relationship is still mainly presented as a win-win scenario, and regional economic impacts are well established through the provision of employment, tax revenues, and services to many peripheral locations that may otherwise have been suffering from declining populations and the loss of natural resource industries (Puhakka & Saarinen, 2013). However, a protected area driven or supported tourism may come with broader trade-offs, even when there is no local population. Indeed, one related concern is the inherent scalar challenge related to tourism stemming from the local-level juridical limits of a protected area's administration. This is often in contrast to the state's economic interests that may even further support tourism development in already crowded areas. Such scalar mismatches are additionally backed-up by reliance on government funding and state power to invest in services outside protected areas' governance needs (e.g., roads, project funding, and airports). In summary, the broader sustainability of protected area tourism is less frequently in the hands of the local administration. Thus, using tourism as a tool for more sustainable development requires increased critical scrutiny.

Temporal and Scalar Complexities of Sustainable Tourism Governance

Protected areas in Finland have become increasingly significant for socio-economic and regional benefits, particularly in remote locations, such as Arctic Finland. This increase can be attributed to various factors, such as advancements in assessing the indirect economic impacts of protected areas and tourism, strategic use of benefits in public relations, and decreased employment opportunities in the forestry sector. Consequently, sustainable tourism within protected areas has emerged as a progressive solution under the broad umbrella of sustainable development (Hardy et al., 2002). Protected area management often involves local-level, place-based activities, such as route zoning, area and temporal restrictions, route planning and monitoring, and engaging local communities in governance (Milman, 2019). This approach transforms overtourism into a point-based managerial challenge, necessitating investment in the infrastructure to maintain environmental integrity and visitor appeal.

The tradition of *everyman's rights* plays a crucial role in Nordic land use, forming the cultural foundation for governance and access-driven development. Protected areas, such as national parks and wilderness areas, hold a prominent position in Finnish nature relationship branding and national destination marketing (Visit Finland, 2023). This prominence underscores the importance of protected areas in tourism, extending their value and influence beyond mere visitor rate metrics. In general, sustainable tourism governance within protected areas seeks to balance ecological and cultural values by implementing appropriate zoning and visitor flow controls. However, increasing visitor numbers may require further infrastructure development to mitigate the negative effects of tourism growth. Moreover, in the Sami homeland, Sápmi, protected area planning employs the Akwé: Kon principles, underlining the need to protect not only ecological biodiversity but also indigenous knowledge in conservation (Metsähallitus, 2020), bringing another set of values to protected area governance compatible with national cultural heritage and export revenues.

The role of protected areas in contemporary tourism is not neutral in terms of time, space, and indirect visitation-related impacts. The long-term dilemmas and implications of visitor growth and the associated challenges can be examined by considering the temporal nature of governance decisions, the interdependency of protected areas and tourism's economic valuation, and changes in visitor profiles and protected area use (Table 13.1). First, the COVID-19 pandemic has led to a surge in domestic tourism, with overcrowding in protected areas prompting investment

Table 13.1 Temporal and scalar complexities of protected area tourism governance solutions and challenges

Sustainability complexity	Short-term solution	Long-term challenge
On-point sustainability actions	ACTION Invest in infrastructure and accessibility to manage domestic overcrowding in protected areas during the COVID-19 pandemic. BENEFIT Immediate management of high visitor numbers to mitigate potential negative environmental impacts and ensure the protected areas continue to serve their conservation and leisure purposes.	• Enhanced infrastructure may attract more foreign tourists, leading to increased tourism growth and larger environmental impact outside the destination. • The built infrastructure may cater to inexperienced tourists and tourism operators, potentially increasing the risk of environmental degradation.
Foreign focus	ACTION Encourage foreign tourism in protected areas to generate higher socioeconomic benefits for local communities and businesses. BENEFIT Foreign tourists typically spend more than domestic visitors, which can significantly contribute to the regional economy and support the development of local communities.	• Increased foreign tourism contributes to higher carbon emissions due to their reliance on air travel. • The environmental impacts of climate change may result in alterations to protected area ecosystems, potentially undermining their original conservation goals.

(*continued*)

Table 13.1 (continued)

Sustainability complexity	Short-term solution	Long-term challenge
Public virtues	ACTION Lower barriers to access and provide low-threshold services in protected areas to promote public virtues such as environmental education, health, and well-being. BENEFIT Ensuring easy access and low-threshold services can help educate visitors about conservation and foster a sense of appreciation and responsibility for the environment.	• Increased accessibility and services may inadvertently promote unsustainable growth in foreign tourism, contributing to global carbon emissions. • The focus on easy access and low-threshold services may prioritise the needs of less experienced tourists over the long-term conservation objectives of protected areas.

in infrastructure and accessibility. However, as international tourism resumes, the improved infrastructure may attract more foreign visitors who rely heavily on carbon-intensive air travel. This could inadvertently exacerbate long-term sustainability challenges. Second, protected areas are valued for their indirect regional and socioeconomic impacts, but the higher spending by foreign visitors comes at the environmental cost of increased emissions due to reliance on air travel. This highlights the need to balance the economic benefits of foreign tourism with the environmental consequences of travel. Finally, the changing visitor profiles in protected areas and the intersection of Nordic and democratic values play pivotal roles in sustainable protected area governance. As the demand for low-threshold services and easy access increases, so does the potential for unsustainable growth in foreign tourism, and its contribution to global carbon emissions. Thus, sustainable tourism growth in protected areas requires a nuanced understanding of the complex, short-, and long-term decisions that shape visitor behaviour and protected area governance.

The three intrinsic sustainable tourism governance challenges shown in Table 13.1 raise critical concerns regarding the temporality and scale of sustainability policies in protected areas as well as at destinations more broadly. Although adaptive management strategies and visitor regulation would ideally provide some answers to overcrowding in the short-term, these managerial interventions would in principle challenge the Nordic ethos of free access to nature. While investment in infrastructure and accessibility may address immediate issues, it could inadvertently attract unsustainable tourism in the long run, leading to greater environmental impacts. Moreover, increasing accessibility may change visitor profiles, produce local displacement, and compete with local use based on solitude and unregulated access. The territorial and place-based focus of management bodies for protected areas reflects the high level of public sector performance and environmental capability in the Nordics (Sachs et al., 2022; EPI, 2022). This is evidenced by standard practices such as zoning, the use of acceptable change frameworks, and participatory approaches.

Second, balancing socioeconomic benefits with environmental conservation creates a scalar dilemma. Here, the crucial question is whether to evaluate the role of protected areas more broadly: For whom and for what purposes are protected areas established? (Saarinen, 2019). Protected area visits in countries with public and legislative protected area funding often emphasise the regional, educational, and health benefits of these areas for a nation and its citizens. However, foreign visitors contribute relatively more to socioeconomic gains and subsequent evaluations of protected areas' regional economic impacts, while their modes of travel generate increased emissions that exacerbate climate change and alter protected area ecosystems.

Finally, protected areas are often and historically framed within the values associated with civic virtues and national modernisation (Niemelä & Ruuskanen, 2019). Thus, traditional domestic tourism in protected areas supports these areas' nationalistic values and nature-based outdoor norms. For example, the establishment of national parks has been used in Finland to celebrate national figures (Urho Kekkonen National Park was established to celebrate the president's 80-year birthday and was named after him) and independence (Hossa National Park was established to mark Finland's 100 years of independence). However, the key protected areas in Finland are now part of broader tourism destinations and industries and are often located nearby ski resorts (see already mentioned Oulanka and Pallas-Ylläs, but also Syöte and Urho Kekkonen national parks). Such areas create synergetic relationships in terms of tourism development, further blurring the divisions between these destinations tourism flows. Here, protected area management focuses on easy access, and low-threshold services that support public virtues but may inadvertently promote unsustainable growth in foreign tourism, thereby contributing to global carbon emissions.

Contrary to sudden events, climate change can, however, be understood as a slow-onset governance crisis for climate-vulnerable protected areas, embedded in longer processes linked with the Anthropocene (Varnajot & Saarinen, 2021). Climate change is expected to cause major disruptions and create new opportunities for tourism, especially in the Arctic (Demiroglu & Hall, 2020). In relation to Arctic protected areas and their use, climate change has already begun to affect elements such as the availability and upkeep of visitor sites. This can be attributed to the uncertain and irregular snow and ice conditions affecting visitor infrastructure in remote areas, which have historically been serviced during winter using snowmobiles. Changes in ice and snow conditions will likely affect routing (e.g., snowmobile routes) and visitor sites within protected areas, as their maintenance can no longer rely solely on historical winter patterns.

As climate change does not affect all regions in the same way and at the same pace, some destinations may benefit from temporary competitive advantages, creating new sets of winners and losers (Varnajot & Saarinen, 2022). For destinations in the European Arctic, such as Lapland, the main concerns relate to the presence of snow and white vistas during the high tourist season (November–December). However, this "new normal" of Arctic tourism has not yet unravelled, and in many cases, Arctic destinations' resilience and growth-focus can be extended by technological and managerial fixes (e.g., production of artificial snow, reinforcing route infrastructure) (Varnajot & Saarinen, 2022; Lunden, 2022). In practice, in the context of Arctic protected areas, climate change has already impacted the maintenance, route, and infrastructure of some areas owing to the lack of snow and ice required for many protected area management activities. Additionally, the phenomenon of "last chance tourism" (Demiroglu & Hall, 2020) has created new incentives for people to learn, witness, experience, and engage with the Arctic cryosphere before and while it is lost for good or irrevocably changed.

Conclusions

The governance of protected area tourism will continue to be a challenging issue owing to climate change and the nexus of tourism growth. From a broader perspective, the issue of protected area tourism becomes an example of the politics of future consumption and how and who (e.g., domestic or foreign visitors) should consume protected areas in a given country. Such political weighting would likely change the geography of nature-based tourism in many countries, question established protected areas' societal legitimacy and destabilise their status. In other words, and paradoxically, protected areas need to extend their sphere of societal legitimacy beyond tourism deduced regional and communal benefits. Although the SDGs provide a framework for addressing these challenges, they have been criticised for their limited ability to change the status quo of tourism and the associated environmental unsustainability (Bianchi & de Man, 2021; Hall et al., 2023). The implementation of SDGs is generally hindered by the lack of political will, limited resources, and a narrow focus on individual goals rather than a systemic approach to sustainability (Fukuda-Parr & McNeill, 2019; Hall et al., 2023).

Therefore, while SDGs provide a framework for addressing environmental sustainability in tourism, they alone are insufficient to address the root causes of environmental degradation and the negative impacts of tourism. A more critical and systemic approach is needed to reconsider the notion of environmental sustainability in tourism, adopt practices that prioritise the long-term viability of protected areas, and minimise the negative environmental impacts associated with tourism activities. This requires a fundamental shift in the way we think about and manage tourism, prioritising the long-term robustness and resilience of the environment and the well-being of both human and non-human communities. Thus, by focusing on the immediate place-based challenges of tourism, visitor management solutions and responsible organisations largely side-line the environmental concerns related to broader tourism growth. In other words, the temporalities and scalar politics of sustainable tourism governance are central questions that require development and further research in the context of climate change and the Anthropocene.

References

Balmford, A., Green, J. M. H., Anderson, M., Beresford, J., Huang, C., Naidoo, R., Walpole, M., & Manica, A. (2015). Walk on the wild side: Estimating the global magnitude of visits to protected areas. *PLOS Biology, 13*(2), e1002074.

Bianchi, R. V., & de Man, F. (2021). Tourism, inclusive growth and decent work: A political economy critique. *Journal of Sustainable Tourism, 29*(2–3), 353–371.

Bolić, T., & Sivčev, Ž. (2011). Eruption of Eyjafjallajökull in Iceland: Experience of European air traffic management. *Transportation Research Record, 2214*(1), 136–143.

Bonham, C., Edmonds, C., & Mak, J. (2006). The impact of 9/11 and other terrible global events on tourism in the United States and Hawaii. *Journal of Travel Research, 45*(1), 99–110.

Brechin, S. R., Murray, G., & Mogelgaard, K. (2010). Conceptual and practical issues in defining protected area success: The political, social, and ecological in an organized world. *Journal of Sustainable Forestry, 29*(2–4), 362–389.

Brouder, P. (2020). Reset redux: Possible evolutionary pathways towards the transformation of tourism in a COVID-19 world. *Tourism Geographies, 22*(3), 484–490.

Budowski, G. (1976). Tourism and environmental conservation: Conflict, coexistence, or symbiosis? *Environmental Conservation, 3*(1), 27–31.

Cheer, J. M., Lapointe, D., Mostafanezhad, M., & Jamal, T. (2021). Global tourism in crisis: Conceptual frameworks for research and practice. *Journal of Tourism Futures, 7*(3), 278–294.

Clark, C., Nyaupane, G. P., Timothy, D. J., & Buzinde, C. (2022). Scenario planning as a tool to manage tourism uncertainties during the era of COVID-19: A case study of Arizona, USA. *Current Issues in Tourism, 25*(7), 1063–1073.

Demiroglu, O. C., & Hall, C. M. (2020). Geobibliography and bibliometric networks of polar tourism and climate change research. *Atmosphere, 11*(5), 498.

Dudley, N., & Stolton, S. (Eds.). (2008). *Defining protected areas: An international conference in Almeria, Spain*. IUCN.

EPI. (2022). *Environmental performance index: Results overview*. Yale Center for Environmental Law & Policy. https://epi.yale.edu/epi-results/2022/component/epi

Fredman, P., Haukeland, J. V., Tyrväinen, L., Stensland, S., & Wall-Reinius, S. (2021). Nature-based tourism in a Nordic context. In P. Fredman & J. V. Haukeland (Eds.), *Nordic perspectives on nature-based tourism: From place-based resources to value-added experiences* (pp. 2–15). Edward Elgar Publishing.

Fukuda-Parr, S., & McNeill, D. (2019). Knowledge and politics in setting and measuring the SDGs: Introduction to special issue. *Global Policy, 10*, 5–15.

Gössling, S., Scott, D., & Hall, C. M. (2020). Pandemics, tourism and global change: A rapid assessment of COVID-19. *Journal of Sustainable Tourism, 29*(1), 1–20.

Hall, C. M. (2006). Tourism, biodiversity and global environmental change. In S. Gössling & C. M. Hall (Eds.), *Tourism and global environmental change* (pp. 211–226). Routledge.

Hall, C. M. (2010). Crisis events in tourism: Subjects of crisis in tourism. *Current Issues in Tourism, 13*(5), 401–417.

Hall, C. M., Seyfi, S., & Koupaei, S. N. (2023). Politics and the sustainable development goals: Tourism agenda 2030 perspective article. *Tourism Review, 78*(2), 314–320.

Hardy, A., Beeton, R. J., & Pearson, L. (2002). Sustainable tourism: An overview of the concept and its position in relation to conceptualisations of tourism. *Journal of Sustainable Tourism, 10*(6), 475–496.

Jacobs, L. A., Blacketer, M., Peterson, B., Levithan, E., Russell, Z., & Brunson, M. (2020). Responding to COVID-19 and future times of uncertainty: Challenges and opportunities associated with visitor use, management, and research in parks and protected areas. *Parks Stewardship Forum, 36*(3), 481–488.

Kaltenborn, B., Haaland, H., & Sandell, K. (2001). The public right of access—some challenges to sustainable tourism development in Scandinavia. *Journal of Sustainable Tourism, 9*(5), 417–433.

Lehtonen, S., Heikkinen, T., & Hirvonen, J. (2007). Jokamiehenoikeuksien tulkintoja ja haasteita luontoliikunnan ja -matkailun kannalta (Interpretations and challenges of everymans' rights from the perspective of nature-based leisure and tourism). *Maaseutupolitiikan yhteistyöryhmän julkaisu*, 1/2007. Ministry of Economic Affairs and Employment of Finland.

Löfgren, O. (2002). *On holiday: A history of vacationing*. University of California Press.

Lunden, A. (2022). The biopolitics of Arctic tourism development and sustainability. *Via. Tourism Review*, (21). https://doi.org/10.4000/viatourism.8084.

Lundmark, L. J., Fredman, P., & Sandell, K. (2010). National parks and protected areas and the role for employment in tourism and forest sectors: A Swedish case. *Ecology and Society, 15*(1), 19.

McGinlay, J., Gkoumas, V., Holtvoeth, J., Fuertes, R. F. A., Bazhenova, E., Benzoni, A., et al. (2020). The impact of COVID-19 on the management of European protected areas and policy implications. *Forests, 11*(11), 1214.

Metsähallitus. (2020). *Procedure for applying the Akwé: Kon principles in the collaboration between Metsähallitus and Sámi Parliament*. Metsähallitus. https://julkaisut.metsa.fi/assets/pdf/lp/Muut/akwekon_toimintamalli.pdf [Website in Finnish].

Metsähallitus. (2022). *Visitor numbers: The popularity of national parks was not just a coronavirus phenomenon*. Metsähallitus. https://www.metsa.fi/en/press-releases/the-popularity-of-national-parks-was-not-just-a-coronavirus-phenomenon/

Metsähallitus. (2023). *Development of visitor rates*. Metsähallitus. https://www.metsa.fi/vapaa-aika-luonnossa/kayntimaarat/kayntimaarien-kehitys/ [Website in Finnish].

Milman, A. (2019). Visitor management in highly-visited attractions: Lessons that practitioners can learn from the US theme park industry and the national parks. In H. Pechlaner, E. Innerhofer, & G. Erschbamer (Eds.), *Overtourism: Tourism management and solutions* (pp. 104–124). Routledge.

Newsome, D. (2020). The collapse of tourism and its impact on wildlife tourism destinations. *Journal of Tourism Futures, 7*(3), 295–302.

Niemelä, J., & Ruuskanen, E. (2019). References to iconic landscapes in the debate surrounding the founding of Finland's national parks, circa 1880–1910. *International Journal for History, Culture and Modernity, 7*(1), 741–764.

OECD. (2022). Norway. *OECD tourism trends and policies 2022*. https://www.oecd-ilibrary.org/sites/c41cef4a-en/index.html?itemId=/content/component/c41cef4a-en

Øian, H., Fredman, P., Sandell, K., Sæþórsdóttir, A. D., Tyrväinen, L., & Jensen, F. (2018). *Tourism, nature and sustainability: A review of policy instruments in the Nordic countries*. Nordic Council of Ministers, Rosendahls.

Puhakka, R., Pitkänen, K., & Siikamäki, P. (2017). The health and well-being impacts of protected areas in Finland. *Journal of Sustainable Tourism, 25*(12), 1830–1847.

Puhakka, R., & Saarinen, J. (2013). New role of tourism in national park planning in Finland. *The Journal of Environment & Development, 22*(4), 411–434.

Ritchie, H. (2020). Google mobility trends: How has the pandemic changed the movement of people around the world? *Our World in Data*. https://ourworldindata.org/covid-mobility-trends

Saarinen, J. (2003). The regional economics of tourism in Northern Finland: The socio-economic implications of recent tourism development and future possibilities for regional development. *Scandinavian Journal of Hospitality and Tourism, 3*(2), 91–113.

Saarinen, J. (2007). Tourism in peripheries: The role of tourism in regional development in northern Finland. In D. K. Müller & B. Jansson (Eds.), *Tourism in peripheries: Perspectives from the far north and south* (pp. 41–52). CABI.

Saarinen, J. (2019). What are wilderness areas for? Tourism and political ecologies of wilderness uses and management in the Anthropocene. *Journal of Sustainable Tourism, 27*(4), 472–487.

Sachs, J., Kroll, C., Lafortune, G., Fuller, G., & Woelm, F. (2022). *Sustainable development report 2022*. Cambridge University Press.

Sandell, K. (2007). Access, tourism and democracy: A conceptual framework and non-establishment of a proposed national park in Sweden. In D. K. Müller & B. Jansson (Eds.), *Tourism in peripheries: Perspectives from the far north and south* (pp. 141–153). CABI.

Searle, A., Turnbull, J., & Lorimer, J. (2021). After the anthropause: Lockdown lessons for more-than-human geographies. *The Geographical Journal, 187*(1), 69–77.

Snyman, S., & Bricker, K. S. (2019). Living on the edge: Benefit-sharing from protected area tourism. *Journal of Sustainable Tourism, 27*(6), 705–719.

Sönmez, S. F., Apostolopoulos, Y., & Tarlow, P. (1999). Tourism in crisis: Managing the effects of terrorism. *Journal of Travel Research, 38*(1), 13–18.

Statistics Finland. (2023). Yearly nights spent and arrivals in Lapland region, 2002–2022. Statistics Finland's statistical databases. https://pxdata.stat.fi/PxWeb/pxweb/fi/StatFin/StatFin__matk/statfin_matk_pxt_11j1.px/table/tableViewLayout1/

Tuulentie, S., & Rantala, O. (2013). Will "free entry into the forest" remain? argumentation analysis of Finnish and Swedish Discussions on everyman's rights. In D. K. Müller, L. Lundmark, & R. H. Lemelin (Eds.), *New issues in polar tourism: Communities, environments, politics* (pp. 177–188). Springer Science & Business Media.

Varnajot, A., & Saarinen, J. (2021). 'After glaciers?' Towards post-Arctic tourism. *Annals of Tourism Research, 91*, 103205.

Varnajot, A., & Saarinen, J. (2022). Emerging post-Arctic tourism in the age of Anthropocene: Case Finnish Lapland. *Scandinavian Journal of Hospitality and Tourism, 22*(4–5), 357–371.

Visit Finland. (2023). *The happiest country in the world.* https://www.visitfinland.com/en/

Volenec, Z., Abraham, J., Becker, A., & Dobson, A. (2021). Public parks and the pandemic: How park usage has been affected by COVID-19 policies. *PLOS ONE, 16*(5), e0251799.

Wearing, S., Stevenson, D., & Young, T. (2009). *Tourist cultures: Identity, place and the traveller.* Sage.

14

Proximity Tourism in the New Normal: Toward a More Sustainable Tourism Behavior in Spain

Adrián Mendieta-Aragón
and Raquel Arguedas-Sanz

Introduction

The COVID-19 pandemic caused a profound change in the interactions between consumers and companies, giving rise to a new context in the markets for goods and services (Mehta et al., 2020; Turoń & Kubik, 2021). The study of the changes in tourist behaviors that characterize the New Normal of the COVID-19 pandemic is an essential avenue of tourism research (Assaf et al., 2022). Indeed, recent academic studies have focused on identifying and understanding these new travel habits and preferences (Gowreesunkar et al., 2022; Huang et al., 2021; Khare & Sathe, 2021; Li et al., 2021; Maingi et al., 2022; Mendieta-Aragón, 2022).

Pre-pandemic tourist behaviors have led to tourism activities being directly responsible for a significant proportion of global greenhouse gas (GHG) emissions, for example, transport-related emissions from tourism

A. Mendieta-Aragón (✉) • R. Arguedas-Sanz
National University of Distance Education (UNED), Madrid, Spain
e-mail: amendieta@cee.uned.es; rarguedas@cee.uned.es

S. W. Maingi et al. (eds.), *Tourist Behaviour and the New Normal, Volume II*,
https://doi.org/10.1007/978-3-031-45866-8_14

account for around 5% of all man-made CO2 emissions (UNWTO, 2019). Therefore, in line with Higgins-Desbiolles (2021), this study wonders whether this period of transformation toward a New Normal could be the perfect opportunity to tackle the main current challenge of the tourism sector, that is, the development of a more ethical, sustainable, and responsible tourism.

Consumers have the option to reorient the tourism industry toward sustainable development by adopting new travel habits, such as proximity tourism (Seidel et al., 2021; Sharma et al., 2021). Proximity tourism promotes local destinations, which can have a positive impact on the role of tourism in the territorial imbalance and economic development of local destinations (Díaz-Soria & Llurdés-Coit, 2013; Romagosa, 2020). In addition, it can contribute to reducing the carbon footprint caused by the tourism industry (Jeuring & Díaz-Soria, 2017). Therefore, proximity tourism allows addressing the most pressing challenges of our time, included in the United Nations Sustainable Development Goals (SDG), such as avoiding overcrowding and uncontrolled urban growth related to the tourism sector (SDG 11), the decoupling of economic growth from environmental degradation and the promotion of sustainable consumption that advances toward green economies (SDG 12) or the reduction of CO2 levels and other greenhouse gases in the atmosphere (SDG 13).

Despite the relevance of proximity tourism for sustainable tourism development, the existing literature is still scarce (Salmela et al., 2021), and, to the best of our knowledge, there are no previous studies that define and identify the proximity tourist profile. Therefore, this chapter analyzes tourist behavior in the New Normal by answering the following research questions: (i) Has tourist behavior changed toward proximity tourism in the New Normal of the COVID-19 pandemic?; (ii) What is the profile of proximity tourists that emerged in the New Normal?; (iii) What are the characteristics of proximity tourism?

This study contributes to understanding the behavior and profile of the proximity tourist, which presents significant research implications for tourism industry companies to promote new models of sustainable tourism to achieve the Sustainable Development Goals. To answer the proposed research questions, this empirical study focuses on the case of residents in Spain. This study uses a set of microdata with 72,232 observations obtained from the Resident Travel Survey (RTS) of the National Statistics Institute of Spain (INE) during the period 2019–2021 to compare the behavior of tourists in the pre-pandemic stage and the behavior of tourists in the New Normal. The data used in this study have been weighted based on the information provided by the RTS and make it possible to guarantee the representativeness of the sample of all leisure trips made by residents in Spain during the study period.

Literature Review

The Concept of Proximity Tourism

Proximity tourism has been conceptualized as a form of tourism that boosts travel to local destinations. However, proximity tourism is not simply defined by the trip's distance but is a form of tourism based on the idea of getting to know and exploring the travelers' everyday surroundings (Höckert et al., 2022; Rantala et al., 2020; Salmela et al., 2021). In fact, the concept of proximity challenges the traditional way of defining tourism according to how close or far it is from the place of origin, rather, tourism and proximity will be linked through the questioning of the "usual environment," and this usual environment is not necessarily close (Díaz-Soria & Llurdés-Coit, 2013).

Previous empirical and conceptual literature have highlighted how proximity tourism could be a new form of sustainable tourism that allows tourists to explore their usual environment (Díaz-Soria, 2017), avoid the overcrowding of tourist activity in certain destinations (Müller et al., 2023), reduce the carbon footprint generated by tourism (Higgins-Desbiolles, 2022; Rico et al., 2019), and contribute to the development of the local economy (Jeuring & Haartsen, 2017). Likewise, proximity tourism promotes a more sustainable tourism by establishing a close connection between tourists and the surrounding environment, reducing tourism activity's negative environmental and social impact. Therefore, based on previous literature, proximity tourism can be defined as a type of travel that allows tourists to rediscover their usual environment, focusing attention on local destinations and characterized by producing a lower carbon footprint.

Despite the relevant importance of the development of proximity tourism, Salmela et al. (2021), in their thematic literature review on proximity tourism, consider that the existing literature is still incipient and scarce. It is, therefore, necessary to advance empirical research that defines the new profile of the proximity tourist and its determinant factors to understand the motivations, needs, and demand for this form of sustainable tourism.

Empirical Review of Tourist Behavior During Pre-Pandemic and Post-Pandemic Tourism

Empirical evidence has shown that the tourism sector is susceptible to external shocks, such as economic crises, terrorist attacks, political instability, or natural disasters (Korstanje, 2023; Mansfeld & Winckler, 2015; Rosselló et al., 2020). In the context of the recent SARS-CoV-2 health crisis, the spread of COVID-19 around the world has meant that its effects on the tourism industry are not concentrated in specific regions but affect all countries. Likewise, consumers around the world were forced to change their purchasing behaviors, causing a significant psychological, social, and economic impact on them (Oana, 2020; Rahmanov et al., 2021).

According to Warde (1997), the behavior of consumers depends on their context. Therefore, unforeseen events, such as the pandemic, or new regulations, such as those established by governments to control the spread of COVID-19, can severely affect consumer habits, especially in the tourism industry, which has been the most affected by these events.

Figure 14.1 presents the four contexts identified by Sheth (2020) that modify consumer habits or preferences, these are: (i) social context, which is affected by alterations in social interactions in the family, work, or friendship environment; (ii) technological context, it is affected by technological improvements, such as the development of e-Commerce, virtual reality, Internet of Things, to mention a few examples; (iii) regulatory context, this is affected by all changes in the legal framework of a country;

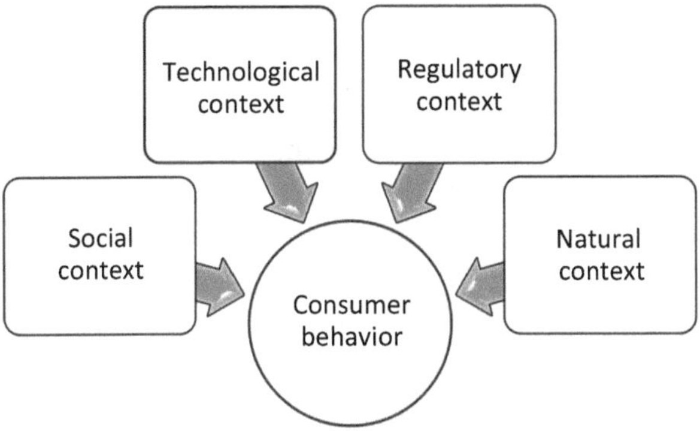

Fig. 14.1 Contexts that govern consumption habits. (Source: Own elaboration from Sheth (2020))

(iv) natural context, natural disasters such as earthquakes, tsunamis, health crises, or wars.

The reduction of social interactions and the increase of social distance to avoid the risks of contagion, the acceleration of the digitalization of the tourism industry to respond to new consumer preferences, the closure of national and regional borders, and the COVID-19 health crisis have severely altered every tourist context, thus justifying the profound change of tourism behavior in the New Normal.

Recent literature has tried to identify the change in tourist habits in the New Normal toward more sustainable tourism. Existing studies indicate that nature tourism has aroused greater interest in the New Normal since the new post-pandemic tourist prefers clean, outdoor, and sparsely crowded destinations (Seraphin & Dosquet, 2020; Wachyuni & Kusumaningrum, 2020). Regarding the type of destination, some studies highlight a change in tourist preferences toward short-stay trips to short-distance destinations (Mendieta-Aragón, 2023; Mirzaei et al., 2021).

Furthermore, the behavior of proximity tourists has shown greater resilience during the pandemic and, like domestic tourism in general, has played a key role in the recovery of the tourism industry (Klaniczay, 2022). Dot Jutglà et al. (2022) analyze the supply and demand of proximity tourism in the case of rural spaces in the region of Catalonia (Spain), noting an increase in proximity tourism in the post-pandemic stage since

2020. Similarly, Lebrun et al. (2022) examine the impact of the COVID-19 pandemic on the travel intention of French tourists. Their findings reveal that the increased risk perceived by tourists due to the pandemic encouraged proximity tourism on vacation trips. Bertacchini et al. (2021) investigate the behavior patterns of proximity tourism and same-day visits to museums and heritage sites. This research extracted tourist information from 76,000 museum card holders in Piedmont (Italy) to map behavior patterns and intra-regional flows. Therefore, although proximity tourism has traditionally remained under-researched, in recent years, it has become increasingly popular in tourism research (Salmela et al., 2021).

In summary, proximity tourism has been established as a way of valuing the environment surrounding individuals, bringing benefits to local communities, and contributing to a more sustainable tourism. Therefore, research that combines consumer behavior in the New Normal, proximity tourism, and its advantages toward a new sustainable tourism model is particularly interesting for society and academic research.

Methodology

This study focuses on defining the profile of the proximity tourist and identifying the changes in the determinants of tourist behavior that have a more significant impact on the development of proximity tourism in the New Normal. In particular, this research uses the set of microdata obtained from the Resident Travel Survey (RTS) prepared by the National Statistics Institute of Spain (INE)[1] to identify and detect changes in the socioeconomic and demographic characteristics of the proximity tourist and the characteristics of the trip since the period before the pandemic until the New Normal. A relevant advantage of this national survey is that it provides information on sample weights. Sample weights in tourism research are essential to guarantee the representativeness of the

[1] https://www.ine.es/dyngs/INEbase/es/operacion.htm?c=Estadistica_C&cid=1254736176990&menu=ultiDatos&idp=1254735576863

population, eliminate sample selection bias, and provide valid inferences about the population (Boto-García, 2023; Loomis, 2007).

The RTS allows us to disaggregate according to three main types of tourism: leisure tourism, business tourism, or visits to friends and family. In this study, we focus on leisure tourism due to two reasons. Firstly, because it is the majority and represents more than half of the trips made by residents in Spain, and secondly, because in this type of tourism, the tourist has a greater capacity to decide the trip's destination than in business tourism or visits to friends and family. This study is carried out for the years 2019–2021. However, it is evaluated and analyzed each year individually because they have different characteristics that can influence tourist behavior. In this way, the year 2019 is taken as the pre-pandemic period. In 2020 the pandemic broke out, and with it, a stage of strong restrictive measures on mobility, confinement, and closure of international and regional borders to prevent its spread. And finally, 2021 (the most recent full year for which information is available) was characterized as the beginning of the recovery of tourism and the New Normal.

Following the conceptualization of previous literature on proximity tourism as the form of tourism that takes place within the tourist's usual environment (Díaz-Soria, 2017; Höckert et al., 2022; Rantala et al., 2020; Salmela et al., 2021), the dependent variable of our empirical study is defined as a binary variable that takes the value 1 if the trip has taken place in the same region of residence of the tourist, or conversely, the dependent variable takes the value 0 if the trip has taken place outside the tourist's region of residence.

Equation (14.1) defines two categories of determinants of proximity tourism, the sociodemographic and economic characteristics of the tourist (gender, age, educational level, population, income, region of residence) and the characteristics of the trip (length of stay, type of accommodation, type of transport, and size of group).

$$\text{Proximity} = \text{f}\begin{pmatrix} \text{gender,age,education,population,} \\ \text{income,region of residence,} \\ \text{length of stay,type of accommodation,} \\ \text{type of transport,group size} \end{pmatrix} \tag{14.1}$$

In this chapter, the logistic regression model is used to identify the profile of consumers most likely to participate in proximity tourism according to the socioeconomic and demographic characteristics of the traveler and the characteristics of the trip. Mathematically, the model would be expressed as follows:

$$P\left(PROXIMITY_i = 1\right) = \frac{\exp(z)}{1+\exp(z)} \tag{14.2}$$

where $P(PROXIMITY_i = 1)$ represents the probability that the leisure, recreation, and vacation trip i takes the value 1 (proximity tourism), while z refers to:

$$\begin{aligned} z = {} & \beta_0 + \beta_1\, GENDER_i + \beta_2\, AGE_i + \beta_3\, EDUC_i \\ & + \beta_4\, HABITAT_i + \beta_5\, INCOME_i + \beta_6\, REG_i \\ & + \beta_7\, L\_STAY_i + \beta_8\, ACCOMM_i + \beta_9\, TRANSP_i \\ & + \beta_{10}\, G\_SIZE_i \end{aligned} \tag{14.3}$$

In the present chapter, all explanatory variables were converted into categorical values to calculate the odd ratio estimates.[2]

Results and Discussions

This section presents the results of the changes in the profile of tourist proximity in Spain during the New Normal. First, we analyze descriptively the penetration rate of proximity tourism. Subsequently, we model tourist behavior and identify the profile and determinants of proximity tourism. Table 14.1 presents the penetration rates of proximity tourism according to the sociodemographic characteristic of the tourist and the

---

[2] The odds ratio estimates indicate the odds of an event compared to the odds of a reference group. Categories with estimated values greater than 1, indicate that the odds have increased, therefore, there is a higher probability of occurrence than the reference group. Conversely, categories with ORs below 1, indicate lower odds, hence, there is a lower probability of occurrence than the reference group.

Table 14.1 Proximity tourism penetration rates and their recent evolution

		2019	2020	2021
GENDER	Female	44.5%	54.1%	50.9%
	Male	44.3%	53.2%	52.1%
AGE	[15–25)	41.6%	49.3%	52.1%
	[25–35)	39.5%	46.1%	45.2%
	[35–45)	45.4%	54.9%	52.0%
	[45–55)	43.2%	54.2%	50.8%
	[55–65)	46.1%	58.9%	53.1%
	[65–75)	49.2%	56.3%	56.7%
	+ 75	51.8%	57.1%	55.3%
EDUCATION	Primary	51.8%	58.8%	62.8%
	Secondary	51.0%	61.1%	57.6%
	Bachelor	44.4%	49.1%	49.1%
	Master or PhD	40.9%	52.4%	49.3%
POPULATION	>100.000	43.8%	52.4%	50.6%
	[50.000–100.000)	40.3%	50.1%	49.9%
	[20.000–50.000)	46.9%	55.5%	55.7%
	[10.000–20.000)	48.9%	56.6%	53.3%
	<10.000	44.4%	57.0%	50.2%
INCOME	[0–1000€)	51.5%	61.9%	61.5%
	[1000–1500 €)	48.5%	57.7%	55.6%
	[1500–2500 €)	45.8%	54.1%	52.6%
	[2500–3500 €)	44.6%	52.7%	48.7%
	[3500–4999 €)	39.7%	49.4%	50.0%
	+ 5000€	36.0%	51.9%	48.1%
REGION OF RESIDENCE	Andalusia	75.1%	86.2%	83.4%
	Aragon	40.9%	46.5%	47.0%
	Asturias	40.9%	60.1%	47.5%
	Balearic Islands	50.3%	67.8%	77.8%
	Canary Islands	77.1%	90.3%	84.6%
	Cantabria	25.8%	38.6%	44.7%
	Castile and León	37.7%	56.7%	49.6%
	Castilla–La Mancha	23.9%	35.4%	26.2%
	Catalonia	64.8%	71.5%	68.0%
	Valencian Community	51.6%	64.9%	67.6%
	Extremadura	20.6%	48.8%	30.8%
	Galicia	64.1%	76.1%	70.3%
	Community of Madrid	10.0%	12.6%	10.3%
	Region of Murcia	34.2%	45.9%	42.4%
	Navarre	26.7%	33.7%	35.0%
	Basque Country	12.6%	13.5%	12.9%
	La Rioja	20.7%	27.6%	29.4%

(continued)

Table 14.1 (continued)

		2019	2020	2021
LENGTH OF STAY	[1–6)	51.6%	60.4%	59.2%
	[6–11)	16.6%	32.7%	24.6%
	[11–16)	20.5%	32.1%	26.1%
	+ 16	29.2%	39.7%	33.1%
ACCOMMODATION	Hotel	25.8%	37.1%	36.5%
	Rented	30.8%	43.0%	36.9%
	Rural	42.5%	54.4%	49.3%
	Other accommodation	59.3%	63.7%	63.7%
TRANSPORTATION	Car	51.9%	57.4%	55.9%
	Bus	32.7%	41.2%	46.5%
	Train	20.3%	39.3%	24.3%
	Air	2.7%	5.4%	4.4%
	Sea	45.1%	61.8%	50.5%
SIZE OF GROUP	1 tourist	39.0%	47.6%	47.0%
	2 tourists	43.9%	56.0%	51.5%
	3 or more tourists	47.7%	55.3%	53.8%
TOTAL		44.4%	53.7%	51.5%

Note: Results of the penetration rates with weighted sample

trip characteristic. The results confirm a significant increase in the penetration rates of proximity tourism in the New Normal, which has important implications for aiding the recovery of domestic destinations and advancing more sustainable tourism development in the long term. However, although proximity tourism has experienced significant growth after the emergence of COVID-19, it should be noted that there is still a wide margin for improvement. On the other hand, the results show that proximity tourism is not evenly distributed among the different groups of tourists but has different penetration rates depending on the characteristics of the tourist and the trip. Therefore, it is convenient to statistically model the profile of the proximity tourist to identify the determining factors that influence the tourist to move toward this emerging form of sustainable tourism.

First, it is verified that there is no multicollinearity between the nominal explanatory variables using the Phi or Cramer's V coefficient. The Wald test also confirms that the determining factors used are valid to explain the behavior of the proximity tourist. In addition, according to Hensher and Stopher (1979), the pseudo R^2 shows excellent goodness-of-fit. Finally, the odds ratios of the logistic regression model obtained through the statistical software STATA v.17 are presented in Table 14.2. Robust estimation was used to avoid problems of heteroscedasticity of the models. The representativeness of the results is guaranteed by weighing the sample.

Table 14.2 Results of the modeling of the profile of proximity tourists in Spain

		2019	2020	2021
Constant		**2.31**	**4.39**	**7.39**
GENDER	Male	1.10	1.05	1.12
Ref. category: Female				
AGE	[25–35)	1.08	0.91	0.83
Ref. category: [15–25)	[35–45)	1.13	1.12	0.97
	[45–55)	1.07	1.24	0.99
	[55–65)	**1.30**	**1.29**	1.22
	[65–75)	**1.82**	1.21	1.11
	+ 75	**2.77**	**1.96**	1.16
EDUCATION	Secondary	**1.22**	**1.37**	0.89
Ref. category: Primary	Bachelor	0.94	1.03	0.74
	Master or PhD	0.92	1.19	0.83
POPULATION	[50.000–100.000)	0.90	0.93	0.92
Ref. category: >100.000	[20.000–50.000)	**0.79**	**0.71**	**0.80**
	[10.000–20.000)	0.93	0.90	**0.85**
	<10.000	**0.80**	0.92	**0.73**
INCOME	[1000–1500 €)	1.14	0.98	**0.77**
Ref. category: [0–1000€)	[1500–2500 €)	1.07	0.87	**0.81**
	[2500–3500 €)	1.00	0.93	0.82
	[3500–4999 €)	1.00	0.92	0.87
	+ 5000€	1.00	**1.69**	1.00
REGION OF RESIDENCE	Aragon	**0.11**	**0.06**	**0.09**
Ref. category: Andalusia	Asturias	**0.10**	**0.11**	**0.09**
	Balearic Islands	**3.78**	1.84	**8.70**
	Canary Islands	**50.63**	**38.15**	**76.28**
	Cantabria	**0.06**	**0.05**	**0.07**
	Castile and León	**0.11**	**0.09**	**0.11**
	Castilla-La Mancha	**0.05**	**0.04**	**0.04**
	Catalonia	**0.53**	**0.33**	**0.35**
	Valencian Community	**0.22**	**0.17**	**0.25**
	Extremadura	**0.05**	**0.07**	**0.05**
	Galicia	**0.40**	**0.34**	**0.32**
	Community of Madrid	**0.01**	**0.01**	**0.01**
	Region of Murcia	**0.08**	**0.06**	**0.07**
	Navarre	**0.05**	**0.03**	**0.05**
	Basque Country	**0.01**	**0.01**	**0.01**
	La Rioja	**0.04**	**0.03**	**0.03**

(continued)

Table 14.2 (continued)

		2019	2020	2021
LENGTH OF STAY	[6–10]	**0.22**	**0.29**	**0.22**
Ref. category: [1–5]	[11–15]	**0.24**	**0.25**	**0.23**
	+ 16	**0.25**	**0.32**	**0.28**
ACCOMMODATION	Apartments	**1.56**	**1.84**	**1.87**
Ref. category: Hotel	Rural	**2.34**	**3.12**	**2.30**
	Others	**6.40**	**6.17**	**5.92**
TRANSPORTATION	Bus	**0.54**	**0.76**	**0.62**
Ref. category: Car	Train	**0.21**	**0.56**	**0.22**
	Air	**0.00**	**0.00**	**0.00**
	Sea	**0.06**	**0.09**	**0.07**
SIZE OF GROUP	2 tourists	0.95	1.07	0.93
Ref. category: 1 tourist	3 or more tourists	**1.25**	1.05	0.98
N. Obs.		32,994	15,458	23,780
Wald Test		4480.73	2239.09	3080.56
Pseudo R^2		0.44	0.41	0.42

Note: Values in bold represent odds ratios with a significance level of 10%

Regarding the socioeconomic and demographic profile of the tourist, the results confirm that gender does not significantly affect proximity tourism. However, a relevant finding is that there was a greater probability of proximity tourists in individuals older than 55 years, compared to the reference group (travelers ages between 16 and 25), in the pre-pandemic stage. In fact, the probability that a tourist over 75 years of age would participate in proximity tourism was 2.77 times higher than a tourist with ages between 16 and 25. Nevertheless, after the pandemic outbreak, the significant differences by age have gradually disappeared, and in the New Normal, there are no statistically significant differences caused by the age of the tourist. In the pre-pandemic stage, our results are in line with Xiang (2013), tourists with a higher level of academic study seek culture shock as part of the travel experience and broaden their knowledge beyond the tourist's usual environment. However, in the New Normal stage, our results reveal that the differences in education have disappeared.

In the New Normal, tourists residing in cities with more than 100,000 inhabitants are more likely to engage in proximity tourism than residents of cities with smaller populations. The income level of individuals does not significantly affect the behavior of proximity tourism. In line with the results obtained in this study, proximity tourism is not related to the purchasing power of individuals but to the idea that tourists get to know the usual environment that surrounds them, the culture of local destinations, and establish a greater connection with the local residents.

The results highlight notable interregional differences in proximity tourism between the regions of Spain. On the one hand, the insular regions (Balearic Islands and Canary Islands) present the highest odds, which confirms that the insularity of the territory is possibly increasing this practice of sustainable tourism among its residents, especially in the post-pandemic stage. On the other hand, the results show that, in general, residents of the most touristic regions of Spain have higher rates of participation in proximity tourism. For example, Andalusia, Catalonia, or the Valencian Community (in addition to the island regions already mentioned) occupy the top positions in terms of their residents' propensity to travel to proximity destinations. A large number of the main tourist attractions in Spain are found in these regions, such as the Alhambra (Andalusia), the Sagrada Familia (Catalonia), and the City of Arts and Sciences (Valencian Community), among others. In line with Bertacchini et al. (2019), the tourist heritage of these regions could benefit the development of proximity tourism.

Our results empirically confirm that proximity tourism is mainly limited to short-stay trips. According to Lebrun et al. (2021), this new form of tourism is gaining importance among national tourists with less than one week vacation periods. In the New Normal, proximity tourism has emerged as a more resilient type of tourism in the face of sudden regulatory changes during the pandemic, as being characterized by short stays has given tourists greater flexibility in planning their trips.

In proximity tourism, the tourist seeks to learn about the culture of local destinations and establish a deeper connection with local residents. Therefore, our results confirm that proximity tourists prefer to stay in tourist apartments, rural accommodations, or second homes instead of hotels. These types of accommodations allow the proximity tourist to get closer to the local culture and the inhabitants of the tourist destinations. Shorter distance trips characterize proximity tourism, which reduces tourism's CO2 emissions into the atmosphere. As expected, our results confirm that the means of transport most likely to be used in proximity tourism is the car, followed by public transport, such as buses and trains. While air transport, which is the main emitter of greenhouse gases generated by tourism, has a very low probability of being used in proximity tourism. Finally, in the pre-pandemic stage, there was a higher probability of doing proximity tourism in groups of more than three tourists, however, our results confirm that the size of the traveling group in proximity tourism has homogenized in the New Normal and the statistically significant differences with respect to the number of travelers have disappeared.

Conclusions

The proximity tourism that has emerged in the New Normal offers new opportunities that could favor the development of a more sustainable tourism model to achieve the SDGs. We agree with Rantala et al. (2020) that we are in a critical situation in which it is necessary to reduce carbon emissions while maintaining the economic balance of the tourism sector. In this chapter, we propose proximity tourism as an emerging form of tourism that contributes significantly to the sustainability of tourism. Proximity tourism not only helps to reduce the carbon footprint generated by tourism but also boosts the development of local economies. Tourism can exacerbate climate change if timely measures are not taken to reduce the environmental footprint and achieve the SDGs. Society could maintain the sustainable tourist habits acquired during the New Normal to make tourism a sustainable industry, which allows reducing the emission of greenhouse gases, as well as the preservation of ecosystems and biodiversity. Therefore, the definition of the new profile of proximity tourists and the components of their trips can be of great value to local tourist destinations to manage and develop marketing strategies focused on the potential tourists that have emerged in the New Normal.

The results of this chapter confirm that there has been a significant increase in the penetration of proximity tourism among residents in Spain since the outbreak of COVID-19, which will have a positive impact on the development of sustainable tourism. One of the most relevant findings is that a significant change in the behavior of proximity tourists in the New Normal is confirmed. Specifically, the results show that, even though there are still significant differences between the sociodemographic and economic profiles of proximity tourists, the differences due to the age and educational level of tourists have been significantly reduced. In this way, proximity tourism is becoming more widespread, and the profile of the proximity tourist is becoming more homogenized.

This study also finds interregional differences in the participation of proximity tourism between the regions of Spain. The results confirm the positive effect of the geographical conditioning of the insular regions in proximity tourism. Likewise, it is found that the regions with a greater cultural heritage in Spain present a greater probability of participating in proximity tourism. Regarding the characteristics of the trip, our results empirically confirm the preference for short stays in this type of tourism and the use of alternative means of transport to air transport. Therefore, the promotion of proximity tourism contributes significantly to sustainable development by reducing air traffic and the carbon footprint it generates. Despite its contributions, this study is not without its limitations. This analysis uses a nationally representative sample of trips made by a resident in Spain, so this study is limited to the information from this survey. In future studies, it would be interesting to analyze the motivation of proximity tourists with primary data. Likewise, this study focuses on the proximity tourism of residents in Spain. The results in other countries must be contrasted in future studies to validate the determining factors and the profile of the proximity tourist defined here. In any case, this chapter provides stakeholders with a comprehensive empirical analysis of a new form of tourism boosted during the pandemic due to regulatory restrictions, but persisting after the pandemic, with relevant implications for sustainable tourism development in the New Normal. Nevertheless, this study opens up new avenues of research on changes in tourist behavior in the New Normal that contribute to the development of sustainable tourism that maintains the balance between the economy, society, and the environment.

References

Assaf, A. G., Kock, F., & Tsionas, M. (2022). Tourism during and after COVID-19: An expert-informed agenda for future research. *Journal of Travel Research, 61*(2), 454–457. https://doi.org/10.1177/00472875211017237

Bertacchini, E., Nuccio, M., & Durio, A. (2019). Proximity tourism and cultural amenities: Evidence from a regional museum card. *Tourism Economics, 27*(1), 187–204. https://doi.org/10.1177/1354816619890230

Bertacchini, E., Nuccio, M., & Durio, A. (2021). Proximity tourism and cultural amenities: Evidence from a regional museum card. Tourism Economics, 27(1), 187–204. https://doi.org/10.1177/1354816619890230

Boto-García, D. (2023). Good results come to those who weight: On the importance of sampling weights in empirical research using survey data. *Current Issues in Tourism*. Advance online publication. https://doi.org/10.1080/13683500.2023.2178394

Díaz-Soria, I. (2017). Being a tourist as a chosen experience in a proximity destination. *Tourism Geographies, 19*(1), 96–117. https://doi.org/10.1080/14616688.2016.1214976

Díaz-Soria, I., & Llurdés-Coit, J. C. (2013). Thoughts about proximity tourism as a strategy for local development. *Cuadernos de Turismo, 32*, 303–305.

Dot Jutglà, E., Romagosa Casals, F., & Noguera Noguera, M. (2022). El incremento del turismo de proximidad en Cataluña en verano de 2020: una oportunidad para la consolidación del turismo rural como una forma de turismo sostenible y segura. *Investigaciones Turísticas, 23*, 162–185. https://doi.org/10.14198/INTURI2022.23.8

Gowreesunkar, V. G. B., Maingi, S. W., & Ming'ate, F.L.M. (2022). *Management of tourism ecosystem services in a post pandemic context: Global perspectives*. Taylor & Francis.

Hensher, D. A., & Stopher, P. R. (1979). Behavioural travel modelling. In D. A. Hensher & P. R. Stopher (Eds.), *Behavioural travel modelling* (pp. 1–42). Croom Helm. https://doi.org/10.4324/9781003156055

Higgins-Desbiolles, F. (2021). The "war over tourism": Challenges to sustainable tourism in the tourism academy after COVID-19. *Journal of Sustainable Tourism, 29*(4), 551–569. https://doi.org/10.1080/09669582.2020.1803334

Higgins-Desbiolles, F. (2022). Subsidiarity in tourism and travel circuits in the face of climate crisis. *Current Issues in Tourism*. Advance online publication. https://doi.org/10.1080/13683500.2022.2116306

Höckert, E., Rantala, O., & Jóhannesson, G. T. (2022). Sensitive communication with proximate messmates. *Tourism, Culture and Communication, 27*(2), 181–192. https://doi.org/10.3727/109830421X16296375579624

Huang, S., Shao, Y., Zeng, Y., Liu, X., & Li, Z. (2021). Impacts of COVID-19 on Chinese nationals' tourism preferences. *Tourism Management Perspectives, 40*, Article 100895. https://doi.org/10.1016/J.TMP.2021.100895

Jeuring, J., & Díaz-Soria, I. (2017). Introduction: Proximity and intraregional aspects of tourism. *Tourism Geographies, 19*(1), 4–8. https://doi.org/10.1080/14616688.2016.1233290

Jeuring, J., & Haartsen, T. (2017). The challenge of proximity: The (un)attractiveness of near-home tourism destinations. *Tourism Geographies, 19*(1), 118–141. https://doi.org/10.1080/14616688.2016.1175024

Khare, A., & Sathe, S. (2021). A study of tourism choices by tourists in India in new normal. *International Journal of Application or Innovation in Engineering & Management, 10*(5), 54–64. http://localhost:8080/xmlui/handle/123456789/9795

Klaniczay, J. (2022). The resilience of proximity tourism during the pandemic: Local walking tours of Budapest. *International Journal of Tour Guiding Research, 3*(1), 19–30. https://doi.org/10.21427/hj3y-9g86

Korstanje, M. E. (2023). The Janus face of terrorism and tourism: Terrorism as a risk, as a danger and as a worry. In C. Seabra & M. E. Korstanje (Eds.), *Safety and tourism* (Tourism security-safety and post conflict destinations) (pp. 103–115). Emerald Publishing Limited. https://doi.org/10.1108/978-1-80382-811-420231006

Lebrun, A. M., Corbel, R., & Bouchet, P. (2022). Impacts of Covid-19 on travel intention for summer 2020: A trend in proximity tourism mediated by an attitude towards Covid-19. *Service Business, 16*(3), 469–501. https://doi.org/10.1007/s11628-021-00450-z

Lebrun, A. M., Su, C. J., & Bouchet, P. (2021). A more sustainable management of domestic tourists in protected natural parks: A new trend in sport tourism after the covid-19 pandemic? *Sustainability (Switzerland), 13*(14), Article 7750. https://doi.org/10.3390/SU13147750

Li, X., Gong, J., Gao, B., & Yuan, P. (2021). Impacts of COVID-19 on tourists' destination preferences: Evidence from China. *Annals of Tourism Research, 90*, Article 103258. https://doi.org/10.1016/J.ANNALS.2021.103258

Loomis, J. (2007). Correcting for on-site visitor sampling bias when estimating the regional economic effects of tourism. *Tourism Economics, 13*(1), 41–47. https://doi.org/10.5367/000000007779784506

Maingi, S. W., Ming'ate, F. L. M., & Gowreesunkar, V. G. B. (2022). Sustainable cultural ecosystem services and Community-based Tourism (CBT) models post COVID-19 pandemic within Aberdares conservation area and National Park, Kenya. In V. G. B. Gowreesunkar, S. W. Maingi, & F. L. M. Ming'ate (Eds.), *Management of tourism ecosystem services in a post pandemic context: Global perspectives* (1st ed., pp. 382–395). Taylor & Francis.

Mansfeld, Y., & Winckler, O. (2015). Can this be spring? Assessing the impact of the "Arab Spring" on the Arab tourism industry. *Tourism: An International Interdisciplinary Journal, 63*(2), 205–223.

Mehta, S., Saxena, T., Purohit, N., University, I., & Dayal Marg, P. (2020). The new consumer behaviour paradigm amid COVID-19: Permanent or transient? *Journal of Health Management, 22*(2), 291–301. https://doi.org/10.1177/0972063420940834

Mendieta-Aragón, A. (2022). Cambios en el comportamiento turístico tras la COVID-19: hacia un nuevo perfil del turista y del viaje de ocio en España. *Revista Electrónica de Comunicaciones y Trabajos de ASEPUMA, 23*(1), 37–51. https://doi.org/10.24309/RECTA.2022.23.1.03

Mendieta-Aragón, A. (2023). Efectos de la COVID-19 en el sector turístico de la Unión Europea: Análisis económico para el caso de España. *Revista Universitaria Europea, 38*, 41–70.

Mirzaei, R., Sadin, M., & Pedram, M. (2021). Tourism and COVID-19: Changes in travel patterns and tourists' behavior in Iran. *Journal of Tourism Futures, 9*(1), 49–61. https://doi.org/10.1108/JTF-01-2021-0017/FULL/PDF

Müller, N., Fletcher, R., & Blázquez-Salom, M. (2023). Convivial tourism in proximity. In F. Higgins-Desbiolles & B. Chew Bigby (Eds.), *The local turn in tourism* (pp. 215–232). Channel View Publications. https://doi.org/10.21832/9781845418809

Oana, D. (2020). The impact of the current crisis generated by the COVID-19 pandemic on consumer behavior. *Studies in Business and Economics, 15*(2), 85–99. https://doi.org/10.2478/SBE-2020-0027

Rahmanov, F., Mursalov, M., & Rosokhata, A. (2021). Consumer behavior in digital era: Impact of COVID 19. *Marketing and Management ofInnovations, 5*(2), 243–251. https://doi.org/10.21272/MMI.2021.2-20

Rantala, O., Salmela, T., Valtonen, A., & Höckert, E. (2020). Envisioning tourism and proximity after the Anthropocene. *Sustainability, 12*(10), Article 3948. https://doi.org/10.3390/SU12103948

Rico, A., Martínez-Blanco, J., Montlleó, M., Rodríguez, G., Tavares, N., Arias, A., & Oliver-Solà, J. (2019). Carbon footprint of tourism in Barcelona. *Tourism Management, 70*, 491–504. https://doi.org/10.1016/J.TOURMAN.2018.09.012

Romagosa, F. (2020). The COVID-19 crisis: Opportunities for sustainable and proximity tourism. *Tourism Geographies, 22*(3), 690–694. https://doi.org/10.1080/14616688.2020.1763447

Rosselló, J., Becken, S., & Santana-Gallego, M. (2020). The effects of natural disasters on international tourism: A global analysis. *Tourism Management, 79*, Article 104080. https://doi.org/10.1016/J.TOURMAN.2020.104080

Salmela, T., Nevala, H., Nousiainen, M., & Rantala, O. (2021). Proximity tourism: A thematic literature review. *Matkailututkimus, 17*(1), 46–63. https://doi.org/10.33351/MT.107997

Seidel, S., Vrenegoor, F., & Cavagnaro, E. (2021). Sustainable behaviour in tourism and hospitality. *Scandinavian Journal of Hospitality and Tourism, 21*(5), 471–474. https://doi.org/10.1080/15022250.2021.1984986

Seraphin, H., & Dosquet, F. (2020). Mountain tourism and second home tourism as post COVID-19 lockdown placebo? *Worldwide Hospitality and Tourism Themes, 12*(4), 485–500. https://doi.org/10.1108/WHATT-05-2020-0027/FULL/PDF

Sharma, G. D., Thomas, A., & Paul, J. (2021). Reviving tourism industry post-COVID-19: A resilience-based framework. *Tourism Management. Perspectives, 37*, Article 100786. https://doi.org/10.1016/J.TMP.2020.100786

Sheth, J. (2020). Impact of Covid-19 on consumer behavior: Will the old habits return or die? *Journal of Business Research, 117*, 280–283. https://doi.org/10.1016/J.JBUSRES.2020.05.059

Turoń, K., & Kubik, A. (2021). Business innovations in the new mobility market during the COVID-19 with the possibility of open business model innovation. *Journal of Open Innovation: Technology, Market, and Complexity, 7*(3), Article 195. https://doi.org/10.3390/JOITMC7030195

UNWTO. (2019). Transport-related CO2 emissions of the tourism sector. In *Transport-related CO2 emissions of the tourism sector—Modelling results*. World Tourism Organization (UNWTO). https://doi.org/10.18111/9789284416660

Wachyuni, S. S., & Kusumaningrum, D. A. (2020). The effect of COVID-19 pandemic: How are the future tourist behavior? *Journal of Education Society and Behavioural Science, 33*(1), 67–76. https://doi.org/10.9734/JESBS/2020/v33i430219

Warde, A. (1997). Afterword: The future of the sociology of consumption. *The Sociological Review, 44*(1_suppl), 302–312. https://doi.org/10.1111/J.1467-954X.1996.TB03446.X

Xiang, Y. (2013). The characteristics of independent Chinese outbound tourists. *Tourism Planning and Development, 10*(2), 134–148. https://doi.org/10.1080/21568316.2013.783740

15

Tourism Degrowth in the New Normal: Exploring the Impacts of COVID-19 on Sustainable and Equitable Tourism Development

Isaac Kimunio, Martin W. Nandelenga, and Steve Makambi

Introduction

The travel and tourist business has been significantly impacted all around the world as a result of the epidemic that was caused by COVID-19. The travel and tourist business was one of the areas that was negatively impacted by the COVID-19 epidemic the most. This was owing to measures that restricted travel, shut down borders, and locked down businesses in an effort to control the spread of the disease. The tourist business suffered massive losses to measures taken to prevent the virus's spread,

I. Kimunio (✉) • S. Makambi
School of Economics, Business and Tourism, Kenyatta University, Nairobi, Kenya
e-mail: kimunio.isaac@ku.ac.ke; makambi.steve@ku.ac.ke

M. W. Nandelenga
African Development Bank, Nairobi, Kenya
e-mail: m.nandelenga@afdb.org

S. W. Maingi et al. (eds.), *Tourist Behaviour and the New Normal, Volume II*,
https://doi.org/10.1007/978-3-031-45866-8_15

including travel restrictions, social isolation, temperature checks, and increased hygiene.

In the years leading up to COVID-19, the tourism and travel industry was a very significant part of the global super economy. It was responsible for 10% of the global GDP and over 320 million employment opportunities across the globe. By the year 1950, there had been 25 million people who had traveled to other countries. By the year 2019, that number had increased to over 1.5 billion as a result of the tremendous growth that the tourism and travel sector had seen in a number of nations. The United Nations World Tourism Organization (UNWTO) estimates that international visitor travel fell by around 72% in 2020 compared to 2019, which resulted in a loss of $4.5 trillion throughout the globe. The first quarter of 2021 saw a worldwide decline in tourism of 87% compared to the same period in 2020, with the biggest effects being seen in Asia, the Pacific Region, and Europe. Over 120 million jobs in micro, small-, and medium-sized sectors, which employed 54% of the tourist workers with the majority of them women, were placed in jeopardy as a result of the one-of-a-kind epidemic that swept the world.

Even if the world economy is beginning to recover, the pandemic's effects on tourism have altered how people see and participate in the industry forever (Rahman et al., 2021). The concept of tourism degrowth, which refers to the intentional reduction of the scale and impact of tourism in order to achieve more sustainable and equitable outcomes, has gained renewed interest (Alexander & Gleeson, 2019).

This section examines the current and prospective effects of the COVID-19 virus outbreak on tourism and business. This chapter addresses a variety of techniques and regulations that may be used to both encourage the growth of tourism in a sustainable manner and lessen the adverse effects that can be caused by tourism. We specifically examine the possibilities of degrowth as a framework for attaining sustainable and fair tourism in the post-COVID-19 era. We also examine the role of governments, tourism industry stakeholders, and local communities in implementing degrowth strategies and promoting sustainable tourism practices.

The following outline constitutes the chapter's structure. In the beginning, a general explanation of the idea of tourist degrowth and its significance on the globe after a pandemic will be provided. Next, an analysis will be conducted to determine how the COVID-19 epidemic has altered people's vacation habits and their perspectives on tourism, as well as the ways in which these shifts may play a role in the growth of tourism practices that are more environmentally friendly and egalitarian. following that, the chapter will investigate the influence that the pandemic had on the experiences that tourists had while visiting popular tourist locations, as well as the possible avenues that may lead to a more sustainable and fair expansion of tourism in the globe following the epidemic. Finally, the chapter provides concluding remarks and suggestions for future research.

Economic Rationale for Tourism Degrowth

The economic rationale for tourism degrowth is based on the negative impacts of traditional approaches to tourism that prioritize growth and expansion. Tourism degrowth offers a sustainable alternative approach that prioritizes quality over quantity and equity. One economic rationale for tourism degrowth is that it can reduce the negative impacts of tourism on the environment and local communities, which can impact the long-term economic growth of a destination. By reducing these impacts, destinations can attract more responsible and sustainable tourists, ultimately supporting the growth of the industry. Another rationale is that tourism degrowth can promote more equitable distribution of economic benefits. Sustainable tourism practices can involve local communities more and help reduce economic leakage. Finally, tourism degrowth can enhance economic resilience to external shocks such as natural disasters or pandemics. This can support long-term economic growth and stability. Overall, the economic rationale for tourism degrowth is based on promoting sustainable and equitable tourism practices to support economic growth while preserving natural and cultural heritage for future generations.

Tourism Degrowth in the Post-Pandemic World

COVID-19 posed a serious danger to the tourist industry's bottom line, as we illustrated in the previous section. The industry has been forced to close down or lay off employees due to the lack of tourism activities. The reality is that the level of financial instability and capital erosion caused by the pandemic may lead to prioritizing economic recovery over sustainable and equitable practices. Furthermore, the drop in economic development due to decreased tourism might make it harder for local communities to recover from the economic disruption created by the epidemic. To this end, the question of Tourism degrowth is one that warrants careful consideration because the need for rebuilding the sector and economic recovery surpasses sustainability concerns particularly in the short-term period.

Furthermore, encouraging tourism degrowth and adopting more responsible and sustainable tourist practices may not be economically viable for many locations, especially those that depend significantly on tourism for their economic success. Several well-liked tourist destinations may not be able to survive the economic damage caused by the COVID-19 outbreak if visitor numbers drastically drop. Implementing sustainable tourism practices may be difficult because certain stakeholders may put short-term economic rewards ahead of long-term sustainability objectives.

The argument that the COVID-19 epidemic has offered a chance for the tourist sector to adopt more sustainable and equitable practices is met with the counterargument that the pandemic has had a significant effect on the financial viability of the tourism industry. Tourism degrowth has become an even more pressing issue as countries and communities seek to build back better and avoid a return to unsustainable tourism practices and patterns. Tourism degrowth is a concept that has gained increasing attention in recent years, particularly as concerns about the sustainability of the global tourism industry have grown. At its core, tourism degrowth is a response to the sector's detrimental effects on local communities' cultures, and environments, and seeks to address these issues by reducing the scale and impact of tourism (Canavan, 2014; Sharpley, 2021).

The goal is to create a tourism model that aligns with principles of environmental and social justice, rather than one that prioritizes economic growth at all costs and perpetuates patterns of inequality and exploitation. Although the idea of degrowth in tourism has gained popularity among scholars and policymakers, it is still a debatable topic for many in the worldwide tourism sector (Sharpley, 2021). Tourism degrowth has faced opposition from critics who argue that it could result in negative economic effects, such as decreased revenues and job losses for destinations that heavily rely on tourism (Higgins-Desbiolles, 2020). Moreover, some critics argue that tourists themselves may not be willing to limit their travel experiences.

Despite these criticisms, supporters of tourism degrowth maintain that transitioning toward a more sustainable and equitable model of tourism is crucial. In an attempt to address concerns related to tourism sector growth and sustainability, proponents of paradigm shifts in global tourism have identified the promotion of alternative forms of tourism. In this regard, many nations are setting green and sustainable goals as they plan their recovery efforts. As a result, politicians, tourist operators, and other stakeholders have begun rethinking the role of tourism in society and how it may be improved to be more sustainable and fair in the wake of the epidemic (OECD, 2020; Oralkan, 2021).

One potential pathway for achieving more sustainable and equitable tourism practices is through the concept of tourism degrowth. Degrowth potential has been looked at vis a vis a variety of zones. Hitherto application of the concept to tourism has been moderately narrow. An early assessment of this potential is presented by Bourdeau and Berthelot (2008) in their paper for the first global degrowth conference, which took place in Paris the same year. To grasp the magnitude of tourism's contribution to sustainable development, we must place it within degrowth processes, which provide a counter-narrative to the prevailing economism paradigm. To put this option into effect, Hall advocated for what he called a "steady-state tourism" paradigm, which prioritizes quality growth above quantity growth at the cost of natural capital. This model predicts that in order for tourism to conform to bigger, more sustainable patterns of consumption and production, it would need to "degrow" significantly (Fletcher et al., 2019). Meanwhile, Panzer-Krause (2019) offers a unique empirical addition to the debate by tracing the emergence of a degrowth-oriented subset of ecotourism operators in rural Ireland. Intentionally, the concept of tourism degrowth advocates for scaling down the impact of mass and/or commercially oriented tourism in favor of possible more sustainable and meaningful tourism frontiers that benefit local communities, cultures, and environments. This may entail a variety of steps, including limiting the number of visitors to specific areas, raising environmental and social standards for the tourism industry, and fostering new types of travel that put preservation and interaction with locals ahead of profit (Andriotis, 2018; Alexander & Gleeson, 2019).

One of the key concepts emerging from these discussions has been that of tourism degrowth. This is a paradigm shift that challenges the prevailing belief in perpetual economic growth and seeks to intentionally reduce the scale and impact of tourism. This concept recognizes that the tourism industry, just like any other economic sector, has limits to its growth and cannot continue to expand indefinitely without negative environmental social consequences. As such, tourism degrowth emphasizes the need to transition toward a more sustainable and equitable model of tourism that prioritizes community well-being and environmental protection over unrestricted economic growth and profit maximization (Markwell et al., 2012; OECD, 2020; Oralkan, 2021).

Promoting slow tourism, ecotourism, cultural tourism, or rural tourism are all examples of alternate types of tourism that may help with the shift to a more sustainable model. These forms of tourism are typically less resource-intensive and have a lower environmental impact than mass tourism. For example, slow tourism and ecotourism are emerging trends that could balance between profitability and sustainable goals. Slow tourism prioritizes on eco-friendly modes of transportation and puts emphasis on fewer destinations. This strategy aims at maximization of quality, as opposed to quantity, and meaningful experiences. On the other hand, ecotourism is characterized by promotion of tourism that focuses on appreciation of natural beauty support environmental conservation. Both strategies are not only mutually reinforcing but also provide strategies that can accentuate tourist's satisfaction, environmental conservation, and community benefits. In addition, they also tend to distribute the benefits of tourism more fairly among local communities, and promote cultural exchange and understanding between tourists and host communities (OECD, 2020; Markwell et al., 2012; Oralkan, 2021).

Promotion of alternative forms of tourism prompts the careful deliberations and formulation of policies that minimize the industry's environmental impact and resource consumption. The implementation of regulations and plans to protect the tourism industry varied across countries, and was influenced by factors such as the significance of the industry, competing demands of other industries, and international relationships with collaborating nations, as indicated by the context (Nyaruwata & Mbasera, 2021). For instance, SADC member states adopted the policies and measures in response to COVID-19's effects on the travel and tourism sector. This is illustrated in Table 15.1 below adapted from Nyaruwata & Mbasera (2021).

Table 15.1 Adopted policies and measures by SADC countries

	Measures and policies taken	Examples from SADC countries
1	State of emergency being declared nationally	March through April 2020 for all member states
2	Nationwide lockdowns are put into effect	Beginning in March of 2020, each and every member state will have done so at different times
3	National airspace and borders are closed	All member nations will have done so at different periods starting in March 2020
4	Strategies to promote tourism	Several member states, such as South Africa, Zimbabwe, Zambia, Namibia, and Mauritius, have developed recovery plans and are presently executing their domestic tourism components as part of their efforts to recover from the impacts of the COVID-19 pandemic

(*continued*)

Table 15.1 (continued)

	Measures and policies taken	Examples from SADC countries
5	Employees cash grants and subsidies	• South African businesses accessed funding for three-month employee wages (March–May 2020) • The government of Mauritius has made US$265 million available to company owners in the form of two schemes in 2020: the salary support scheme and the self-employment assistance program • Businesses in Seychelles were able to get grants for the period of March to December 2020. Businesses in Botswana were able to receive 50% of employee wage costs through the COVID-19 relief fund • Several grant and subsidy programs available to businesses in Lesotho, Namibia, Tanzania, and Zambia
6	Tax rebates and relief	Provided by all participating governments; specifics may vary by country
7	Liquidity support	The governments of Seychelles, Mauritius, and South Africa supplied funds to businesses at reduced interest rates (e.g., Seychelles funds had an interest rate of 1.5% while the South African government contributed funds totaling R1.6 billion)
8	Loan guarantee schemes	Available in the majority of member states
9	Incentives for training and professional development within the industry	This was available in various African countries, including South Africa, Botswana, Namibia, Mauritius, Seychelles, and Zambia
10	Monetary assistance to employees who have been laid off	The government of Seychelles offered financial help to its laid-off workers, including a portion of their monthly salary, assistance finding new employment, and training and education options
11	Moratorium on loan repayment	Several countries such as Mauritius, Botswana, Namibia, and Seychelles offered this facility
12	Relief on payment of rent	During the COVID-19 epidemic, rent payment discounts were made available to businesses in some nations, such as Mauritius and Botswana, that operated on sites that were held by the government or leased property from local councils and government ministries

Source: Nyaruwata and Mbasera (2021, pp. 11–13)

Impacts of the COVID-19 Pandemic on Travel Patterns and Attitudes Toward Tourism

Both traveler habits and public perception have been profoundly altered by the COVID-19 outbreak (Luković & Stojković, 2020). Many people were forced to scrap or postpone their vacations as a direct consequence of governments throughout the world enacting travel restrictions and social distancing measures. The consequence has been a drop in demand for tourism-related services and a serious economic disruption for many businesses. At the same time, the epidemic has caused many civilizations to reevaluate the role that tourism plays in their economies. Many individuals have recently gained greater awareness of the detrimental effects that tourism may have on the local communities, cultures, and habitats, and many have voiced a wish for tourist practices that are more sustainable and fair. The aftermath of the epidemic has sparked an increasing interest in the concept of tourism degrowth. Governments, tourist operators, and various stakeholders are now striving to build a more sustainable and equitable tourism economy on a global scale (OECD, 2020; Oralkan, 2021; Babii & Nadeem, 2021).

Before the pandemic, the travel and tourism industry was a significant contributor to global employment and GDP (World Travel & Tourism Council, 2023). From 2014 to 2019, it accounted for one in five new jobs created worldwide and 10.3% of all jobs (334 million) and 10.4% of global GDP (US$ 10 trillion) in 2019. Additionally, international visitor spending was at US$ 1.9 trillion in 2019 (World Travel & Tourism Council, 2023). The travel and tourism sector showed signs of recovery in 2022. It contributed 7.6% to global GDP, a 22% increase from 2021, and only 23% below 2019 levels (World Travel & Tourism Council, 2023). In terms of employment, there were 22 million new jobs created in 2022, representing a 7.9% increase from 2021 and only 11.4% below 2019 levels. Domestic visitor spending increased by 20.4% in 2022, which is only 14.1% below 2019 levels. While international visitor spending rose by 81.9% in 2022, it is still 40.4% behind 2019 numbers (World Travel & Tourism Council, 2023).

The pandemic has also led to changes in travel patterns and preferences. As people have become more cautious about their health and safety, many have opted for domestic or regional travel instead of international travel. Anecdotal reports show that domestic tourism played a significant role in softening the impact of the pandemic. According to Arbulú (2021) despite the decline of domestic tourists by 43% in Spain, the economy realized 33% increase in overnight stays by domestic tourisms in 2020, compared to 2019, since more than half of outbound tourists preferred to stay and visit local destinations. It was reported that countries such as Solomon Island, Fiji, Samoa, and Vantau introduced policies targeting promotion of domestic tourism and promotion of cultural and rural tourism with great success (ADB, 2021).

Furthermore, the COVID-19 outbreak has significantly altered the ways in which visitors to famous tourist destinations remember those places. To ensure the safety of visitors and residents alike, governments throughout the world have instituted social distancing rules and other safety measures that have forced many famous tourist destinations to alter their business methods. Because of this, the ways in which visitors engage with their travel locations and the kinds of experiences that are offered have both undergone significant transformations as a result (Sharpley, 2021; Babii & Nadeem 2021; ADB, 2021).

For example, many tourist destinations have had to control the number of visitors allowed at certain attractions, or have implemented timed entry systems to prevent overcrowding. This has led to a more personalized, meaningful, and intimate visitor experience, as visitors have more space and time to explore destinations without feeling rushed or focusing on visiting as many destinations as possible. Additionally, there has been a shift toward more outdoor and nature-based tourist experiences, as people seek to avoid enclosed spaces and crowded areas.

Overall, the pandemic played a significant role in shifting tourists' attitude and behavior toward more sustainable and responsible tourism practices as people seek to minimize their carbon footprint and support local economies. Additionally, there has been a trend toward more experiential and authentic tourism, as people seek to engage with local cultures and communities in a more meaningful way.

Economic Sustainability and Governance Trends

Sustainable tourism is a concept that encompasses wider-than tourism response. It should be approached as a process that is aimed at delivering long-term social economic goals (Canavan, 2014). According to Asian Development Bank report (ADB, 2021) decline in general economic activity during the pandemic led to annual carbon dioxide (CO_2) emission reduction by 2.4 billion tons representing the largest annual CO_2 reduction ever recorded since World War II. The degree to which social and economic policies across all sectors are oriented toward sustainable development will be a major factor in determining the pandemic's long-term effect.

An investigation conducted after the event reveals that the influence of the COVID-19 epidemic offers a catalytic effect that moves tourism away from mass tourism and toward a kind of tourism that is more sustainable. The COVID-19 pandemic has established a "new normal" by bringing to light the vulnerabilities of the tourist industry. These vulnerabilities include the industry's overreliance on mass tourism, which may result in overtourism, economic leakage, and adverse effects on both the environment and the people that are directly affected by tourism. In reaction to this, there has been rising emphasis placed on the creation of sustainable tourism that places quality over quantity, community engagement, and environmental conservation at the forefront of its goals (Benjamin et al., 2020; Babii & Nadeem, 2021).

The economic sustainability and governance trends that have emerged in the new normal post-COVID-19 have had a significant impact on tourist behavior. According to ADB (2021), 76% of consumers take into consideration environmental issues when making purchasing decisions after the COVID-19 pandemic. Around 14% of tourists reported that they are in favor of sustainable tourism after the COVID-19 pandemic. Tourists are becoming more conscious of their impact on local communities and the environment, and are seeking out more responsible and sustainable tourism experiences. This includes a growing interest in ecotourism, community-based tourism, and sustainable tourism practices (Benjamin et al., 2020; Babii & Nadeem 2021; ADB, 2021).

Tourists are also becoming more interested in cultural and nature-based tourism experiences that provide opportunities for learning and engagement with local communities. Included in this is the promotion of indigenous tourism and cultural heritage tourism, both of which can aid in the economic growth of nearby communities while maintaining their natural and cultural legacy.

Governments and industry players now recognize the need for more effective governance frameworks that can aid in the development of sustainable tourism, and governance trends have also arisen. This comprises laws and rules that support environmentally friendly, socially conscious, and locally focused forms of travel. Increasing stakeholder participation in decision-making processes for tourism development also entails local communities and indigenous peoples.

Partnerships for Sustainable Tourism Development

In the context of developing sustainable tourism, the sustainable development goal (17) on renewing the global cooperation for sustainability is especially pertinent. Collaboration between governments, business partners, local communities, and non-governmental organizations (NGOs) is crucial to fostering the creation of a sustainable tourist industry that can promote social stability, economic stability, and the reduction of poverty.

In addition, partnerships can help with capacity building, the exchange of knowledge and best practices, and access to financial and technical support. This includes public-private partnerships, which can encourage the growth of sustainable tourism by bringing together stakeholders from the public and private sectors. Furthermore, achieving sustainable development goals 1, 16, and 17 may require significant resources and collaboration, which may be difficult to achieve in some destinations due to political, social, or economic factors.

Pathways for More Sustainable and Equitable Tourism Development

The COVID-19 epidemic has provided the tourist sector with a chance to reassess its approach to sustainable practices and equitable distribution of benefits. By embracing the concept of tourism degrowth and implementing more sustainable and responsible tourism practices, it may be possible to create a tourism industry that is more equitable, resilient, and beneficial for all stakeholders.

One pathway for achieving this goal is through balancing between social and economic goals and introduction of more stringent environmental and social standards for tourism businesses. This could involve measures such as the adoption of sustainable tourism regulations, the formulation and implementation of sustainable tourism planning and management frameworks, and the promotion of responsible tourism practices among local populous, businesses fraternity, and tourists alike (Benjamin et al., 2020; Babii & Nadeem 2021; ADB, 2021).

One possible approach would be to encourage alternative types of tourism, such as those that put an emphasis on environmental responsibility and cross-cultural interaction rather than mass tourism. The development of community-based tourism projects, the promotion of agrotourism and ecotourism activities, and the merging of traditional and indigenous knowledge and practices into tourism operations might all be part of this process.

Changes in travel habits, perspectives on tourism's value to society, and the epidemic's overall influence may be seen all over the world as a result of the worldwide COVID-19 pandemic. Despite the colossal economic and social disruption witnessed in the tourism sector, the pandemic presented a rare opportunity for the reinvention of the global tourism industry from mass tourism to sustainable tourism. The extent to which the benefits of reinvention of the tourism sector outweigh the cost will largely depend on strategies and policies adopted in promotion of alternative forms of tourism that support both commercial and sustainable interests. To this end, there is need to embrace more sustainable and equitable practices, and to promote effective tourism degrowth strategies as a means of achieving these long-term socio-economic goals.

The catalytic effect of the pandemic led to paradigm shift in the tourism sector with desirable effects on business practices, tourists' attitude and behavior, environmental conservation, and cultural and community focus. Even though economic-oriented policies are critical for crisis management and economic recovery in the tourism sector, they may be counterproductive in pursuit and achievement of sustainability development in the long run. To this end, in the post-pandemic environment tourism degrowth strategies should be carefully designed to balance both economic and commercial concerns and sustainability in the sector (Oralkan, 2021; ADB, 2021).

Tourism degrowth and economic sustainability have significant implications for the tourism industry and the broader economy. The concept of tourism degrowth aims to produce more equitable and sustainable outcomes by reducing the industry's footprint. Economic sustainability refers to the ability of an economy to support long-term economic growth while also promoting social and environmental well-being.

One implication of tourism degrowth is a shift toward more sustainable forms of tourism, such as ecotourism and community-based tourism, which are more environmentally friendly and socially responsible. This shift could help support local economies and communities and create a more sustainable tourism industry focused on quality over quantity. Reducing the number of tourists could also lessen the negative impacts on the environment and local resources.

Promoting sustainable tourism can also help support long-term economic growth. Sustainable tourism can generate income for local communities, create jobs, and attract more tourists to a destination by promoting its natural and cultural heritage. Governments and industry stakeholders can promote tourism degrowth and economic sustainability by developing policies and regulations that encourage sustainable tourism practices, investing in sustainable infrastructure and services, and supporting research and development.

In conclusion, promoting sustainable tourism practices is essential to achieve economic sustainability and minimize the negative impacts of tourism on the environment and local communities. Governments and industry stakeholders can work together to create a more sustainable tourism industry that supports long-term economic growth and social and environmental well-being.

Policy implications from the economic sustainability and governance trends in sustainable tourism development include the need for effective governance frameworks and partnerships to promote economic growth, poverty reduction, social stability, and environmental protection. To achieve these goals, governments and industry stakeholders should prioritize the development of sustainable tourism practices and policies that encourage ecotourism, responsible tourism, and community-based tourism while discouraging mass tourism. They should also invest in infrastructure and services that support sustainable tourism development, such as public transportation, cycling and walking, and renewable energy sources.

Stakeholders in the sector should develop marketing plans that highlight the advantages of sustainable tourism for the economy, the environment, and society, including how it helps local communities, promotes environmental conservation, and engages with local cultures. To guarantee that tourist development is equitable and sustainable, governance frameworks should consider the viewpoints of local communities, indigenous peoples, and other stakeholders. Additionally, governments and industry stakeholders should support research and development of sustainable tourism practices by collaborating with academic institutions, NGOs, and other stakeholders to identify best practices and develop innovative solutions to sustainability challenges. The sustainable development goals of No Poverty (SDG 1), Peace, Justice, and Strong Institutions (SDG 16), and revitalizing the global partnership for sustainable development (SDG 17) are important for promoting sustainable tourism growth and ensuring that tourism contributes to reaching the sustainable development goals.

Moving ahead, it will be essential for decision-makers, firms in the tourist industry, and other stakeholders to collaborate on the development of tourism practices that are more sustainable and fair, and that provide benefits to the local people, cultures, and surroundings. By embracing tourism degrowth and implementing more responsible and sustainable tourism practices, it may be possible to create a tourism industry that is more resilient, equitable, and beneficial for all.

References

Alexander, S., & Gleeson, B. (2019). *Degrowth in the suburbs: A radical urban imaginary*. Palgrave Macmillan.

Andriotis, K. (2018). *Degrowth in tourism: Conceptual, theoretical and philosophical issues*. CABI.

Arbulú, I., Razumova, M., Rey-Maquieira, J., & Sastre, F. (2021). Can domestic tourism relieve the COVID-19 tourist industry crisis? The case of Spain. *Journal of Destination Marketing & Management, 20*, 100568.

Asian Development Bank. (2021). Sustainable tourism after Covid-19 insights and recommendations for Asia and the Pacific. Retrieved April 2023, from https://www.adb.org/sites/default/files/publication/761511/sustainable-tourism-after-covid-19.pdf

Babii, A., & Nadeem S. (2021). Tourism in a post-Covid world. Retrieved April 2023, from https://www.imf.org/en/News/Articles/2021/02/24/na022521-how-to-save-travel-and-tourism-in-a-post-pandemic-world

Benjamin, S., Dillette, A., & Alderman, D. H. (2020). "We can't return to normal": Committing to tourism equity in the post-pandemic age. *Tourism Geographies, 22*(3), 476–483.

Bourdeau, P., & Berthelot, L. (2008, April 18–19). *La décroissance pour repenser le tourisme?* [Degrowth to rethink tourism?] [Paper presentation]. *First International Conference on Economic De-growth for Ecological Sustainability and Social Equity*, Paris.

Canavan, B. (2014). Sustainable tourism: Development, decline and de-growth. Management issues from the Isle of Man. *Journal of Sustainable Tourism, 22*(1), 127–147.

Fletcher, R., Murray Mas, I., Blanco-Romero, A., & Blázquez-Salom, M. (2019). Tourism and degrowth: An emerging agenda for research and praxis. *Journal of Sustainable Tourism, 27*(12), 1745–1763.

Higgins-Desbiolles, F. (2020). The "war over tourism": Challenges to sustainable tourism in the tourism academy after COVID-19. *Journal of Sustainable Tourism, 29*(4), 551–569.

Luković, S., & Stojković, D. (2020). Covid-19 pandemic and global tourism. *Менаџмент у хотелијерству и туризму, 8*(2), 79–87.

Markwell, K., Fullagar, S., & Wilson, E. (2012). Reflecting upon slow travel and tourism experiences. *Slow Tourism: Experiences and Mobilities, 227*, 233.

Nyaruwata, S., & Mbasera, M. (2021). A critique of contribution of tourism to jobs in Southern African Development Community (SADC): Implications for post COVID19 pandemic. *International Tourism and Hospitality Journal, 4*(5), 1–18.

OECD. (2020). Rebuilding tourism for the future: COVID-19 policy responses and recovery. Retrieved April 2023, from https://www.oecd.org/coronavirus/policy-responses/rebuilding-tourism-for-the-future-covid-19-policy-responses-and-recovery-bced9859/

Oralkan, A. (2021). Paradigm shift of tourist behavior in Eurasian countries under pandemic conditions. *On Eurasian Economies, 2021*, 175.

Panzer-Krause, S. (2019). Networking towards sustainable tourism: Innovations between green growth and degrowth strategies. *Regional Studies, 53*(7), 927–938.

Rahman, M. K., Gazi, M. A. I., Bhuiyan, M. A., & Rahaman, M. A. (2021). Effect of Covid-19 pandemic on tourist travel risk and management perceptions. *Plos one, 16*(9), e0256486.

Sharpley, R. (2021). On the need for sustainable tourism consumption. *Tourist Studies, 21*(1), 96–107.

World Travel & Tourism Council. (2023). https://wttc.org/research/economic-impact#:~:text=WTTC's%20latest%20annual%20research%20shows,and%20only%2011.4%25%20below%202019.

16

Glamping in Nature-Based Destinations: A Product for Tourism Recovery

Riza Mae Mirandilla, Chloe Marie Novo, and Maria Criselda Badilla

Introduction

The COVID-19 pandemic has greatly affected the tourism industry due to the travel restrictions. With tourism destinations slowly reopening, glamping is regarded as one of the suitable products in the new normal since it combines nature, the outdoors, and social distancing. Nature-based tourism products are now being developed as a response to tourist's desires for safer alternatives in selecting destinations (Rocamora, 2020). The activities people engage in when they visit natural areas that are not their usual surroundings are included in nature-based tourism products (Fredman et al., 2009 as cited in Fossgard & Fredman, 2019). A CBI (2020) article states that activities in nature-based tourism include nature tours, observation of flora and fauna, cultural and local heritage tourism, volcano tourism, fishing tourism, cycling tourism, bird watching

R. M. Mirandilla • C. M. Novo • M. C. Badilla (✉)
Asian Institute of Tourism, University of the Philippines,
Quezon City, Philippines
e-mail: ramirandilla@up.edu.ph; cbnovo@up.edu.ph; mgbadilla2@up.edu.ph

S. W. Maingi et al. (eds.), *Tourist Behaviour and the New Normal, Volume II*,
https://doi.org/10.1007/978-3-031-45866-8_16

tourism, kayaking tourism, bush walking, trekking, camping, hiking tourism, cold water tourism, hunting, visiting parks, scenic driving, photography tourism, beach experiencing, relaxing as well as self-improvement travel.

One of the tourism products that can be experienced with nature is glamping. Glamping, which combines the words 'glamour' and 'camping,' is a new camping idea that combines comfort with direct contact with nature (Brochado & Pereira, 2017). It retains the outdoor experience but has pre-setup comfortable accommodation and modern amenities. Glamping can be in the form of cabins, treehouses, tents, yurts, and tipis (Brooker & Joppe, 2013). The outdoors, however, has good air circulation that decreases the risk of transmission (Morgan, 2021) of diseases. Glamping enables travelers to stay and engage in activities outdoors, be surrounded by nature, and avoid crowds.

The chapter aimed to know the viability of glamping (glamorous camping) in nature-based destinations that can assist in sustainable tourism recovery in the Philippines, post-pandemic. The pandemic brought about a new set of preferences for travelers. Tourists seek to experience memorable moments that enable them to be in touch with nature but not lose the essential comfort of luxury accommodations (Fernandes et al., 2021). This focused on analyzing the glamping attributes from both the supply and demand side based on interviews with managers of Your Brother's House Tribal Village in Legazpi City, Albay, Misty Heights in Marilog District, Davao City and Camp Bernardino de Lavigan in Davao Oriental as well as the perception of tourists interested in glamping through an online survey.

Literature Review

People have become cautious in traveling during and after the COVID-19 pandemic. A study by Zheng et al. (2021), showed that people were inclined to feel frightened, nervous, and anxious when they travel in the post-COVID-19 period. As tourism opens post-pandemic, the development of COVID-19 adaptive products is recommended to influence people to be cautious to travel as coping strategies would lead to resilience of the tourism industry. Williams et al. (2021) explored the COVID-19 vaccine confidence in Italian residents and their interest in tourism activities. The findings identified two segments, the 'High Confidence Group (HCG)' and 'Low Confidence Group' (LCG). The HCG was willing to be vaccinated as quickly as possible. The confident respondents wanted to restart their traveling on national territory earlier, suggesting that higher vaccine confidence might lead to a willingness to travel, at least regionally, as soon as possible. With vaccination campaigns gaining ground, tourists are gaining higher travel confidence. The Traveler Confidence Index developed by Travel Again revealed that business travelers were optimistic about the future of business travel (Schutte, 2021). As of September 2021, data shows that 67.2% felt that they are already traveling or will travel as often as they did by the end of 2022, while 8.8% believed they will never travel as often as they did before the pandemic. Additionally, 22.4% thought they will hit the same level of pre-pandemic business travel in 2023 and 4% in 2024. A survey conducted by the European Travel Commission (2021) in July 2021 revealed that the top three factors that boost travelers' confidence were getting vaccinated (37%), pre-travel COVID-19 tests (20%), and destination's effectiveness in managing COVID-19 (20%).

In 2019, the World Travel & Tourism Council and Global Rescue prepared a crisis readiness report where strategies on tourism recovery are recommended when a destination experiences a crisis. The COVID-19 pandemic crisis has taught destinations many lessons in destination development. During this recovery phase, it is critical for a destination to understand the nuances of travelers and their motivations and to invite them back when the destination is ready. The crisis may be an opportunity to rebuild sustainable and efficient infrastructure and rethink or change tourism products and target markets. The goal for the destination should be to overcome the crisis by designing and building a more resilient system with thorough crisis preparedness plans.

In designing nature-based tourism (NBT) products, six dimensions were presented by Fossgard and Fredman (2019). The six dimensions included in the framework were adapting to guests' needs, access to resources, experience facilitation, risk management, crowding management, and connection to place. The findings of the study showed that the products offered by NBT providers were linked to the resources of nature which connect the physical, social, and cultural aspects of the tourist experience. Still, the results suggested that the most important feature of any NBT product was the intangible resources including the sights, scents, and sounds.

Camping has evolved into a highly fragmented niche tourism product from the traditional use of tents for outdoor recreation, it has evolved into a holistic activity and accommodation choice, ranging from recreational vehicles (RVs), caravans, cabins, or other types of temporary shelter where the comforts of home can be enjoyed (Brooker & Joppe, 2013, as cited in Rogerson & Rogerson, 2020). Mikulić et al. (2017) investigated the decisive campsite attributes that influence both campsite choice and experience of tourists in the coastal campsites in Croatia and results showed that sanitary facilities, electricity, and water provision were the most relevant and important campsite attributes. Meanwhile, O'Neill et al. (2010) explored nature-based car camping and visitor satisfaction. The researchers developed a four-factor structure for visitor satisfaction with camping services which are people, service, tangible, and washroom availability and cleanliness.

Brochado and Pereira (2017) described glamping as a subsector of camping that merges 'glamorous' and 'camping.' It is an upscale version of camping that gives tourists an outdoor experience but removes the discomforts and inconveniences of tent camping and replaces them with pre-erected, homelike accommodation. Tourists would not need to pitch tents, use sleeping bags, or build fires. Glamping can be in tents, villas, huts, yurts, teepees, airstreams, treehouses, eco- and safari lodges, cabins, campervans, cabooses, or barns. The facilities and amenities it can offer can include bathrooms, kitchens, televisions, and Wi-Fi service.

Literature revealed different attributes affecting an individual's glamping choices. Lee et al. (2019) in their study of glamping in South Korea, identified five attributes affecting individuals' glamping choices: price (glamping price), atmosphere (provision of convenience tools), cleanliness (personal shower or room condition), congestion, and security (the presence of a security guard). The findings revealed that security was regarded as the most important attribute. Further, the dominant factor that established the intention to revisit the destination was the physical environment or physical quality which includes the design, decoration, temperature, and lighting (Realino et al., 2021). Craig (2020) examined the influence of COVID-19 on future camping and glamping considerations of US-based leisure travelers, of which, 36.3% were more likely to consider tent camping and 34.8% would prefer glamping. Regarding camping plans, 38.6% had plans for 2020 which was slightly more than the 35.6% in 2019. Furthermore, the results suggested that travelers who had previous camping experience were more likely to engage in the same experience post-COVID-19. The recent travelers who camped were 4.29 times more likely to consider tent camping and 2.70 times more likely to consider glamping. Overcrowding was also highly significant with or without COVID-19. Travelers regardless of the pandemic preferred locations that provide social distance.

Another study by Craig and Karabas (2021) surveyed American and Canadian active leisure travelers to address current and future impacts of COVID-19 on glamping. The findings indicated that active leisure travelers had more post-COVID-19 glamping trip plans (45.9%) than hotel/resort trip plans (24.7%). The results also showed 16% more post-COVID-19 glamping trip plans than pre-COVID-19 trip plans; while the percentage of post-COVID-19 hotel/resort trip plans (24.7%) was lower than 2019 hotel/resort trip plans (35.5%) and pre-COVID-19 trip plans (40.2%). The data suggested an increased viability of glamping to a broader market of leisure travel during and after the COVID-19 pandemic. Although pre-pandemic, the accessibility, popularity, and interest in glamping were already increasing. During the pandemic, however, where proximity to home, social distancing, and outdoor recreation is sought after, glamping is a desired accommodation and activity.

There were also some challenges to glamping. In Portugal, Filipe et al. (2018) discussed the main obstacles of customers in choosing glamping were the high cost, the lack of knowledge, the limited offers, and the non-authenticity compared to camping as well as the shortage of offers, high cost, and lack of awareness. Boiko (2021) investigated glamping as a new type of ecologically friendly business in Ukraine where data suggests that the risks of glamping as a business include choosing the wrong location, equipment breakdown, an increase in the share of fixed costs, unfavorable climate, and stronger regulation of the activity. Adamovich and Nadda

(2021) wrote about glamping providers' opinions on glamping quality and standards in the UK where all participants have stated that glamping had quality issues, therefore managing customer expectations is crucial for the success of the business.

The characteristics of glamping were mostly described and understood with its association to luxury, nature, and quality whereas travel motivations were split into a tourist's motivation for nature-based tourism and luxury tourism (Sakáčová, 2013). The Socio-psychological Motivations to Travel developed by Crompton identified seven socio-psychological motives and two cultural motives which urged tourists to travel to a specific destination (Yousaf et al., 2018). The socio-psychological motives or push motives included escape, exploration of self, relaxation, prestige, regression, enhancement of kinship relationships, and facilitation of social interaction while novelty and education were the cultural motives or pull motives (Katsikari et al., 2020). Additionally, Crompton suggested that the socio-psychological motives must be considered by the destinations in developing products and promotional strategies. Dann's Theory of Push and Pull Motivations stated the multiple factors that motivate tourists to travel can be categorized as either push or pull motivations (Yousaf et al., 2018). Push factors were the socio-psychological needs that urged an individual to travel. On the other hand, pull factors were associated with the services and facilities offered at the destination. This framework was based on anomie which is the desire to transcend feelings of isolation and ego-enhancement that are derived from personal needs.

Hennessey et al. (2016) explored what influenced the intentions of potential first-time and repeat visitors to visit a destination. The researchers designed a model hypothesizing that intention to travel was influenced directly by two major elements of tourism marketing: responses to advertising and the respondent's use of the official tourism website for a destination. Findings showed that the two major elements had direct influence on the respondents' intention to travel while travel intentions directly influenced actual visitation. Moreover, advertising recall was the most powerful predictor of the first-time visitors' intention to visit. The number of times the respondent visited the website, however, was the most powerful predictor for the repeat visitors' intention to visit. Nguyen et al. (2021), on the other hand, investigated the travel intention to visit tourism destinations of Vietnam's gen Z. The results indicated that gen Z valued the usefulness of social media and utilized it for seeking information. If they want to experience what they found in social media, it becomes a motivation and leads to visit intention. Furthermore, gen Z used social media to examine the destinations' accessibility. They were highly motivated to visit destinations that had clear and easy access, for example, visa-free destinations.

Study Framework

The diagram shows the framework of the study. Service providers craft their products by considering the different factors, in this case the glamping attributes, that would ensure quality and utmost experience for the tourists. The framework of Sakáčová (2013) (Figure 16.1) showed that the travel motivations for nature-based glamping included the push and pull factors. Dann's Theory of Push and Pull Motivations stated that the factors that pull tourists to travel include the services and facilities of the destination (Yousaf et al., 2018). These glamping attributes were assessed by the tourists which may influence their intention to visit or revisit the site. But because of the fear of contracting the COVID-19 virus when traveling, which was seen in the framework of Zheng et al. (2021), the

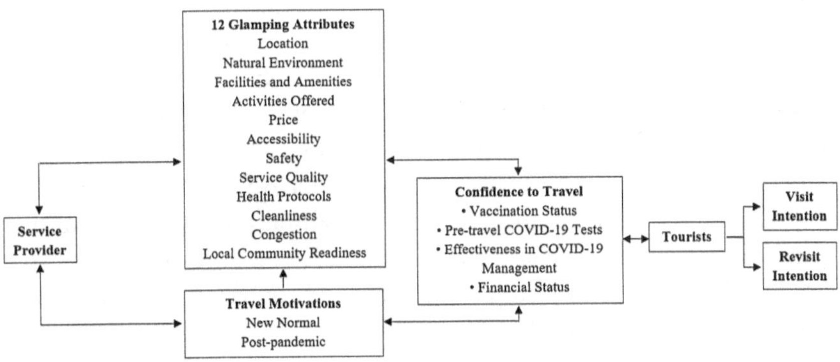

Fig. 16.1 Study framework

travel confidence of the tourists affects how they assess the glamping attributes of the site and the motives that push them to travel. The assessment of the tourists about the glamping attributes of the product offered while considering their travel confidence must be observed and integrated by the service provider in the product offered (Fig. 16.1).

Tourists also have needs as well as socio-psychological and cultural motives based on the theories of Maslow, Crompton, and Dann (Yousaf et al., 2018). These motives may be taken into consideration by tourists in the new normal, post-pandemic, or in both periods. The theories discussed by Yousaf et al. (2018) on travel motivation revealed that the service providers or destinations must understand the behavior, needs, and motives of tourists when traveling to a certain place in order to provide the utmost experience that is expected from them. Since the service provider wants to attract as many visitors as they can, they will incorporate the needs and travel motivations of tourists in the product offered. These factors will influence the intention of the tourists to visit or revisit a certain destination.

Literature points to twelve (12) glamping attributes (Table 16.1) namely location, natural environment, facilities and amenities, activities offered, price, accessibility, safety, service quality, health protocols, cleanliness, congestion, and local community readiness. Location is defined as a place where something happens or exists or is the position of something (Oxford University Press, 2022b). Tourists preferred to travel within the provinces (Zheng et al., 2021) or in places near their homes (Craig & Karabas, 2021) because of the fear of being infected by the virus, thus the location of the destination may be a concern to them.

Since glamping is considered as a nature-based tourism and is also a subsector of camping, the natural environment is one of the factors that will influence the experience of tourists (Fossgard & Fredman, 2019; Rogerson & Rogerson, 2020; O'Neill et al., 2010; Sakáčová, 2013; Filipe et al., 2018). Natural environment is the natural physical surroundings where human life takes place (Lauesen, 2013) which ranges from structured and well-planned gardens to the huge outdoors which include unspoiled mountains and large forest regions (Van Den Bosch, 2017). It covers the access and connection to natural resources (Fossgard & Fredman, 2019; Filipe et al., 2018; Naidoo et al., 2011), the quality of

Table 16.1 Summary of the 12 glamping attributes

Glamping attributes	Meaning and scope
Location	The place where the site is
Natural environment	Includes access and connection to natural resources, the quality of the place and its ecological standards, the beauty and appeal of the natural attractions, as well as the sights, scents and sounds of nature
Facilities and amenities	Includes the availability of electricity, water and sanitary facilities including bathrooms, kitchens, television, Wi-Fi services, and clear signages, the physical condition and quality of the site like the designs, decoration, temperature and lighting, as well as the luxury, privacy, comfort, and convenience
Activities offered	Includes sports, leisure, and entertainment activities, as well as outdoor recreation
Price	Refers to rates of accommodation, facilities, and amenities and activities offered
Accessibility	Includes the ability to locate the site, infrastructure of the destination, wayfinding, capacity to accommodate guests with disabilities, and ease of access to off-the-beaten-path attractions
Safety	Includes risk management and security
Service quality	Includes adapting to the needs of the guests, experience facilitation, customer service, reception service as well as the character of the staffs including their courtesy, responsiveness, communication, assurance, empathy, and care
Health protocols	Includes sanitizing efforts, use of protective equipment like masks, limiting the number of customers to be accommodated, employee training on health and safety protocols as well as observance of social distancing
Cleanliness	The cleanliness of the site, rooms, and facilities
Congestion	The number of visitors in a certain area
Local community readiness	The preparedness of the community in taking action to address a specific health issue

the place and its ecological standards (Mikulić et al., 2017), the beauty and appeal of the natural attractions (Wachyuni & Kusumaningrum, 2020; O'Neill et al., 2010), as well as the sights, scents, and sounds of nature (Fossgard & Fredman, 2019).

The facilities and amenities available in the sites are also evaluated by the travelers. Upen (2018) defined facilities as places or equipment that are built to help facilitate the specific needs of the guests. Amenities, on the other hand, are the things that are designed to give guests comfort and enjoyment. The availability of electricity, (Mikulić et al., 2017; O'Neill et al., 2010) water and sanitary facilities including bathrooms (O'Neill et al., 2010; Brochado & Pereira, 2017), kitchens, television, Wi-Fi services (Brochado & Pereira, 2017), and clear signages (Mikulić et al., 2017), the physical condition and quality of the site like the designs, decoration, temperature, and lighting (O'Neill et al., 2010; Realino et al., 2021), as well as the luxury (Filipe et al., 2018; Sakáčová, 2013), privacy (Filipe et al., 2018), comfort and convenience (Brochado & Pereira, 2017; Lee et al., 2019; Sakáčová, 2013) are all classified under the facilities and amenities. Front desk or concierge service, spas, restaurants, and bars may also be available on the site (Dangel et al., 2020).

Tourists also consider the available activities offered in the sites like the sports, leisure, and entertainment activities (Mikulić et al., 2017), as well as outdoor recreation (Craig & Karabas, 2021). Price is also another factor that affects the decision and satisfaction of the consumers (Lee et al., 2019; Naidoo et al., 2011; Sakáčová, 2013) because Filipe et al. (2018) and Lee et al. (2019) considered price to be an obstacle in experiencing glamping. Price is the amount of money that someone must pay to acquire something (Oxford University Press, 2022c).

When access is addressed regarding destinations, it usually relates to how easy it is for visitors to travel to a place. Accessibility refers to the infrastructure of the destination, wayfinding, capacity to accommodate guests with disabilities, and ease of access to off-the-beaten-path attractions (Belton, 2021). Tourists also think about accessibility (Craig & Karabas, 2021) and the ability to locate the site (O'Neill et al., 2010). Safety is a state of controlling the threats and circumstances that may lead to physical, psychological, or material harm to protect the health and well-being of individuals and the community (INSPQ, 2018). Tourists also assess the safety of the destination (Wachyuni & Kusumaningrum, 2020; Mikulić et al., 2017; Yousaf et al., 2018) which includes risk management (Fossgard & Fredman, 2019) and security (Lee et al., 2019).

The service quality also influences how satisfied the travelers are with their experience (O'Neill et al., 2010). Service quality is a product of the effort that every member of the organization invests in satisfying customers (MBA Knowledge Base, 2021). Adapting to the needs of the guests, experience facilitation (Fossgard & Fredman, 2019), customer service (Brochado & Pereira, 2017), reception service (Mikulić et al., 2017) as well as the character of the staffs including their courtesy, responsiveness, communication, assurance, empathy, and care (Mikulić et al., 2017; O'Neill et al., 2010; Naidoo et al., 2011) are all part of the service quality.

The health protocols implemented in the sites are also examined. The sanitizing efforts, use of protective equipment like masks, limiting the number of customers to be accommodated, employee training on health and safety protocols (Gursoy et al., 2020) as well as observance of social distancing (Gursoy et al., 2020; Craig, 2020; Craig & Karabas, 2021) are placed under the health protocols. According to the Oxford University Press (2022a), cleanliness is 'the state of being clean or the habit of keeping things clean.' The cleanliness of the site, rooms, and facilities are also demanded by tourists (Wachyuni & Kusumaningrum, 2020; O'Neill et al., 2010; Mikulić et al., 2017; Lee et al., 2019). Congestion is the amount of visitors in a certain area. During and even before the pandemic, several studies claimed that congestion or crowds affect the experience and preferences of the tourists when traveling (Park et al., 2021; Lee et al., 2019; Craig, 2020; Fossgard & Fredman, 2019; Falk et al., 2021).

Community readiness is the preparedness of the community in taking action to address a specific health issue (Rural Health Information Hub, 2018). This must also be looked into as tourists are confident to travel when destinations manage COVID-19 effectively (European Travel Commission, 2021). World Travel and Tourism Council and Global Rescue (2019) also acknowledged that destinations must invite tourists back when they are ready, recovered, and have rebuilt their confidence. It was also stated that creating policies to facilitate travel will help in enticing tourists to visit a place. The study of Kim et al. (2021) revealed that preparedness of a place in responding to a health crisis and the willingness of the people to address the problem significantly helped in quickly decreasing the cases.

Due to the fear of contracting COVID-19, people consider their health and safety before deciding to travel. This was evident in the framework of Zheng et al. (2021) where the pandemic travel fear may cause protection motivation of the people which may then lead to cautious travel. Hassan and Soliman (2021) also stated that COVID-19 fear arousal affects the perception of holidaymakers with regard to the social responsibility and reputation of the destination as well as the perceived trust of the visitors (Table 16.1).

The term confidence to travel is also a factor that affects the assessment of the different attributes of the site as well as the push factors that motivate tourists to travel. To measure the confidence of the tourists in traveling, the vaccination status, pre-travel COVID-19 tests, effectiveness in COVID-19 management (European Travel Commission, 2021), and financial status (Department of Tourism et al., 2020) were evaluated. Vaccination status refers to the traveler being vaccinated with 1st dose and 2nd dose of COVID-19 vaccine as well as the booster shot. On the other hand, pre-travel COVID-19 tests include RT-PCR and antigen tests for COVID-19. The effectiveness in COVID-19 management includes the availability of treatment, reduced rate of infections, and the absence of new cases in the area (Department of Tourism et al., 2020). The financial status of the tourists was also considered since the pandemic generally reduced the income of the Filipinos (Department of Tourism et al., 2020). Further, tourist travel behavior can be determined on their visit and revisit intention which is defined as the plan of the tourist for future travel in the glamping site.

Using convergent mixed methods design through an online interview and survey form was utilized to assess tourists' perceptions and the case study approach was used to evaluate the three sites. The study was conducted in March 2022 while still on lockdown, hence, online interviews (Zoom) and online surveys (Google Docs) were conducted.

Data Discussion and Analysis

On the supply side, the three glamping sites ranked the prioritization of the glamping attributes. The ranking was based on the level of importance that the sites have given to the attributes (Table 16.2). The table

Table 16.2 Degree of importance of glamping attributes (supply side)

Rank	Your Brother's House Tribal Village	Misty Heights	Camp Bernardino de Lavigan
1	Facilities and amenities	Location	Facilities and amenities
2	Health protocols	Safety	Activities offered
3	Activities offered	Natural environment	Service quality
4	Service quality	Local community readiness	Health protocols
5	Safety	Accessibility	Safety
6	Natural environment	Service quality	Natural environment
7	Cleanliness	Cleanliness	Accessibility
8	Local community readiness	Health protocols	Cleanliness
9	Congestion	Price	Price
10	Accessibility	Facilities and amenities	Location
11	Location	Congestion	Local community readiness
12	Price	Activities offered	Congestion

below shows the ranking of the three sites in which 1 is the most prioritized glamping attribute and 12 is the least prioritized. Both Your Brother's House Tribal Village and Camp Bernardino de Lavigan prioritized facilities and amenities the most while location is the first priority of Misty Heights. The three sites differ with the least prioritized attribute. It was price for Your Brother's House Tribal Village, activities offered for Misty Heights and congestion for Camp Bernardino de Lavigan (Table 16.2).

In all the sites, service quality, safety, and natural environment were in the upper half of prioritized attributes while cleanliness and congestion were in the lower half. Price and congestion did not matter very much mainly due to the lack of visitors in the sites during the pandemic. The data indicates that the supply side is adjusting to the needs of the market.

Figure 16.2 shows the degree of importance of the glamping attributes from the demand side. Using a Likert-type scale, the survey revealed that the top three attributes of glamping tourists were cleanliness (4.9), safety (4.89), and natural environment (4.85) while price (4.62), location (4.36), and activities offered (4.27) were at the bottom three contrary to

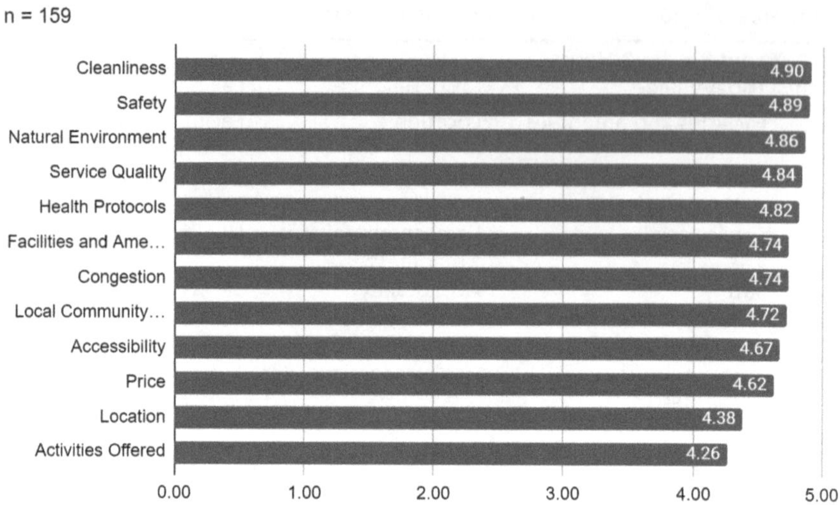

Fig. 16.2 Degree of importance of the glamping attributes of all the respondents (demand side)

pre-pandemic where price, location, and activities have shown to be the top travel motivators.

Travel confidence is influenced by vaccination rate, pre-travel COVID-19 tests, effectiveness in COVID-19 management, and financial resources. Data implies that glamping can be regarded as a viable product for tourism recovery and demand for this special-interest tourism activity has grown with Fig. 16.3 showing an increase in the willingness of people to engage in glamping post-pandemic with only 1.3% of respondents not willing to engage in glamping.

While glamping is relatively new in the market, there was already a demand for it as the majority of the respondents have already glamped or considered glamping before the pandemic. Additionally, almost all the respondents who have visited the sites have revisit intentions. Even at the time of the pandemic, the willingness to engage in glamping plus those that have already glamped in this period increased. There is also a large increase in the willingness to travel during the pandemic to after the pandemic. The data implies that glamping can be regarded as a viable product during and after the pandemic.

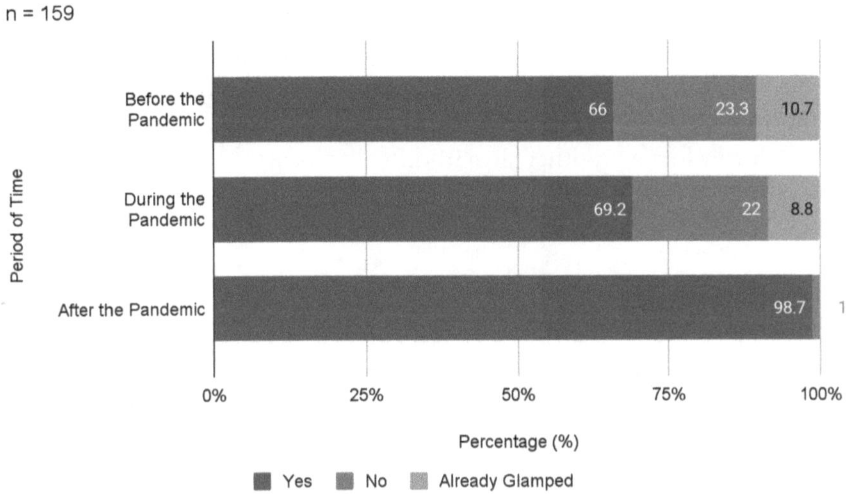

Fig. 16.3 Willingness to engage in glamping before, during, and after the pandemic

Given the findings, business owners need to prioritize safety and cleanliness of its sites to ensure customer satisfaction. The second most important attribute is safety due to the fear of traveling during the pandemic and the expectation for the site to have risk management policies and plans. Natural environment is the third most prioritized attribute which suggests that people prefer to travel to sites with direct contact to nature to de-stress from the tension brought by the pandemic. It is also worth noting that price and activities offered, which are typically mainly considered when traveling, did not rank at the top. This implies a shift in the priorities when traveling toward safety rather than affordability. Glamping sites must incorporate the 12 glamping attributes in their operations as these affect the needs and wants of tourists most especially with regard to health and safety protocols.

Glamping sites need to consider caring for the natural environment through sustainable practices to ensure long-term viability. The sites should also consider access to the elderly and persons with disabilities to ensure inclusivity. Destination Management Organizations and local government units should provide assistance in the sites' business operations and promotions since glamping is a relatively new tourism product. Tourists should observe responsible care for the environment. There needs to be a concerted effort among all stakeholders to ensure that glamping will help redefine sustainable travel in the new normal. Future research can focus on other sites in different locations in the Philippines to examine the long-term viability of glamping.

References

Adamovich, V., & Nadda, V. (2021). Quality issues in glamping tourism from providers perspective in the UK. *International Journal of Entrepreneurship Management Innovation and Development*, *4*(1), 99–115. https://www.researchgate.net/publication/349338034_Quality_Issues_In_Glamping_Tourism_From_Providers_Perspective_In_The_UK

Belton, V. (2021). *Key elements of a destination, part 1: Access.* Stamp. https://stampdestinations.com/good-work/insights/key-elements-of-a-destination-part-1-access/

Boiko, V. (2021, December 20). *Glamping as a promising niche in tourism and hotel business.* DSpace Repository. http://dspace.ksau.kherson.ua/handle/123456789/7492?show=full

Brochado, A., & Pereira, C. (2017). Comfortable experiences in nature accommodation: Perceived service quality in glamping. *Journal of Outdoor Recreation and Tourism, 17*, 77–83. https://doi.org/10.1016/j.jort.2017.01.005

Brooker, E., & Joppe, M. (2013). Trends in camping and outdoor hospitality—An international review. *Journal of Outdoor Recreation and Tourism, 3–4*, 1–6. https://doi.org/10.1016/j.jort.2013.04.005

CBI. (2020, January 8). *The European market potential for nature and ecotourism | CBI.* https://www.cbi.eu/market-information/tourism/nature-tourism/nature-eco-tourism/market-potential

Craig, C. A. (2020). Camping, glamping, and coronavirus in the United States. *Annals of Tourism Research, 89*, 103071. https://doi.org/10.1016/j.annals.2020.103071

Craig, C. A., & Karabas, I. (2021). Glamping after the coronavirus pandemic. *Tourism and Hospitality Research, 21*(2), 251–256. https://doi.org/10.1177/1467358421993864

Dangel, S., LaRocca, M., & Jaeger, J. (2020, February 21). *Sleeping under the stars in style: An overview of glamping*. Boston University. https://www.bu.edu/bhr/2020/02/21/sleeping-under-the-stars-in-style-an-overview-of-glamping/

Department of Tourism, Dr. Andrew L. Tan Center for Tourism & Guide to the Philippines. (2020, June). *The Philippine travel survey report: Insights on Filipino travelers' sentiments on the new normal*. Department of Tourism, Asian Institute of Management, Guide to the Philippines. https://philippines.travel/data/PHTravelSurveyReport2020.pdf

European Travel Commission. (2021, September 22). *Monitoring sentiment for domestic and intra-European travel—Wave 8*. European Travel Commission. https://etc-corporate.org/reports/monitoring-sentiment-for-domestic-and-intra-european-travel-wave-8/

Falk, M., Hagsten, E., & Lin, X. (2021). High regional economic activity repels domestic tourism during summer of pandemic. *Current Issues in Tourism, 25*(8), 1209–1225. https://doi.org/10.1080/13683500.2021.1968805

Fernandes, S., Ferreira, D., Alves, T., & de Sousa, B. M. B. (2021). Glamping and the development of sustainable tourism: A Portuguese case study. In A. Lubowiecki-Vikuk, B. M. B. de Sousa, B. M. Đerčan, & W. Leal Filho (Eds.), *Handbook of sustainable development and leisure services. World sustainability series*. Springer. https://doi.org/10.1007/978-3-030-59820-4_14

Filipe, S., Santos, C. A., & Barbosa, B. (2018). Tourists' motivations and obstacles for choosing glamping: An exploratory study. *CBU International Conference Proceedings, 6*, 113–119. https://doi.org/10.12955/cbup.v6.1142

Fossgard, K., & Fredman, P. (2019). Dimensions in the nature-based experiencescape: An explorative analysis. *Journal of Outdoor Recreation and Tourism, 28*, 100219. https://doi.org/10.1016/j.jort.2019.04.001

Gursoy, D., Chi, C. G., & Chi, O. H. (2020). *Restaurant and hotel customers' sentiment analysis*. The University of West Indies. https://uwi.edu/covid19/sites/covid19/files/Covid-19%20Summary%20Report%20-%20Restaurant%20and%20hotel%20customers%E2%80%99%20sentiment%20analysis.pdf

Hassan, S. B., & Soliman, M. (2021). COVID-19 and repeat visitation: Assessing the role of destination social responsibility, destination reputation, holidaymakers' trust and fear arousal. *Journal of Destination Marketing & Management, 19*, 100495. https://doi.org/10.1016/j.jdmm.2020.100495

Hennessey, S., Yun, D., & MacDonald, R. (2016). Influencing the intentions to visit a destination: The case of potential first-time and repeat visitors. *Travel And Tourism Research Association: Advancing Tourism Research Globally, 26*.

https://scholarworks.umass.edu/cgi/viewcontent.cgi?article=1503&context=ttra

INSPQ. (2018, August 17). *Definition of the concept of safety*. https://www.inspq.qc.ca/en/quebec-collaborating-centre-safety-promotion-and-injury-prevention/definition-concept-safety

Katsikari, C., Hatzithomas, L., Fotiadis, T., & Folinas, D. (2020). Push and pull travel motivation: Segmentation of the Greek market for social Media Marketing in Tourism. *Sustainability, 12*(11), 4770. https://doi.org/10.3390/su12114770

Kim, J. H., An, A. R., Oh, S. J., Oh, J., & Lee, J. K. (2021, March 5). *Emerging COVID-19 success story: South Korea learned the lessons of MERS*. Our World in Data. https://ourworldindata.org/covid-exemplar-south-korea

Lauesen, L. M. (2013). Natural environment. In S. O. Idowu, N. Capaldi, L. Zu, & A. D. Gupta (Eds.), *Encyclopedia of corporate social responsibility*. Springer. https://doi.org/10.1007/978-3-642-28036-8_368

Lee, W. S., Lee, J. K., & Moon, J. (2019). Influential attributes for the selection of luxury camping: A mixed-logit method. *Journal of Hospitality and Tourism Management, 40*, 88–93. https://doi.org/10.1016/j.jhtm.2019.05.004

MBA Knowledge Base. (2021, August 14). *Service quality*. https://www.mbaknol.com/marketing-management/service-quality/

Mikulić, J., Prebežac, D., Šerić, M., & Krešić, D. (2017). Campsite choice and the camping tourism experience: Investigating decisive campsite attributes using relevance-determinance analysis. *Tourism Management, 59*, 226–233. https://doi.org/10.1016/j.tourman.2016.07.020

Morgan, O. (2021, May 27). *Episode #40—Transmission indoors and outdoors*. World Health Organization. https://www.who.int/emergencies/diseases/novel-coronavirus-2019/media-resources/science-in-5/episode-40%2D%2D-transmission-indoors-and-outdoors

Naidoo, P., Munhurrun, P. R., & Seegoolam, P. (2011). An assessment of visitor satisfaction with nature-based tourism attractions. *International Journal of Management and Marketing Research, 4*(1), 87–98. https://www.researchgate.net/publication/228312598_An_Assessment_of_Visitor_Satisfaction_with_Nature-Based_Tourism_Attractions

Nguyen, V. H., Truong, T. X. D., Pham, H. T., Tran, D. T., & Nguyen, P. H. (2021). Travel intention to visit tourism destinations: A perspective of generation Z in Vietnam. *The Journal of Asian Finance, Economics and Business, 8*(2), 1043–1053. https://doi.org/10.13106/JAFEB.2021.VOL8.NO2.1043

O'Neill, M. A., Riscinto-Kozub, K. A., & Van Hyfte, M. (2010). Defining visitor satisfaction in the context of camping oriented nature-based tourism—The driving force of quality! *Journal of Vacation Marketing, 16*(2), 141–156. https://doi.org/10.1177/1356766710364541
Oxford University Press. (2022a). Cleanliness. In *Oxford learner's dictionary at OxfordLearnersDictionaries.com.* Oxford University Press. https://www.oxfordlearnersdictionaries.com/us/definition/english/cleanliness?q=cleanliness
Oxford University Press. (2022b). Location. In *Oxford learner's dictionary at OxfordLearnersDictionaries.com.* Oxford University Press. https://www.oxfordlearnersdictionaries.com/us/definition/english/location
Oxford University Press. (2022c). Price. In *Oxford learner's dictionary at OxfordLearnersDictionaries.com.* Oxford University Press. https://www.oxfordlearnersdictionaries.com/us/definition/english/price_1#:~:text=%2Fpra%C9%AAs%2F,have%20to%20pay%20for%20something
Park, I. J., Kim, J., Kim, S. S., Lee, J. C., & Giroux, M. (2021). Impact of the COVID-19 pandemic on travelers' preference for crowded versus non-crowded options. *Tourism Management, 87*, 104398. https://doi.org/10.1016/j.tourman.2021.104398
Realino, D., Mugiono, & Moko, W. (2021). The effect of customer experiential quality on revisit intention with positive emotion and perceived value as mediation variables. *The International Journal of Social Sciences World, 3*(1), 245–258. https://doi.org/10.5281/zenodo.5064043
Rocamora, J. A. L. (2020, December 3). Boom in nature-based tourism seen under new normal. *Philippine News Agency.* https://www.pna.gov.ph/articles/1123771
Rogerson, C. M., & Rogerson, J. M. (2020). Camping tourism: A review of recent international scholarship. *Geojournal of Tourism and Geosites, 28*(1), 349–359. https://doi.org/10.30892/gtg.28127-474
Rural Health Information Hub. (2018, April 30). *Community readiness model—Rural health promotion and disease prevention toolkit.* RHI Hub. https://www.ruralhealthinfo.org/toolkits/health-promotion/2/program-models/community-readiness#:%7E:text=Community%20readiness%20refers%20to%20how,These%20stages%20are%3A&text=Denial%20or%20resistance%20%E2%80%93%20there%20is,members%20about%20the%20health%20issue
Sakáčová, K. L. (2013). *GLAMPING—Nature served on silver platter.* Aalborg University. https://projekter.aau.dk/projekter/files/76860450/Glamping.pdf

Schutte, C. (2021, October 12). Business travelers optimistic about long-term future for recovery. *Travel Again Project.* https://www.travelagainproject.org/post/business-travelers-optimistic-about-long-term-future-for-recovery

Upen, B. (2018, June 29). Difference between amenities and facilities. *Pediaa.Com.* https://pediaa.com/difference-between-amenities-and-facilities/#:%7E:text=The%20difference%20between%20amenities%20and%20facilities%20is%20that%20the%20amenities,the%20guests%20in%20their%20needs

Van Den Bosch, M. (2017). Natural environments, health, and well-being. *Oxford Research Encyclopedia of Environmental Science.* https://doi.org/10.1093/acrefore/9780199389414.013.333

Wachyuni, S. S., & Kusumaningrum, D. A. (2020). The effect of COVID-19 pandemic: How are the future tourist behavior? *Journal of Education, Society and Behavioural Science,* 67–76. https://doi.org/10.9734/jesbs/2020/v33i430219

Williams, N. L., Nguyen, T. H. H., Del Chiappa, G., Fedeli, G., & Wassler, P. (2021). COVID-19 vaccine confidence and tourism at the early stage of a voluntary mass vaccination campaign: A PMT segmentation analysis. *Current Issues in Tourism, 25*(3), 475–489. https://doi.org/10.1080/13683500.2021.1963216

World Travel & Tourism Council & Global Rescue. (2019). *Crisis readiness.* World Travel & Tourism Council. https://wttc.org/Portals/0/Documents/Reports/2019/Crisis%20Preparedness%20Management%20Recovery-Crisis%20Readiness-Nov%202019.pdf

Yousaf, A., Amin, I., & Santos, J. A. C. (2018). Tourists' motivations to travel: A theoretical perspective on the existing literature. *Tourism and Hospitality Management, 24*(1), 197–211. https://doi.org/10.20867/thm.24.1.8

Zheng, D., Luo, Q., & Ritchie, B. (2021). Afraid to travel after COVID-19? Self-protection, coping and resilience against pandemic 'travel fear'. *Tourism Management, 83,* 104261. https://doi.org/10.1016/j.tourman.2020.104261

Part IV

Socio-cultural Sustainability, Host Perspectives and Tourist Behavior in the New Normal (SDG 1,3,5,8 and 11)

17

Community-Led Sustainable Tourism in the New Normal: Case Study from Purulia, India

Sneha Bhattacharyya

Introduction

Travel and tourism can be truly considered as the only expenditure, which makes one richer. Tourism acquaints one with the natural and cultural assets of different places and contributes in experience enhancement. Travel and tourism industry also contributes immensely to the national and global economy. Tourism is responsible for 9% of global GDP and accounts for 29% of export in services worldwide (UNWTO, 2013). In developing countries, tourism often acts as a significant source of foreign exchange earnings. Given the importance of tourism in social and economic context, it is important to make the industry resilient and responsible so that holistic development can be facilitated through tourism. United Nations has identified tourism as one of the major sectors to drive change towards a green economy and was included in the Rio+ 20 Outcome Document as one of the sectors capable of making

S. Bhattacharyya (✉)
Contact Base, Kolkata, India
e-mail: sneha@banglanatak.com

S. W. Maingi et al. (eds.), *Tourist Behaviour and the New Normal, Volume II*,
https://doi.org/10.1007/978-3-031-45866-8_17

contributions in Sustainable Development Goals (SDGs) and can create jobs and trade opportunities.

The tourism industry has been one of the majorly hit sectors by the COVID-19 pandemic. International tourism decreased by 80% in post-pandemic times (OECD, 2020). According to UNWTO estimates, the impact of COVID-19 crisis on tourism led to a reduction in global GDP by 1.5% to 2.8%, putting 100 million direct tourism jobs at risk (UNWTO, n.d.); 54% of tourism workforce comprising of women and youth workers are among the most at-risk categories. Small businesses accounting to around 80% of global tourism became particularly vulnerable, striving to survive in the new normal. While the pandemic impacted the tourism industry in insurmountable amount, it also offers innovative avenues to revive tourism. The COVID-19 crisis has provided a watershed moment to align efforts and make tourism more resilient, inclusive, carbon-neutral and resource efficient. Given the unprecedented impact of the pandemic on tourism, new and disruptive adaptations and innovations may find a very fertile ground in this crisis.

This chapter is dedicated to demystify how sustainable tourism made its way in India in the new normal. By citing the instance of Purulia district in the state of West Bengal in India, the chapter highlights the efficacy of community-led cultural tourism in changing the fate of the district and bringing grass-root development in post-pandemic times. This chapter is divided into two sections: the first section talks about the shift in tourism landscape and the opportunities post-pandemic times offer in reorienting the sector towards a sustainable future. The latter section of the chapter validates this theoretical proposition by highlighting the case study of Purulia in juxtaposition. Development of community-led cultural tourism in Purulia has not only altered the fate of the district but has served as a major backbone to support the local economy and bringing socio-economic prospects for the inhabitants. Following a qualitative methodology, the research paradigm provides a redeeming pathway as to how community-led cultural tourism has the potential in making tourism resilient and equipped to cater to the changing demands of tourism in the new normal.

Tourism Landscape in Post-Pandemic Times—The Urgency to Shift Towards Responsible and Sustainable Tourism

The COVID-19 crisis has completely altered the tourism landscape globally. Several reports suggest that while international tourism suffered a major hit, domestic tourism is significantly contributing in softening the blow. Preferences for coastal, regional and rural areas are increasing as opposed to city and congested destinations in post-pandemic tourism (OECD, 2020). Kock et al., 2020 identified that only those destinations and services that understand the new equilibrium at which tourism will possibly settle will be able to position themselves well in the post-COVID-19 era, thereby paving the way for competitiveness and further growth. The COVID-19 pandemic has radically transformed the way socio-economic operations take place. Given the drastic change, it becomes mandatory for the tourism industry to align itself in ways suited to adjust and prosper in the new normal.

Assaf et al. (2021) identifies COVID-19 crisis to be catalytic in providing an opportunity to launch new endeavours that can have a transformative impact on tourism research and practice. In terms of consumer behaviour, post-pandemic tourism witnesses growing importance for safety, risk reduction, cleanliness, reliability, threat anticipation, and changing travel patterns with increased liking for less congested destinations. There is also added focus on coming up with innovative ways to develop resilient destinations and the need to develop trust-based relations among the host and guest communities. The transformed expectations of tourism include within its purview optimal usage of digital media and aspects of e-tourism in shaping holiday experience and larger adoption of Artificial Intelligence (AI) technology to enhance interaction with consumers and facilitate virtual touristic experiences. Finally, quality of life and ensuring sustainability feature to be some of the crucial agendas, which become mandatory for post-pandemic tourism to cater to. Importance of creating safe destinations for both host and guest communities needs to balance the interests of holiday makers, tourists and local communities, and promoting sustainable and climate-friendly tourism products can rightly be understood as the need of the hour in the new normal.

UNWTO (n.d.) advocates for cultivating purposive networking, coordination and cooperation among different tourism stakeholders to make the sector resilient in post-pandemic times. Only collective action has the potential to transform tourism, advance the sector's contribution to 2030 agenda and make a shift towards an inclusive, carbon-neutral tourism industry that has the ability to harness innovation, embrace local and communitarian values and create jobs for all. Fostering sustainability has been identified as crucial by the organization to promote green growth and pave the path for a resilient, competitive, resource efficient and carbon-neutral tourism sector. Green investments on protected areas, local community, smart buildings and circular economy along with advance innovation and digitization of tourism have been identified as mandatory prerequisites in transforming the fate of tourism in post-pandemic times.

Following these theoretical advocacies, several practical initiatives have been undertaken globally post-COVID-19 to reignite the tourism sector. Costa Rica has moved its national holidays to Mondays to boost domestic tourism by extending weekends (Babii & Nadeem, 2021). Barbados has introduced 'Welcome Stamp' visa—a one-year residency permit that allows remote employees to live and work from the country. Fiji launched its Blue Lane initiative that allows yachts to berth in its marinas after meeting strict quarantine measures. To woo foreign tourists, the Indian government announced that the first 500,000 e-visas issued until March 2022 will be free of cost, with the e-visa facility extended to 169 countries (Philip, 2021). In general, the global tourism scenario has seen a major shift towards eco-tourism in post-pandemic times. However, these initiatives will fall insufficient in bringing positive impact if the changed ethos of tourism fails to recognize and acknowledge local values in transforming the fate of the sector in the new normal. UNWTO (n.d.) rightly identified making people the nucleus of tourism development, instead of bystanders, as one of the major critical factors contributing in making tourism resilient in post-pandemic times. Community-led responsible tourism is the primary factor in making tourism sustainable in the changed times.

G20 tourism ministers in Diriyah Communique (G20 Research Group, 2020) recognized the pandemic to cause a major shift in the tourism sector, which requires a jump towards responsible and sustainable tourism to counter the crisis. Rio+20 Outcome Document 'The Future We Want' includes sustainable tourism as contributing to green growth (UNWTO, 2013). UNWTO and UNEP (2005) define sustainable tourism as the tourism that takes full account of its current and future economic, social and environmental impacts, addressing the needs of the visitors, the industry, the environment and the host communities. The organizations have identified 12 aims for sustainable tourism, which include: (1) economic viability, (2) local prosperity, (3) employment quality by creating local jobs, (4) promoting social equity, (5) visitor fulfilment, (6) facilitating local control by endowing in local communities' decision-making abilities, (7) community well-being, (8) promoting cultural richness and pluralism, (9) physical integrity, (10) biological diversity, (11) resource efficiency and (12) environmental purity.

Elaboration of sustainable tourism by UNWTO refers to the need for it to: firstly, make optimal use of environmental resources, that constitute a key element in tourism development, maintaining essential ecological processes and helping to conserve natural heritage and biodiversity. Secondly, promoting respect for socio-cultural authenticity of host communities, conserving their tangible and intangible heritage and traditional values and practices and contributing to cultural pluralism by facilitating inter-cultural understanding and tolerance. Thirdly, ensuring long-term economic operations, providing socio-economic benefits to all stakeholders, including stable employment and income-earning opportunities and social services to host communities, and contributing to poverty alleviation by making the host community the nucleus of tourism development. UNWTO's conceptualization of sustainable tourism is premised on the three pillars of promoting social, economic and environmental sustainability and makes tourism directly a contributing sector in furthering the agenda of SDGs. It is only this responsible and sustainable tourism that has the potential to make tourism resilient in the new normal.

The next section of the chapter is dedicated to demystify how the ethos of sustainable tourism by making community the nucleus has been implemented in the Purulia district of West Bengal, India. The case study highlights how the prosper of community-led cultural tourism in Purulia has boosted responsible tourism in the post-pandemic times and positively transformed the socio-economic fate of the local residents. The next section also bears reference to how responsible and sustainable tourism development in Purulia has altered the fate of the district and contributed in making a progress towards attaining several SDGs.

Community-Led Cultural Tourism Development at Charida Village in Purulia Serving as the Backbone in Pandemic and Post-pandemic Times: A Case Study

Purulia is a district in the western-most part of West Bengal in India. The district presents a classic confluence of natural and cultural assets. A scenic district with several dams, Purulia is rich in its cultural offerings. Chau dance, Jhumur folk song and dance, and basketry weaving tradition are some of the traditional cultural practices innate to the district.

Chau is an acrobatic mask dance form. Colourful masks, rhythmic drum beats, powerful acrobatic movements and somersaults (locally known as ulfa) are characteristics of Chau dance as practised in the Purulia district of West Bengal. Apart from Purulia's Chau, there are two other variants found in India—Seraikella Chau and Mayurbhanj Chau, which are practised in the states of Jharkhand and Odisha respectively.

Purulia's Chau dancers enact stories from the epics of Ramayana, Mahabharata, Puranic texts and others. In addition, the dance is being used to disseminate social messages on burning social issues and promoting public schemes. The dancers wear ornate costumes and elaborately crafted masks that portray the character of the dancer. The masks are made by a community of rural mask makers living in Charida village in Baghmundi block of Purulia, who make masks of deities, epic and also of tribal characters. The masks are of different sizes starting from small to large ones. The tradition of making Chau masks started in Charida around 150 years back during the rule of King Madan Mohan Singh Deo of Baghmundi. Chau dance is inscribed in the UNESCO Representative

List of Intangible Cultural Heritage (ICH) of Humanity in 2010 and Chau mask has been certified with Geographical Indication (GI) in 2018.

Even a decade back, the district was popularly identified as a politically disturbed area infused with Maoist insurgencies. Even Chau as a folk tradition was not as vibrant as its current form. During those days, the master dancers formed a team of Chau dancers and while the script of the Chau act was known to the leading dancers, the remaining members of the troupes were ignorant of the storyline to which they were performing. The traditional steps emulating nature were also gradually getting lost and being replaced by acrobatic movements. To revive the folk tradition of Chau, senior dancers organized several workshops and skill transmission sessions to disseminate knowledge to young practitioners on traditional Chau steps. Workshops were also organized to document traditional Chau steps and to fix a uniform terminology for the steps. Chau dance now occupies a central position as an Intangible Cultural Heritage (ICH) not just of the state of West Bengal but of overall India.

Enhanced prospects from practising Chau dance have attracted the youths in the folk tradition, thereby enriching the folk heritage and curbing out migration from the district. Purulia, once a hotspot for Maoist insurgency, has now transformed into a vibrant cultural tourism destination. Providing a conducive learning ecosystem has not only rejuvenated the folk tradition and the practiced locale but has also steered the fate of Chau in unimaginable ways. Once a majorly male-dominated art form, Purulia's Chau in today's times has also attracted female performers. There are around 13 female Chau dance troupes in Purulia, who are competing with their male counterparts and performing all-round the year in different locales.

Methodology

This study is based on cultural tourism development at Charida, the Chau mask makers' village and how the same served as the backbone of the socio-economy in pandemic and post-pandemic times. Instances of cultural tourism development in the village and its impact on local socio-economic conditions have been captured through qualitative interactions. Unstructured qualitative interviews have been conducted with both senior mask makers and their younger counterparts. The interviews have been conducted individually, as well as through Focused Group Discussions. Every respondent was asked to trace the evolution of Chau dance and the art of mask making with time and the impact of cultural tourism development on the same and vice versa. They were also asked to reflect on the impact of COVID-19 and their surviving strategies to overcome the tough times. After conducting the interviews, the experiences of the respondents have been analysed keeping in mind their local context and the findings have been documented descriptively in the essay.

Developing Community-Led Cultural Tourism in Charida: Changing the Tourism Landscape in the New Normal

Our organization, Contact Base, is a social enterprise having special consultative status with UN ECOSOC, advisory status to UNESCO's 2003 Intangible Cultural Heritage Committee and a global partner of UNWTO for promoting responsible tourism. Our flagship initiative, Art For Life (AFL) (https://youtu.be/4GgNhQ77pO4), is a culture-based development model focusing on holistic development of marginalized cultural practitioners by making their traditional cultural practices the premise for inclusion and poverty alleviation, as well as strengthening art form and the process of heritage transmission to the next generation. AFL promotes village, artist, and art together, both individually and collectively, so that each when strengthened can work towards accrediting pride, recognition and visibility to the others. Thus, adhering to AFL methodology not only improves prospects of cultural practitioners and their aesthetic pursuits by enabling safeguarding and transmission of otherwise dying heritage components, but also develops village or the in-situ space as cultural destinations, so that local community members can serve to be the nucleus of local tourism development (Fig. 17.1). This tripartite focus makes AFL a catalytic methodology in paving the path for

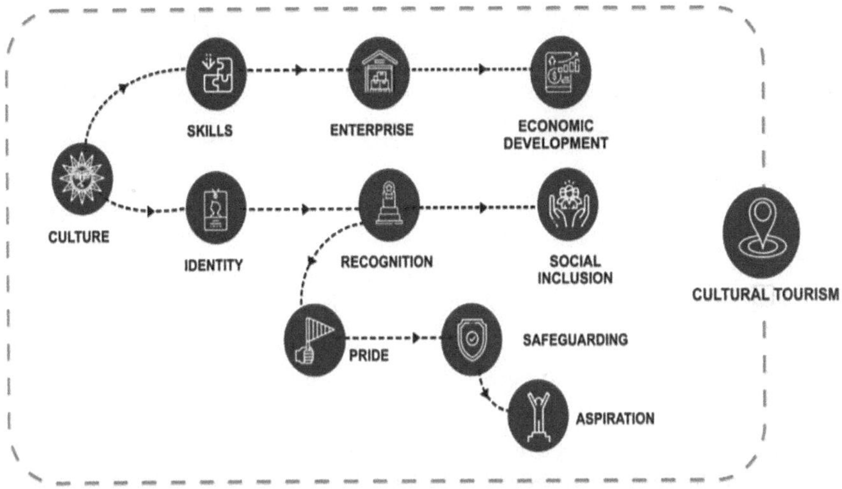

Fig. 17.1 AFL model: Strengthening art, artists and village (The model and the diagrammatic representation have been created by the organization, Contact Base (trading style: banglanatak dot com) (www.banglanatak.com))

inclusive tourism, which makes local communities active stakeholders in the process of tourism development.

Cultural tourism development at Charida has adhered to the AFL methodology. In recent times, the village has grown to be a tourist destination, popularly known as 'the mask makers' village'. The GI tag accredited to Charida's Chau mask has been optimally used in the branding of the destination, keeping its unique cultural offering integral to its tourism offerings. While traditionally the mask makers used to make the masks as crucially supporting props to Chau dance, particularly supplying to the Chau dancers, now the Chau mask has been identified as an independent aesthetic offering and has gained a market value in itself, centring which cultural tourism has prospered over the last decade at Charida. Open shops, adorned with Chau masks of various sizes and characters, line Charida's alleys, which has now grown to be a favoured destination for tourists to witness traditional cultural practices and take back something of authentic handcrafted value.

In pandemic and post-pandemic times, as the literature review above suggests, tourism has taken a different turn. The new normal has witnessed enhanced tourist preferences for rural areas. Charida thus served as the perfect serene setting, which has unique offerings in terms of culture. Our organizations' commitment to develop community-led cultural tourism at Charida premised on the following aspects:

- Safeguarding the heritage of Chau mask making by enabling skill transmission sessions from experienced artists to young practitioners. Coupled with the same, establishing direct market linkage for the mask makers increased the socio-economic prospects associated to mask making. This aspired the youth to pursue the tradition professionally, thereby curbing vulnerable out-migration from the village.
- Capacity building of mask makers to diversify their products by retaining the heritage components of their traditional craft practices. This has enabled them to come up with products of high market value, without compromising on their traditional aesthetic intricacies.
- Conducting exchange and collaboration so that the rural mask makers derive the privilege to travel to different places to present their creative offerings and to interact and co-create with other creative producers. The objective is to provide exposure opportunities to these mask makers so that they can meaningfully contribute in promoting and safeguarding their cultural heritage.
- Developing the village as an in-situ cultural destination. Thus, the intervention strategy has not only focused on directly connecting the mask makers with the market but also bringing the market to their village. Annual village festivals get organized to promote the destination, and celebrate the local cultural heritage. Organizing festivals to celebrate local cultural traditions not only made the art of Chau mask making and Charida visible, it also contributed in initiating a dialogue between folk practices, practitioners and the practiced region and provided exposure opportunities to the inhabitants. It also changed the perception of the neighbouring communities and local governing bodies, who have now started identifying Charida in terms of the practiced handcrafted tradition of mask making.

Community-led cultural tourism development centring traditional practices have immense potential in changing mainstream tourism landscape. Stories of traditional practices, the intricacies of the process and history of these tradition conveyed by the practitioners to the guests have the ability to facilitate experiential tourism in the true sense. Furthermore, community being at the nucleus of tourism development promotes the ethics of responsible tourism and is catalytic in attracting socio-economic benefits for grass-root community members.

It is this community-led cultural tourism that sustained Charida and her inhabitants during pandemic and post-pandemic times. In the new normal, Charida has witnessed significant footfalls, where visitors have come to experience the cultural heritage of the place. Chau mask making has grown to be the central attraction of the tourism offerings. More and more members from the community are showing enhanced enthusiasm to pursue the tradition. While traditionally, women of the household were majorly involved in adorning the masks with beads and jewelleries, now many women are directly related to the art of mask making, where they do everything from moulding the clay, to shaping and designing the mask. During the pandemic, while the entire world was striving to make ends meet, Charida and her residents were welcoming tourists, albeit in small numbers and sustaining itself through community-led cultural tourism. The growth of responsible and experiential tourism is not only catalytic in changing the tenets of mainstream tourism and improving prospects of local community but also significant in addressing several SDGs, thus making an important mark in the larger developmental frame.

Addressing SDGs

Agenda 2030, adopted by the United Nations (UN) in 2015, identifies 17 interconnected SDGs and 169 associated targets to address the most pressing challenges of our times, ranging from the fight against poverty, hunger and climate change to the reduction of inequalities. Inclusive and ambitious, these goals are intended to leave no one behind. Heritage is directly mentioned in only one of the 169 targets (Target 11.4), which aims to 'strengthen efforts to protect and safeguard the world's cultural and natural heritage', which is part of Goal 11: to 'make cities and human settlements inclusive, safe, resilient and sustainable'. While some see this brief mention of culture and heritage in the SDGs as a major victory compared to the complete absence of the topics in earlier international developmental frameworks, it needs to be remembered that the focus mainly lies in safeguarding heritage components. The topics in practical agenda have seldom been linked to addressing broader concerns like how protection of culture and heritage can facilitate poverty alleviation, gender and environmental justice (Labadi, 2022).

Our intervention attempts to further a praxis (theory + practice) based charter showcasing how practical roadmap to safeguard heritage components and community-led cultural tourism development centring on the same can have a positive spillover effect in addressing broader concerns of poverty, gender discrimination, and economic inequality, thereby enhancing community resilience to cope up with tough times. This section discusses at length how community-led cultural tourism development at Charida has not only served as the backbone in sustaining local socio-economy in pandemic and post-pandemic times but has also served as the premise in addressing five major SDG agendas, namely SDG 1 (No Poverty), SDG 5 (Gender Equality), SDG 8 (Decent Work and Economic Growth), SDG 11 (Sustainable Cities and Communities) and SDG 12 (Sustainable Consumption and Production).

SDG 1: No Poverty

Poverty alleviation requires a holistic focus and is not simply restricted to ensuring economic betterment. It includes within its broad purview attributes necessary for bringing overall enhancement of quality of life. While our intervention to facilitate community-led cultural tourism development at Charida has explicitly contributed in income enhancement of the Chau mask makers, it has also paved the path for better living standards within the community. Now, the village is 100% electrified with improved hygiene and sanitation facilities. Education level has also improved, where majority of the community members are now convinced of the importance of education and are sending their children to schools. Health parameters and preventive health awareness have also seen a drastic rise. While initially, over 40% of the community suffered from lingering illness, their general health and awareness has significantly improved now, thereby equipping local communitarian members to cope up in pandemic and post-pandemic times. Our intervention has also been successful in addressing other social maladies like child marriage and drug abuse. Networking and social capital of ICH communities have also significantly contributed in building their resilience and developing coping up mechanisms in them to survive in the new normal.

SDG 5: Gender Equality

Gender equality, fifth agenda of the SDGs, has been addressed and achieved through our intervention. While traditionally, Chau mask making was primarily a male dominated art form, at present, many women from Charida are involved in the art of mask making. They are also breaking gender restrictions and travelling to locales far and wide to promote and exhibit their creative offerings. The women also serve to be the nucleus of local tourism development, which in turn has enhanced their decision-making abilities, freedom and agency.

SDG 8: Decent Work and Economic Growth

Community-led cultural tourism development at Charida has significantly contributed in ensuring decent work and economic growth of overall village community members, not only previously but also in the new normal. Local tourism development has contributed in enhancing visibility and accrediting recognition to the traditional craft practice, their practitioners and the in-situ village locales. Growth of local tourism has also brought economic benefits to the overall village community. The community members, who are not artists, have derived several opportunities to provide tourism services and make it a source of livelihood. Treating creative pursuits as cultural enterprise and the practitioners as cultural entrepreneurs has also enhanced avenues of local employment generation in the new normal premised on the cultural offerings. This highlights the efficacy of our intervention in addressing the major concern of economic inequality, enlisted as the eighth agenda of the SDGs.

SDG 11: Sustainable Cities and Communities

The impact of community-led cultural tourism development at Charida has significantly contributed to enhancing community resilience in the pandemic and post-pandemic times. The community has devised mechanism to collectively work towards practising and promoting their cultural heritage and village locale. In the new normal, they have also started purposive usage of digital and social media to promote and propagate about their creative offerings to a pool of global audience. Enhancement of both physical and virtual purposive exchange within and outside the community of tradition bearers has improved inter- and intra-communitarian linkage and contributed in cultivating and enhancing bridging and bonding social capital, an important parameter to achieve sustainability and resilience (Putnam, 2000). While community-led cultural tourism development has its positives, the commercialization of the otherwise virgin rural locale has also significant risk factors associated to it. Developing communication strategy and training community members on pros and cons of tourism and hospitality and building their awareness on the changed landscape and expectations of tourism in post-pandemic times have equipped them in coming up with strategies to handle the negatives of tourism development and prosper even in tough times.

SDG 12—Sustainable Consumption and Production

Community-led tourism development at Charida has upheld the motto of responsible and sustainable tourism. Optimal usage of local resources by local community to further the cause of experiential tourism has barred undue investments for tourism growth. Charida's tourism is premised on people and experiences and developing strategies to package local cultural offerings in a way that it suits changed tourism needs in the new normal. Usage of local resources to further the cause of tourism and keeping local community as the nucleus of tourism growth is destined to promote the ethics of sustainable consumption and production, enlisted as the twelfth agenda of the SDGs.

Conclusion and Way Forward

The chapter highlights the efficacy of developing community-led cultural tourism in altering the socio-economic fate of the destination as well as the inhabitants. Local tourism development, making community and their creative offerings the nucleus, has the potential to develop collective resilience, endowing coping up mechanisms to deal with difficult times. The research paradigm presented above suggests how local tourism development at Charida served as the backbone of the socio-economy in pandemic and post-pandemic times. It also contributed immensely in changing the landscape of conventional tourism and reflects a shift towards sustainable and responsible tourism. However, it needs to be remembered that while cultural tourism development at Charida has immense positive effects, it is not free from vices. Community resilience through tourism development can only be sustained if the intervention takes into regard certain crucial elements. Firstly, the uniqueness of Charida as a destination lies in its practiced cultural heritage that is Chau mask. Thus, it needs to be remembered that Charida will only cater to a specialized market, not any general market. Hence, to sustain this specialized market, it becomes crucial to safeguard the traditional practice of Chau mask making. Chau is a living heritage and dynamically changes with time. Thus, the art of mask making should not be seen as an aggregate of the past and its safeguarding measures must allow room for diversification and innovation in the creative produce by retaining the heritage components.

Secondly, conducting capacity building training for the hospitality stakeholders in and around Charida like the hoteliers to make them understand the need to shift towards sustainable tourism practices like waste management, optimal utilization of space and hard infrastructure and avoiding unnecessary constructions, which have adverse environmental effects, can be identified as the need of the hour. Thirdly, promoting green transportation and facilitating green investment for tourism development is crucial. Fourthly, to generate sustainable impact, there is immense need to develop entrepreneurial capacities of local community members so that the cause of tourism can get furthered through local self-sustaining ventures. Finally, the GI accreditation of Charida's Chau mask needs to be included in the destination branding strategy so that cultural tourism development is premised on the unique offerings of the destination. Community-led cultural tourism development by keeping into account the above-mentioned aspects has the potential to build community resilience and endow in them coping up abilities to deal with difficult times. The above research material thus shows practical evidence as to how tourist behaviour shifted in the new normal and gave birth to cultural tourism destinations like Charida, both of which symbiotically nurtured each other and how in the background safeguarding of community cultural practices acted as a catalytic force in achieving collective resilience through tourism practices.

References

Assaf, G., Kock, F., & Tsionas, M. (2021). Tourism during and after COVID-19: An expert-informed agenda for future research. *Journal of Travel Research, 61*(2). https://doi.org/10.1177/00472875211017237

Babii, A., & Nadeem, S. (2021). *Tourism in post pandemic world.* International Monetary Fund. Retrieved April 3, 2023, from https://www.imf.org/en/News/Articles/2021/02/24/na022521-how-to-save-travel-and-tourism-in-a-post-pandemic-world

G20 Research Group. (2020). *G20 tourism ministers meeting: Diriyah communiqué.* University of Toronto. Retrieved April 3, 2023, from http://www.g20.utoronto.ca/2020/2020-g20-tourism-1007.html

Kock, F., Nørfelt, A., Josiassen, A., Assaf, A. G., & Tsionas, M. G. (2020). Understanding the COVID-19 tourist psyche: The evolutionary tourism paradigm. *Annals of Tourism Research, 85, 103053.*

Labadi, S. (2022). *Rethinking heritage for sustainable development: International frameworks, LocalImpacts.* UCL Press.

OECD. (2020). *Mitigating the impact of COVID-19 on tourism and supporting recovery*. OECD. ISSN: 23071672 (online). https://doi.org/10.1787/23071672

Philip, L. (2021, October 31). Tourism looks to spread its wings post pandemic upheaval. *Economic Times*. Retrieved April 3, from https://economictimes.indiatimes.com/industry/services/travel/tourism-looks-to-spread-its-winds-post-pandemic-upheaval/articleshow/87413808.cms?from=mdr

Putnam, R. D. (2000). *Bowling alone: The collapse and revival of American community*. Simon & Schuster.

United Nations World Tourism Organization (UNWTO). (2013). *Sustainable tourism for development guidebook: Enhancing capacities for sustainable tourism for development in developing countries*. UNWTO. Retrieved April 3, 2023 from https://www.e-unwto.org/doi/pdf/10.18111/9789284415496

UNWTO. (n.d.). *Tourism and covid-19-unprecedented economic impacts (secretary-general's policy brief on tourism and COVID-19)*. Retrieved April 3, 2023, from https://www.unwto.org/tourism-and-covid-19-unprecedented-economic-impacts

UNWTO and UNEP. (2005). *Making tourism more sustainable: A guide for policy makers*. UNWTO, Madridand UNEP.

18

Ecotourism and Grand Challenges: A Gen-Z Perspective in the Context of Lebanon

Alain Daou, Leila El Zeenni, Anna Hourani, and Salma N. Talhouk

Introduction

Ecotourism refers to "responsible travel to natural areas that conserves the environment, sustains the well-being of the local people, and involves interpretation and education" (TIES, 2015). This type of responsible tourism aiming to minimize the negative impacts on the environment is

A. Daou (✉)
School of Business, American University of Beirut – Mediterraneo, Pafos, Cyprus
e-mail: daou@aubmed.ac.cy; ad73@aub.edu.lb

L. El Zeenni • A. Hourani
American University of Beirut – Nature Conservation Center (AUB-NCC), Beirut, Lebanon
e-mail: lse08@mail.aub.edu; ah267@aub.edu.lb

S. N. Talhouk
Department of Landscape Design and Ecosystem Management (LDEM), American University of Beirut, Beirut, Lebanon
e-mail: ntsalma@aub.edu.lb

S. W. Maingi et al. (eds.), *Tourist Behaviour and the New Normal, Volume II*,
https://doi.org/10.1007/978-3-031-45866-8_18

growing in popularity due to its benefits for local communities and tourists (Xu et al., 2022). Ecotourism is also seen as a way to further learn about different cultures and support local economies, leading to more people wanting to experience it (Ibnou-Laaroussi et al., 2020). Despite the increasing popularity of this type of tourism, it still faces many challenges as some critics believe that current ecotourism policies neither benefit environmental conservation nor local communities (Banerjee, 2010). For instance, the increase in the number of tourists could cause environmental degradation when there is a lack of environmental consciousness among them and a lack of proper management in ecotourism sites. Negative consequences that could happen to local communities include getting labor from the urban sector instead of focusing on training less-skilled locals, as well as inequitable distribution of income among locals which results in tensions. The introduction of ecotourism in some cases also requires enforced displacement to create national parks, which leads to large-scale loss of land and homelessness for local communities. In addition, the concept of "ecotourism" remains poorly understood and could become much abused and simply used as a "buzzword" to attract customers (Das & Chatterjee, 2015).

The literature has shown the importance of focusing on the new generation aged between 18 and 25, or Gen-Z, regarding the ecotourism sector. This generation is more conscious about sustainability concepts and inclined toward pro-environmental behaviors (Budac, 2014; Šaparnienė et al., 2022). Gen-Z is a digital generation focused on social media, which makes it better exposed to environmental issues worldwide, eco-conscious, and more inclined to act to save its future (Wu et al., 2020). Hence, tackling the growing ecotourism sector from a Gen-Z perspective is crucial. In addition, with the rise of extreme events around the globe (Bertello et al., 2022), it is essential to study how environmental and social challenges impact Gen-Z's awareness and practices regarding ecotourism.

This chapter aims to understand Gen-Z's perspective on the environment in general and ecotourism in particular in the case of countries undergoing grand challenges, taking Lebanon as a case study and using a mixed-methods approach. The findings revealed that Lebanese Gen-Z has high pro-environmental awareness that is not always translated into pro-environmental behaviors. The study also shows that while Lebanese youth's ecotourism awareness and practices have improved during the COVID-19 pandemic, this progress got hampered by the severe socio-economic crisis that hit Lebanon. These findings show the direct connection between a country's context and the youth's general awareness and practice regarding the environment and ecotourism.

Literature Review

Generation Z and Sustainability

The generational theory explains that the era in which a person was born affects the development of their view of the world. People from the same generation are more likely to have similar underlying value systems, regardless of their country or community of birth, since they face similar issues, are impacted by the same events, and share similar experiences. These value systems drive behavior and attitudes and are good predictors of behavior and expectations (Codrington, 2012). Generational cohorts include baby boomers, generation X, generation Y, generation Z or Gen-Z (year of birth between 1995 and 2009), and generation alpha (Bulut et al., 2017).

According to Goh and Okumus (2020), Gen-Zers are a pivotal emerging workforce, taking up over 20% of total jobs. This generation is more oriented toward sustainability than other generations. For instance, research has reported that younger employees embrace green and sustainable practices and prefer to work in a hotel that adopts sustainable business decisions and culture, such as company initiatives on minimizing food wastage and increased social responsibility. Also, Budac (2014) specified that Gen-Z consumers consider the environmental impact and carbon footprint of products; hence, recycling and buy-back programs can influence this generation's consumers. Among the present generation of consumers, Gen-Z members are the most interested in incorporating sustainability into their activities. They are interested in sustainable development and social responsibility and tend to participate in environmental protection activities. These values have prompted Gen-Zers to adopt sustainable behavior while sticking to sustainability principles. For instance, Šaparnienė et al. (2022) mention that positive attitudes toward sustainable tourism, such as ecotourism, are highly accepted by Gen-Z.

Grand Challenges

A grand challenge is a "specific critical barrier that if removed would help solve an important societal problem with a high likelihood of global impact through widespread implementation" (George et al., 2016). According to Ferraro et al. (2015), there is an emerging interest in these large unresolved problems, which mainly affect large populations, with their impact extending beyond the boundaries of a single community while adversely affecting human welfare and well-being at the same time. By nature, these challenges are complex, uncertain, and evaluative.

Some examples of grand challenges include climate change, water scarcity, poverty alleviation, and the safeguarding of human rights (George et al., 2016). Another prominent example of a grand challenge was the financial crisis of 2008 which was avoidable (Meléndez Hernández & Ladrón de Guevara Cortés, 2011). Finally, the most recent grand challenge is, according to Howard-Grenville (2021), the COVID-19 pandemic, classified as a global problem scientists have been warning about for decades. Bertello et al. (2022) also mention that the COVID-19 pandemic is a grand challenge since its scale is global. Finding a solution requires collective and collaborative efforts from various actors to enable scientific advancement and progress.

From Mass Tourism to Ecotourism

Mass tourism is a term used to describe the movement of a large number of organized tourists to popular holiday destinations for recreational purposes. Mass tourists enjoy the convenience of not having to make travel arrangements and not being obliged to learn a foreign language to buy goods and enjoy services in their destination, thus abandoning genuine, authentic local cultural contact (Ei & Karamanis, 2017). Mass tourism has mainly economic benefits such as job creation, increased income, and foreign exchange earnings. However, its sociocultural and environmental consequences outweigh the benefits. These include the loss of culture, staged experiences, overexploitation of cultural resources for tourism purposes, overcrowding, air and water pollution, depletion and degradation of natural resources, and litter.

Between the 1970s and early 1980s, new forms of tourism, such as sustainable tourism, emerged as an alternative to traditional tourism (Ei & Karamanis, 2017). Sustainable tourism forms are numerous, one of them being ecotourism. Industry consensus is that ecotourism focuses on ecological conservation and educating travelers on local environments and natural surroundings. In contrast, sustainable tourism focuses on travel that has minimal impact on the environment and local communities. The ecotourism spectrum lies between a low human responsibility pole, where the behavior of tourists is mainly passive, and a high human responsibility pole where the behavior of tourists is primarily active, as shown in Fig. 18.1.

The evolving ecotourism definition includes concepts such as preservation/conservation, distribution of benefits, and ethics/responsibility (Donohoe & Needham, 2006).

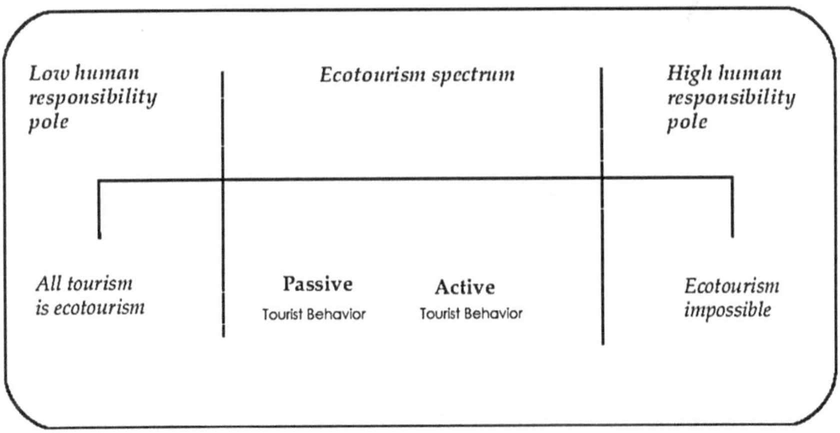

Fig. 18.1 The ecotourism spectrum (Diamantis, 1999)

Grand Challenges and Ecotourism

There is little research on ecotourism and grand challenges with emerging literature tackling the COVID-19 pandemic as the primary focus (Gössling et al., 2020). An exception is Scott (2021) who looks at sustainable tourism and the grand challenge of climate change, advocating for a sector awakening and transition. The ecotourism industry faced severe challenges due to the COVID-19 pandemic after the World Health Organization issued recommendations against traveling and public space crowding. One of these challenges was the significant economic damage, ranging from microeconomic to wide-ranging effects (Hosseini et al., 2021). Also, demand for ecotourism dropped as this activity is considered non-essential in the vast majority of countries, according to Fernández-Bedoya et al. (2021), which has hurt the economy of the villages that receive large incomes from ecotourism due to the closure of borders and access routes. In Morocco, the pandemic has motivated local people to engage in illegal activities that threaten the environment, such as poaching, wildlife trafficking, and forest logging, which resumed due to the country's lockdown and ecotourism collapse (Cherkaoui et al., 2020). These practices were also seen in other parts of Africa, more precisely in Zimbabwe, where a study by Mudzengi et al. (2022) showed that the overall impacts of the pandemic on Mahenye ecotourism elements have been negative. These include increased poaching and reduced ecotourism visitation that caused a severe economic downturn due to decreased income in entrance and lodge fees, selling of crafts, and performing cultural dances.

The Case of Lebanon

Lebanon is a small Middle Eastern country stretching along the Mediterranean Sea's eastern shore. Tourism has traditionally been one of Lebanon's leading economic sectors, representing a major source of income and employment. According to the Investment Development Authority of Lebanon (IDAL), the sector's total contribution was estimated at USD 10.4 billion or 19.1% of GDP in 2018, the second-highest ratio in the Middle East. The tourism sector is also one of the largest employers in the country, as it generated 144,300 direct jobs (or 6.7% of total employment) in 2018. In addition, this sector is one of the fastest-growing sectors in Lebanon, with an average annual growth rate of 10% since 2014 (IDAL, 2022).

In 2021, The United Nations Office for the Coordination of Humanitarian Affairs (UN-OCHA) stated that Lebanon is in survival mode. It published a timeline of the crises that hit the country one after the other, starting with the October 2019 uprising, where people gathered across Lebanon to protest against corruption and poor economy. This uprising which lasted several months was then affected by the emergence of the first reported cases of COVID-19 in February 2020. A few months later, on the 4th of August 2020, one of the biggest non-nuclear explosions in history took place in the port of Beirut, killing over 200 people and injuring more than 6000 (ReliefWeb, 2021). Lebanon was simultaneously going through an economic and financial crisis where the country witnessed a massive collapse in its currency. Lebanese people had their money trapped in the banks, with their minimum wage becoming one of the lowest in the world, with little more than 2$ per day. This pushed the World Bank to say in June 2021 that Lebanon is experiencing one of the most severe economic crises globally since the mid-nineteenth century (ReliefWeb, 2021). In July 2021, the UN estimated that 77% of households now do not have enough food or enough money to buy food, with the cost of food soaring by 700% over the past two years. The UN also announced that the water supply in Lebanon was on the verge of collapse, with over 71% of the population at risk of losing access to water (ReliefWeb, 2021).

With all these consecutive crises hitting Lebanon, the country context presents a series of grand challenges. However, despite the many collapsing sectors, tourism remains a resilient (Daou et al., 2019) sector and a scarce source of growth, where tourist arrivals surged by 101.2% over the first seven months of 2021 (World Bank Group, 2022).

In sum, the literature informs us that Gen-Z is more environmentally conscious than previous generations due to several factors. However, does environmental awareness necessarily translate into pro-environmental behaviors? On the other hand, the literature shows that the COVID-19 pandemic had a negative impact on the ecotourism industry, but can this situation be generalized? And finally, what effect does a grand challenge context have on Gen-Z's perspective of the environment in general and ecotourism in specific?

Methodology

Data Collection Tools and Target Population

This research used a mixed-methods approach combining quantitative and qualitative methods. The quantitative part consisted of an online structured and standardized questionnaire based on the Knowledge-Attitudes-Practices (KAP) model complemented by a socio-demographic data section (KAP, 2014). The questionnaire aimed to assess the knowledge, perceptions, and habits of university students aged between 18 and 25. The knowledge part consisted of questions such as *Do you believe natural resource protection and tourism can go well together? Do you believe the protection of local heritage/culture and tourism can go well together?* The attitude part was inspired by the attitude toward the ecotourism scale developed by Cini et al. (2015), measured on a 5-point Likert scale. For the practices section, participants had to answer several close-ended questions such as *What is your favorite type of tourism? How do you hear about ecotourism activities in Lebanon? How often do you take an ecotourism trip in Lebanon?* They had to choose one or several answers from a provided list of options.

To better understand the above-mentioned results, we conducted 20 qualitative semi-structured interviews targeting university students aged between 18 and 25. The interviews were divided into four parts: socio-demographic information, knowledge concerning ecotourism, the evolution of holiday habits before and after the Lebanese crises, and the relationship of Gen-Z with sustainability concepts (Table 18.1).

Table 18.1 Interviewee characteristics

Interviewee	Gender	Age	Nationality	Major
INT-1	Male	20	Lebanese	Computer and communication engineering
INT-2	Female	21	Palestinian	Graphic design
INT-3	Female	18	Lebanese	Biology
INT-4	Male	18	Lebanese	Biology
INT-5	Male	19	Lebanese	Medical lab
INT-6	Male	19	Lebanese	Computer science and engineering
INT-7	Female	18	Palestinian	Computer and communication engineering
INT-8	Female	19	Lebanese	Mechanical engineering
INT-9	Female	19	Palestinian	Chemical engineering
INT-10	Male	19	Lebanese	Mechanical engineering
INT-11	Female	19	Lebanese	Computer and communication engineering
INT-12	Female	20	Lebanese	Computer science and engineering
INT-13	Male	22	Lebanese	Physics
INT-14	Female	20	Lebanese	Archeology
INT-15	Male	18	Lebanese	Industrial engineering
INT-16	Female	19	Lebanese	Medical lab
INT-17	Female	19	Lebanese	Economics
INT-18	Female	18	Lebanese	Business
INT-19	Female	19	Lebanese	Business
INT-20	Male	18	Lebanese	Business

Data Analysis

The online questionnaire's results were analyzed through descriptive statistics, where the results of each question were summarized in different charts showing percentages for every selected answer. The semi-structured interviews were, for their part, transcribed verbatim, and a thematic analysis was performed to extract the main themes from the given answers.

Results

Survey Results

Of the 208 respondents who participated in the online questionnaire, 67.3% were females, while 32.7% were males. All participants were university students aged between 18 and 25, with the majority being full-time students (86.1%). The majority of the respondents (85.6%) were Lebanese with 14.4% being from other nationalities, mainly Syrian or Palestinian.

- *Knowledge and Attitudes Toward Ecotourism*

Half the participants (54.3%) indicated that they understood the concept of ecotourism. A similar percentage of respondents (53.8%) believed there is demand for such tourism in and around Lebanon. They were more inclined toward "agreeing" with the positive ecotourism attitudes which included statements such as "it allows adventurous activities" and "it gives the possibility to experience silence and rest." On the other hand, most respondents "disagreed" when presented with negative ecotourism attitudes which included "ecotourism is mainly boring and includes activities that are tiring," "it means loneliness," "ecotourism and fun are contrasting realities," "it implies abandoning comforts and services," and "it can damage a fragile natural environment."

- *Ecotourism Practices*

Respondents indicated that they learn or hear about ecotourism mainly through social media (70.7%), and through word of mouth (63.5%). Regarding the frequency of their ecotourism trips in Lebanon, the majority chose the "once a year" option with 28.4%, the second highest option chosen was the "once every three months" with 26%, and 19.7% of participants reported never going on an ecotourism trip. The duration of such trips was mainly a full day (58.1%) with no accommodation (40.3%), while the others who spent more than a day (16.8%) chose guesthouses as their preferred accommodation (28.9%). Regarding how they organize such trips, almost half (48.7%) ask friends and family, while 19.7% hire a guide, a tour operator, or go with professional groups and organizations. Finally, the most selected activity practiced during the ecotourism trips is "hiking, trekking, snowshoeing" (78%), followed by "going on tours" (44%), "bird/nature watching" (38.4%), and "cycling/kayaking" (34.6%).

Thematic Analysis

Two themes emerged from the thematic analysis performed. The first theme sheds light on Lebanese Gen-Z's high environmental awareness, which needs to be better translated into pro-environmental practices. The second theme tackles the hindered positive impact of COVID-19 on Gen-Z's domestic ecotourism awareness and practice.

- *Disconnect Between Knowledge and Action*

When it comes to awareness, many interviewed students believed that the technology-centered life of Gen-Z had a positive effect on making them more conscious about sustainability concepts, as social media for instance could help shed light on environmental issues happening around the world:

> *I really believe that the new generation is more aware of sustainability concepts than previous generations. You know, when the new businesses started minimizing the use of plastic, along with social media awareness campaigns about climate change...this started to happen in our generation. And you know the new generation is strongly attached to their phones so this might have positively affected sustainability awareness within the new generation.* (INT-7)

Access to quality education was another aspect that helped improve Gen-Z's environmental consciousness according to the students, as their parents did not benefit from such an advanced educational level:

> *I believe that the new generation is more aware of sustainability concepts because of their improved education that emphasizes taking care of the environment, not polluting, using alternative energy sources, etc.* (INT-4)

Students mentioned that the increased environmental awareness of Gen-Z is caused by the fact that their generation will undergo the effects of all the environmental damage that could harm our planet in the near future:

We as a new generation are more aware than our parents when it comes to sustainability concepts and more interested. Because we are the ones that will see and undergo the effects of pollution in the near future. (INT-9)

However, the interviewees also focused on how Gen-Z is not making enough effort to implement and translate this gained environmental knowledge into practice in their everyday life:

I believe that our generation has more knowledge when it comes to sustainability concepts, we are more aware, but I don't think we are making enough effort. (INT-18)

From my experience, the new generation is much more aware than their parents when it comes to sustainability concepts. We pay more attention to such things. But it is still not enough. We use normalized things, but that pollute which is bad. (INT-8)

- *A Hindered Positive Impact*

Most interviewed students thought there was a link between the COVID-19 pandemic and their domestic ecotourism awareness. The pandemic had a positive effect on local ecotourism consciousness as people (in particular Gen-Z) were subject to many videos on social media during lockdowns that shed light on the beauty of the Lebanese natural landscapes:

In general, I believe that the pandemic made people more aware of the beauty of the natural destinations in Lebanon, mainly because of social media. (INT-18)

Also, they felt a need to seek nature as a means to reconnect with the world:

In general, I believe that COVID-19 positively affected ecotourism awareness as people needed to leave the city during the pandemic, so they started focusing more on activities that happen in the wild to feel that they are still connected to the world. (INT-9)

A pattern can also be observed from the collected answers whereby before 2019, the year when the economic crisis started to become evident, many of the interviewees used to spend their holidays abroad, while after the beginning of the crisis, they started discovering domestic destinations instead:

> *Before the crisis, I used to travel more abroad; after the crisis, I started discovering my own country, and tourism became more domestic.* (INT-17)

In addition, some students mentioned that the type of activities they henceforth focus on are simpler and nature-related:

> *Due to the economic crisis, it is now hard to go abroad as we used to before the Revolution so before 2019-2018. We focus more on activities in Lebanon because it is less costly, so we focus more on local tourism inside Lebanon. I believe that people after the crisis are more into doing simple outdoor activities like planting and farming; they are more into nature, playing games in nature, doing sports in nature rather than doing expensive activities.* (INT-5)

These findings align with the results of the online questionnaire where most of Gen-Zers confirmed that they were familiar with ecotourism, with positive attitudes toward it. However, they also stated that this new local ecotourism awareness in Gen-Z was consecutively hindered by the severe economic crisis that hit Lebanon after the COVID-19 lockdown period, in that the large increase in gas prices notably decreased their domestic trips within the country in general, and ecotourism trips more specifically:

> *We used to go hiking, especially during the COVID-19 pandemic. However, when gas prices started to rise, the trips to the mountain started to drastically decrease until it only reached three times a year.* (INT-1)

Discussion and Conclusion

This chapter sheds light on Gen-Z's perspective regarding environmental and ecotourism awareness and practice in a context riddled with consecutive grand challenges. First, this study aligns with Gen-Z's environmental awareness literature (Budac, 2014; Šaparnienė et al., 2022). The students interviewed in our research were highly concerned about the environment, primarily due to their increased pro-environmental awareness. However, it is essential to note that the same respondents admitted that this consciousness is not translated into pro-environmental behaviors or actions. This can be explained by the Value-Action gap concept, which occurs when a person's actions do not align with their previous intentions regarding a particular subject. A study by Kollmuss and Agyeman (2002) discussed barriers to pro-environmental behaviors, which include interlinked external factors (such as economic factors intertwined with social, infrastructural, and psychological factors) and internal factors (such as motivation), which are said to be very complex to identify and analyze in general. Therefore, we can say that in a country where the youth is undergoing consecutive extreme conditions, such as in our case study, these barriers may become even more salient, explaining the poor pro-environment behavior in Lebanese Gen-Z.

Second, this study provides a more nuanced perspective regarding what the literature had to say about grand challenges and their impact on ecotourism. In particular, we look at the effects of multiple grand challenges on the ecotourism industry. The COVID-19 pandemic considered a more recent grand challenge (Howard-Grenville, 2021), negatively affected worldwide ecotourism (Hosseini et al., 2021). Our research shows that this is not necessarily the case in Lebanon, where the pandemic led Gen-Z to become more familiar with ecotourism and to resort particularly to domestic ecotourism. According to the interviewed youth, they needed to seek nature to reconnect with the world during confinement. The 2019 economic crisis also increased domestic ecotourism practice among Lebanese people, as the reduced capacity to travel abroad during holidays made domestic ecotourism an active and effective alternative. However, the 2019 economic crisis, which caused a sudden increase in gas prices and devaluation of the local currency, hindered the increased domestic ecotourism practice in Gen-Z. According to Sarraf (2019), generations get affected differently depending on their context. Therefore, it is important to look at intra-generational differences when conducting such research, as labeling people according to the year they were born puts them in a box, preventing us from seeing them as distinctive individuals, which would, as a consequence, increase the stereotyping phenomenon (Jauregui et al., 2020).

This research shows that although Lebanese Gen-Z has a good general environmental awareness, the latter is not fully translated into pro-environmental behaviors. More specifically and focusing on ecotourism, it was seen that Lebanese youth had positive knowledge and attitudes toward this type of tourism. In addition, domestic ecotourism practices by Gen-Z in Lebanon witnessed an improvement, especially during the COVID-19 pandemic. It was then slowed down due to the socioeconomic crisis that hit the country. For that reason, and since such external factors cannot be easily ruled out, we propose to focus on adapting domestic ecotourism packages in a way that offers value for money. This can be done by developing offers that cater to the ambitions and needs of this generation. At the same time, it is important to work on youth motivation, by highlighting the numerous benefits of ecotourism (whether on oneself or the community) through promotion campaigns. These can be conducted in educational contexts such as schools and universities, or other institutions such as non-governmental organizations or local municipalities.

Acknowledgments The authors would like to thank Dr. Maia Sievering from the Faculty of Health Sciences and the S. Olayan School of Business (OSB) Impact fund for their support.

References

Banerjee, A. (2010). Tourism in protected areas: Worsening prospects for tigers? *Economic and Political Weekly*, 27–29.

Bertello, A., Bogers, M. L., & De Bernardi, P. (2022). Open innovation in the face of the COVID-19 grand challenge: Insights from the pan-European hackathon 'EUvsVirus'. *R&D Management, 52*(2), 178–192.

Budac, A.-C. (2014). Strategic considerations on how brands should deal with generation Z. *Revista Economică, 66*(5).

Bulut, Z. A., Kökalan Çımrin, F., & Doğan, O. (2017). Gender, generation and sustainable consumption: Exploring the behaviour of consumers from Izmir, Turkey. *International Journal of Consumer Studies, 41*(6), 597–604.

Cherkaoui, S., Boukherouk, M., Lakhal, T., Aghzar, A., & El Youssfi, L. (2020). Conservation amid COVID-19 pandemic: Ecotourism collapse threatens communities and wildlife in Morocco. In E3S *Web of Conferences* (Vol. 183, p. 01003). EDP Sciences.

Cini, F., Van der Merwe, P., & Saayman, M. (2015). Tourism students' knowledge and tenets towards ecotourism. *Journal of Teaching in Travel & Tourism, 15*(1), 74–91.

Codrington, G. (2012). *Mind the gap: Own your past, know your generation, choose your future*. Penguin Random House South Africa.

Daou, A., Joseph, J., Yousif, D. S., Fathallah, R., & Reyes, G. (2019). Intellectual capital and resilience in torn societies. *Journal of Intellectual Capital.*

Das, M., & Chatterjee, B. (2015). Ecotourism: A panacea or a predicament? *Tourism Management Perspectives, 14*, 3–16.

Diamantis, D. (1999). The concept of ecotourism: Evolution and trends. *Current Issues in Tourism, 2*(2–3), 93–122.

Donohoe, H. M., & Needham, R. D. (2006). Ecotourism: The evolving contemporary definition. *Journal of Ecotourism, 5*(3), 192–210.

Ei, T., & Karamanis, K. (2017). The evolution of alternative forms of tourism: A theoretical background. *Business & Entrepreneurship Journal, 6*(1), 1–4.

Fernández-Bedoya, V. H., Meneses-La-Riva, M. E., & Suyo-Vega, J. A. (2021). Ecotourism in times of Covid-19: A systematic review from the five continents on how this activity is being carried out and what proposals they have for the near future. *Academic Journal of Interdisciplinary Studies*, 1–10.

Ferraro, F., Etzion, D., & Gehman, J. (2015). Tackling grand challenges pragmatically: Robust action revisited. *Organization Studies, 36*(3), 363–390.

George, G., Howard-Grenville, J., Joshi, A., & Tihanyi, L. (2016). Understanding and tackling societal grand challenges through management research. *Academy of Management Journal, 59*(6), 1880–1895.

Goh, E., & Okumus, F. (2020). Avoiding the hospitality workforce bubble: Strategies to attract and retain generation Z talent in the hospitality workforce. *Tourism Management Perspectives, 33*, 100603.

Gössling, S., Scott, D., & Hall, C. M. (2020). Pandemics, tourism and global change: A rapid assessment of COVID-19. *Journal of Sustainable Tourism, 29*(1), 1–20.
Hosseini, S. M., Paydar, M. M., Alizadeh, M., & Triki, C. (2021). Ecotourism supply chain during the COVID-19 pandemic: A real case study. *Applied Soft Computing, 113*, 107919.
Howard-Grenville, J. (2021). Grand challenges, Covid-19 and the future of organizational scholarship. *Journal of Management Studies, 58*(1), 254.
Ibnou-Laaroussi, S., Rjoub, H., & Wong, W. K. (2020). Sustainability of green tourism among international tourists and its influence on the achievement of green environment: Evidence from North Cyprus. *Sustainability, 12*(14), 5698.
IDAL. (2022). Sectors in focus—Tourism. *IDAL.COM*. https://investinlebanon.gov.lb/en/sectors_in_focus/tourism
Jauregui, J., Watsjold, B., Welsh, L., Ilgen, J. S., & Robins, L. (2020). Generational 'othering': The myth of the millennial learner. *Medical Education, 54*(1), 60–65.
Kollmuss, A., & Agyeman, J. (2002). Mind the gap: Why do people act environmentally and what are the barriers to pro-environmental behavior? *Environmental Education Research, 8*(3), 239–260.
Meléndez Hernández, F., & Ladrón de Guevara Cortés, R. (2011). Analysis of the financial crisis inquiry report.
Mudzengi, B. K., Gandiwa, E., Muboko, N., & Mutanga, C. N. (2022). Innovative community ecotourism coping and recovery strategies to COVID-19 pandemic shocks: The case of Mahenye. *Development Southern Africa, 39*(1), 68–83.
ReliefWeb. (2021). A timeline of crisis in Lebanon. https://reliefweb.int/report/lebanon/timeline-crises-lebanon
Šaparnienė, D., Mejerė, O., Raišutienė, J., Juknevičienė, V., & Rupulevičienė, R. (2022). Expression of behavior and attitudes toward sustainable tourism in the youth population: A search for statistical types. *Sustainability, 14*(1), 473.
Sarraf, A. R. A. (2019). Generational groups in different countries. *International Journal of Social Sciences & Humanities, 4*(1), 41–52.
Scott, D. (2021). Sustainable tourism and the grand challenge of climate change. *Sustainability, 13*(4), 1966.
The International Ecotourism Society (TIES). (2015). https://ecotourism.org/what-is-ecotourism/

The KAP Survey Model. (2014). https://www.spring-nutrition.org/publications/tool-summaries/kap-survey-model-knowledge-attitudes-and-practices

World Bank Group. (2022, April 14). Lebanon's economic update—April 2022. *World Bank*. World Bank Group. https://www.worldbank.org/en/country/lebanon/publication/economic-update-april-2022

Wu, J., Snell, G., & Samji, H. (2020). Climate anxiety in young people: A call to action. *The Lancet Planetary Health, 4*(10), e435–e436.

Xu, L., Ao, C., Liu, B., & Cai, Z. (2022). Ecotourism and sustainable development: A scientometric review of global research trends. *Environment, Development and Sustainability*, 1–27.

19

Conclusion: Tourist Behavior in the New Normal and Its Implications on Sustainable Tourism Development: Emerging Realities, Tensions and Prospects

Maximiliano E Korstanje, Vanessa GB Gowreesunkar, and Shem Wambugu Maingi

We live in a world of low mobility after the COVID-19 pandemic. Over decades, the theory of mobilities has discussed the necessary changes that passed the end of industrialism to a new era, the mobile culture or the expansion of the capitalist system. The theory of mobilities has been of paramount importance in helping us understand how transport and the

M. E Korstanje
University of Palermo, Buenos Aires, Argentina

V. GB Gowreesunkar (✉)
Indian Institute of Management, Sirmaur, Himachal Pradesh, India

S. W. Maingi
Kenyatta University, Nairobi, Kenya
e-mail: maingi.shem@ku.ac.ke

S. W. Maingi et al. (eds.), *Tourist Behaviour and the New Normal, Volume II*,
https://doi.org/10.1007/978-3-031-45866-8_19

explosion of travel have been accompanied by a new stage of capitalism where cultures and landscapes are commoditized and exchanged (Vannini, 2010; Sheller, 2014; Sheller & Urry, 2006). Notwithstanding this fact, the mobile culture reaches only a small portion of the global population in the Global South. The capitalist system, which kept some material inequalities above all among classes, has blurred the geopolitical national borders as never before (Korstanje, 2018). Several interesting studies have brought attention to the challenges associated with globalizing the right to travel. At first glimpse, the tourism industry has witnessed the proliferation of risks which include natural disasters, political violence, and even terrorism. Raoul Bianchi argues convincingly that international tourism associates with the zenith of consumer capitalism. Anyway, the capitalist expansion has been accompanied by military interventions and surveillance tech applied to monitor undesired agents. This expansion is conducive to a climate of hostility against Western tourists which is channelized by terrorist groups (Bianchi, 2006). What is equally important, the turn of the century calls attention to the so-called right to travel as a liberalized discourse of capital exchange. In this respect, the right to travel should be seen as an invention of the neoliberal agenda. While recent legislation aimed at empowering tourists as cosmopolitan citizens, the persistence of xenophobic expressions, particularly towards strangers linked to travel restrictions like travel bans, has been noted (Bianchi & Stephenson, 2014). It's essential to acknowledge this phenomenon, including the recent COVID-19 pandemic, which temporarily suspended the right to move freely for everyone (Bianchi et al., 2020).

As the previous argument is given, this second volume—which is vested in a political tone—is formed by a set of chapters aimed at deciphering the connection between geopolitics and sustainable tourism (as well as degrowth tourism). In this context, the first chapter, which is authored by Shem W. Maingi & Vanessa Gowreesunkar, reviews the effects of COVID-19 and geopolitical tensions in the decolonized East Africa. Per their viewpoint, in the Global South—some underdeveloped economies burdened by unregulated consumption—had taken the lead in adopting some austerity measures to promote degrowth tourism. Citing Butcher, the authors use the term *moralization of tourism* as a new merging pattern to reconsider global tourism and its environmental degradation. Following this, Maximiliano Korstanje explores the qualitative changes in travel behavior accelerated by the COVID-19 pandemic. The chapter interrogates philosophically on some issues such as the biological passports, the travel bans, the neglect of the "Other" as well as the decline of Western hospitality. Maraísa de Silva Soares Costa, Cláuda A. Avelar Ferreira & Mahendar R Gavinolla fill a gap left by the lack of specialized literature. Authors analyze the possibilities of disabled people accessing tourism after the COVID-19 pandemic. Having said this, accessible tourism occupies a central position, as a new merging opportunity in the new normal to fasten the industry recovery timeframe. Sweety Mishra and Nimit Chowdhary write on *mindful consumption in the new normal.* What does it really mean? The tourism industry—though less than other services—has shown some adaptable capacity to embrace sustainable forms of consumption. The Covid-19 pandemic prompted a state of crisis that opened the doors to degrowth tourism and new sustainable practices, at least in the Indian market. Meaningful tourism corresponds with an emerging niche of visitors worried about the environment and authentic experiences—even in urban contexts. Rami Isaac researches the impacts of COVID-19 on Dutch travelers' patterns compatible with eco-friendly practices. The obtained outcomes suggest that COVID-19 is not seen necessarily as a close threat in the short run, but travelers are more worried about long-lasting health consequences. Manuel G. Herrera, Silvia Giralt Escobar, and Julian A. Alvarez Hernandez bring an interesting study case based on Majalca National Park, Mexico. The chapter is aimed at sharing information on tourist behavior practices in protected areas. Complementarily, Deborah Kangai, Eliyas E. Aman and Árpád Papp-Váry discuss the future of ecotourism as a vehicle for environmental preservation and development tourism. The new normal is a real Pandora's Box exhibiting great opportunities for new forms of tourism consumption as never before. In this direction goes Zeynep G. Hashmi who holds the thesis that sustainable tourism—after COVID—should be reconsidered in the paradigm of BEST (beyond sustainable tourism) which means a new mindset working in a mixed balance between demand and supply. At the same time, the concept of well-being is helpful to overcome the barriers left by overcrowding and over-tourism. From Lithuania, Darius Liutikas and Viktorija Baranauskienė conduct interesting research centered on the assessment of primary data on a focus group of tourism experts in the country. The consulted persons alerted on the importance to adopt the new model in tourism adaptation. Local tourism is pondered as a less degradable option while it lays the foundations for the placating of social maladies such as riots, terrorism or social upheavals. Francis Boadu and Silvia Fernandes coordinate efforts to integrate digital instruments to enhance digital marketing. Based on a qualitative approach, the authors conclude that tourism still remains the main contributor to economic growth in Ghana even if these projects failed to use digital marketing. Aapo Lundén and Alix Varnajot (who come from the University of Oulu, Finland) assess the current domains of sustainable tourism in the post-COVID-19 context. The chapter outlines and pros and cons (contradictions) of protected areas management in the Arctic. The case of Kenya, which is shared by Isaac Kimunio, Martin Nandelenga and S. Makambi, evinces an urgency to reconsider the function of the tourism industry in society. In fact, the tourism industry has been devastated by COVID-19 altering the ways people travel as well as their habits. Tourism marking should be reformulated according to the restrictive measures—during the lockdowns—that ignited fear in society. M. Korstanje puts critically the lens of the tourist gaze—with his attention on the Russia-Ukraine war. Travel behavior has notably shifted according to the disposition of new surveillance techniques now designed to "control" what the author dubbed as undesired guests. Korstanje calls attention to discussing *the shadows of hospitality.* Lastly, proximity tourism in the fields of sustainable consumption for Spain is analyzed by Adrián Mendieta Aragón and Raquel Arguedas Sanz who reflect on the tourist profile for proximity tourism to impact positively in the community. A Gen-Z perspective approach in the context of Lebanon is explained by Alain Daou, Leila El Zeenni, Anna Hourani and Salma Talhouk. These authors give an innovative snapshot reminding us that Gen-Z (a generation ranging between 18 and 25 years old) is not supposedly pro-environmental in many aspects. In some developing countries, high pro-environmental awareness in Gen Z is not correlated directly to pro-environmental practices. This happens simply because there is a dissociation between the cognitive and the emotional sphere which needs to be investigated.

All chapters gathered in this second volume successfully worked on the problem of proximity tourism, digital technologies, geopolitical tensions as well as the importance of sustainable practices in developing economies. Complementarily to the first volume—which was oriented to personal travel behavior—this second volume is conducive to studying the effects of COVID-19 from a catch-all approach where all stakeholders' perspectives are taken into serious consideration.

References

Bianchi, R. (2006). Tourism and the globalisation of fear: Analysing the politics of risk and (in) security in global travel. *Tourism and Hospitality Research, 7*(1), 64–74.

Bianchi, R., & Stephenson, M. (2014). *Tourism and citizenship: Rights, freedoms and responsibilities in the global order*. Routledge.

Bianchi, R. V., Stephenson, M. L., & Hannam, K. (2020). The contradictory politics of the right to travel: Mobilities, borders & tourism. *Mobilities, 15*(2), 290–306.

Korstanje, M. E. (2018). *The mobilities paradox: A critical analysis*. Edward Elgar Publishing.

Sheller, M. (2014). The new mobilities paradigm for a live sociology. *Current Sociology, 62*(6), 789–811.

Sheller, M., & Urry, J. (2006). The new mobilities paradigm. *Environment and Planning A, 38*(2), 207–226.

Vannini, P. (2010). Mobile cultures: From the sociology of transportation to the study of mobilities. *Sociology Compass, 4*(2), 111–121.

Correction to: Tourist Behavior for Sustainable Development in the Cumbres de Majalca National Park, Mexico: Challenges in a Post-pandemic Context

Manuel Ramón González Herrera,
Silvia Giralt Escobar, and
Julián Alberto Álvarez Hernández

Correction to:

Chapter 7 in: S. W. Maingi et al. (eds.), *Tourist Behaviour and the New Normal, Volume II,* https://doi.org/10.1007/978-3-031-45866-8_7

The original version of the chapter was inadvertently published with incorrect affiliations of the author group. The affiliations of the authors has been corrected as below

M. R. González Herrera (✉) • J. A. Álvarez Hernández
Department of Tourism, Autonomous University of Ciudad Juárez, Ciudad Juárez, Mexico
e-mail: manuel.gonzalez@uacj.mx; jualvare@uacj.mx

The updated version of this chapter can be found at
https://doi.org/10.1007/978-3-031-45866-8_7

S. W. Maingi et al. (eds.), *Tourist Behaviour and the New Normal, Volume II*,
https://doi.org/10.1007/978-3-031-45866-8_20

S. Giralt Escobar
Department of Economics and Business Management, University of Alcala, Alcalá de Henares, Spain
e-mail: silvia.giralt@uah.es

Index

S. W. Maingi et al. (eds.), *Tourist Behaviour and the New Normal, Volume II*,
https://doi.org/10.1007/978-3-031-45866-8

N

O

P

R

S

T

FSC
www.fsc.org
MIX
Papier | Fördert gute Waldnutzung
FSC® C083411

Kolibri 360 GmbH
Ferdinand-Jühlke-Straße 7
99095 Erfurt, Deutschland
produktsicherheit@kolibri360.de